Progress in Mathematics
Vol. 10

Edited by
J. Coates and
S. Helgason

Birkhäuser
Boston · Basel · Stuttgart

Ergodic Theory and Dynamical Systems I

Proceedings
Special Year, Maryland 1979-80

A. Katok, editor

1981

Birkhäuser
Boston · Basel · Stuttgart

Editor

A. Katok
Department of Mathematics
University of Maryland
College Park, Maryland

Library of Congress Cataloging in Publication Data
Ergodic theory and dynamical systems.
(Progress in mathematics ; 10)
Bibliography: p.
Includes index.
1. Ergodic theory--Addresses, essays, lectures.
2. Differentiable dynamical systems--Addresses, essays, lectures. I. Katok, A. II. Series: Progress in mathematics (Cambridge, Mass.) ; 10.
QA611.5.E73 1981 515.4'2 81-1859
ISBN 3-7643-3036-8 AACR2

CIP - Kurztitelaufnahme der Deutschen Bibliothek
Ergodic theory and dynamical systems: proceedings, special year, Maryland 1979-80 - edited by A. Katok.
Boston ; Basel ; Stuggart: Birkhauser, 1981.
(Progress in Mathematics ; 10)
NE: Katok, Anatole (Hrsg.)

ISBN: 3-7643-3036-8
Printed in USA

TABLE OF CONTENTS

PARTICIPANTS OF THE SPECIAL YEAR IN ERGODIC THEORY AND DYNAMICAL SYSTEMS

(Other than permanent faculty of University of Maryland)

University of Maryland
College Park
1979-80

J. Aaronson - IHES, Bures-sur-Yvette - August 1979

R. Adler - IBM - November-December 1979

S. Alpern - London School of Economics - January 1980

M. Brin - University of Maryland - the whole year

C. Conley - University of Wisconsin - Madison, March 1980

E. Coven - Wesleyan University - April 1980

R. Devaney - Tufts University - Spring semester 1980

J. Feldman - University of California, Berkeley - Fall semester 1979

S. Foguel - Hebrew University - September 1979

J. Ford - Georgia Institute of Technology - March 1980

J. Franks - Northwestern University - November 1979

H. Furstenberg - Hebrew University, Jerusalem - November 1979, January-February 1980

M. Gerber - NSF Postdoctoral Fellow - the whole year

A. del Junco - Ohio State University - October 1979

S. Kakutani - Yale University - November 1979

Y. Katznelson - Hebrew University, Jerusalem - August-September 1979

M. Keane - University of Rennes - April 1980

A. Lasota - Sylesian University, Poland - April-May 1980

D. Lind - University of Washington, Seatle - Fall semester 1979

B. Marcus - University of North Carolina - Fall semester 1979

H. Masur - University of Illinois, Chicago Circle - April 1980

J. Mather - Princeton University - January 1980

J. Moser - Courant Institute, NYU - February 1980

S. Newhouse - University of North Carolina - November 1979

Z. Nitecki - Tufts University - April 1980

D. Ornstein - Stanford University - November 1979

J. Palmore - University of Illinois, Urbana - May 1980

K. Peterson - University of North Carolina - November 1979

P. Rabinowitz - University of Wisconsin, Madison - March 1980

M. Ratner - University of California, Berkeley - September 1979

C. Robinson - Northwestern University - February 1980

D. Rudolph - Stanford University - Fall semester 1979, March 1980

D. Ruelle - IHES, Bures-sur-Yvette - December 1979

J.-M. Strelcyn - University of Paris - North - January - February 1980

M. Stuart - Northeastern University - March 1980

L. Swanson - Texas A & M University - Fall semester 1979

W. Szlenk - University of Warsaw - September 1979

W. Veech - Rice University - March-May, 1980

R. Williams - Northwestern University - January 1980

P. Winternitz - University of Montreal, April 1980

Graduate Students

A. Fisher - University of Washington, Seattle - Fall semester 1979

B. Kitchens - University of North Carolina - Fall semester 1979

S. Williams - Yale University - November 1979

PROGRAM OF THE SPECIAL YEAR IN ERGODIC THEORY AND DYNAMICAL SYSTEMS

A. SPECIAL COURSES AND SEMINARS

Fall semester 1979

J. Feldman - Orbit structure in Ergodic Theory.

A. Katok - Constructions in Ergodic Theory.

D. Lind - *D. Rudolph* - Finitary Isomorphism of Measure-preserving transformations.

B. Marcus - Horocycle Flows.

Spring semester 1980

R. Devaney - Non-integrable Classical Dynamical Systems.

W. Veech - Interval Exchange Transformations.

B. SERIES OF LECTURES

H. Furstenberg - Mildly Mixing Transformations.

M. Gerber, A. Katok - Smooth Models for Thurston's Pseudo-Anosov maps.

D. Ruelle - Characteristic Exponents and Invariant Manifolds in Hilbert space.

C. TALKS

J. Aaronson - Infinite Measure Preserving Transformations.

R. Adler - Cross Section for Geodesic Flows and Continuous Fractions.

S. Alpern - Ergodic Properties of Measure-preserving Homeomorphisms.

M. Brin - Topology and Spectrum of Anosov Diffeomorphisms.
Topology and Ergodicity of Frame Flows.

C. Conley - On Limiting for Directal Families of Flows with no Limit Flow and a New Way of Codifying Algebraic Properties of Flows. (2 talks)

J. Feldman - Amenable and Anti-Amenable Group Actions.

S. Foguel - Asymptotic Behaviour of Iterates of a Harris Operator.

J. Ford - Computer Models of Dynamical Systems. (2 Talks)

J. Franks - Symbolic Dynamics and Knot Theory. Anomalous Anosov Flows.

H. Furstenberg - Some Dynamical Systems Connected with Number Theory.

M. Gerber - A Zero-entropy Mixing Transformation Whose Cartesian Products are Loosely Bernoulli.

A. del Junco - Finitary Coding for $\mathbb{Z}^n$-actions.

S. Kakutani - Proof of Maximal Ergodic Theorem.

Y. Katznelson - An Ergodic Approach to the Szemeridi Theorem and its Generalizations.

M. Keane - Continuous Homomorphisms of Bernoulli Shifts.

A. Lasota - Exactness for One-dimensional Maps.

An Application of a Fixed Point Theorem to Ergodic Theory.

H. Masur - Strict Ergodicity of Measured Foliations.

J. Mather - Caratheodory Theory of Prime Ends.
Invariant Subsets for Area-preserving Homeomorphisms.

J. Moser - Isospectral Deformations and Separation of Variables.

S. Newhouse - Modern Bifurcation Theory for Smooth Dynamical Systems.

Z. Nitecki - Non-wandering Points for Maps of the Interval.

D. Ornstein - Automorphism Non-equivalent to the Direct Product of Bernoulli and Zero Entropy.

Orbit Equivalence with Additional Structures.

J. Palmore - New Stable Configurations and Relative Equilibria of Interacting Vortices by a Topological Method.

P. Rabinowitz - Periodic Solutions of Hamiltonial Systems.

Critical Points of Indefinite Functionals.

M. Ratner - Horocycle Flows and the Loose Bernoulli Property.

C. Robinson - Non Absolutely Continuous Invariant Foliations.

D. Rudolph - Measure-preserving Maps with Minimal Self-joinings and Some Applications.

Non-isomorphic Flows with Isomorphic Automorphisms. (2 talks)

J.-M. Strelcyn - Dynamical Systems with Singularities and Billiard Systems. (2 talks)

L. Swanson - Minimal Self-joinings and an Example of Chacon.

W. Szlenk - Dynamics of One-dimensional Maps.

R. Williams - Lorentz Attractors and Knotted Periodic Orbits. (2 talks)

P. Winternitz - Quadratic Hamiltonians in Phase Space and Their Eigenfunctions.

A CHARACTERIZATION OF THOSE PROCESSES FINITARILY ISOMORPHIC TO A BERNOULLI SHIFT

Daniel J. Rudolph

I. Introduction

Those stationary stochastic processes measurably isomorphic to an independent process have been characterized by Ornstein and Weiss (6), (7), (8), as those satisfying a condition called "very weak Bernoulli". A measure preserving transformation is often also continuous with respect to some natural topology. The topology which will most interest us here is that, on a countable state stochastic process, generated by the cylinder sets, i.e., if the outputs of the system are the symbols $R_1, R_2, \ldots,$ we can view the underlying probability space as $\{R_1, \ldots\}^Z$ with its product topology.

In such a case one can ask whether the isomorphism of Ornstein's theory might also be a homeomorphism. Complications to this question arise from periodic points, dimension and topological entropy which are of interest in topological dynamics but not so much to measure theory. Hence a slightly different notion has been introduced of an "almost" continuous isomorphism. We allow ourselves to remove from each space a set of measure zero and then attempt the continuous isomorphism on what remains.

What we shall do here is characterize those processes with their process topology which are almost continuously isomorphic to an independent process.

Keane and Smorodinsky (2), (3), (4) have shown that any two finite state mixing Markov chains with the same entropy are almost homeomorphic in this sense. Their argument is via a very elegant coding, mirroring in this context the machinery of Ornstein's measure theoretic isomorphism. Our argument here is a refinement of theirs, modifying or eliminating those parts which depend too explicitly on knowledge of the process.

One can view these notions of isomorphism in the following more constructive way. We wish to build an isomorphism between two processes (T,P) and (T',P'). Our problem is to construct within the σ-algebra $\bigvee_{i=-\infty}^{\infty} T^i(P)$ a partition $\overline{P}'$ so that $(T,\overline{P}')$ has the same finite distribution as (T',P'). (That this is an isomorphism will come from requiring $P \subset \bigvee_{i=-\infty}^{\infty} T^i(\overline{P}'))$. We can constructively describe $\overline{P}'$ by saying that for each N we look at the atoms of $\bigvee_{i=N}^{N} T^i(P)$ and make a choice for which set in $\overline{P}'$ each cylinder will lie. Call this approximation $\overline{P}(N,\omega)$. As our P cylinder sets get ever smaller we refine the accuracy of our approximation.

Let $E(N)$ be the set of points which, for some $n \geq N$, lie in differently labeled sets of $\overline{P}'(n,\omega)$ and $\overline{P}'$. In Ornstein's isomorphism construction (6) all we get is that $\mu(E(n)) \xrightarrow{n} 0$, which only gives us that $\overline{P}'$ is measurable. To get an almost continuous isomorphism we must also have $E(N)$ an open set, and it is easily seen w.l.o.g. that we can also ask that $E(N) \subset \bigvee_{i=-N}^{N} T^i(P)$. Thus not only must we be able to make a good guess within this algebra, but we must also be able to tell, within this algebra, where our errors will arise. For this reason Smorodinsky and Keane call such an isomorphism a "finitary code". As we read out more and more of the P-name of a point ω, at some point almost surely, the cylinder is in will no longer intersect $E(N)$, at which point we will know which set in $\overline{P}'$ to place ω in. See (12) for a more detailed discussion of these ideas.

II. Finitarily Bernoulli

We will call our condition on a process (T,P) "finitarily Bernoulli". It will appear rather involved and as we shall see a stronger and simpler condition is necessary. This more involved sufficient condition arose out of consideration of certain examples, and is an attempt to give a more easily verified condition.

Fix a process (T,P), $P = \{R_1 \ldots\}$ and let τ be the topology generated by finite cylinder set of $\bigvee_{i=-\infty}^{\infty} T^i(P)$. We will indicate, as is becoming standard, $\bigvee_{i=n}^{m} T^i(P)$ by P_n^m.

For any map $a; \{n,n+1,\ldots,m\} \to \{R_1,\ldots\}$ we will identify a with $\{\omega | T^{-i}(\omega) \in a(i),\ i=n,\ldots,m\}$, i.e., we identify a T,P-name with the points possessing it. We will call such a map an "atom" of P_n^m. Similarly, we let $P_i(\omega)$ be that R_j with $T^{-i}(\omega) \in R_j$ and $P_n^m(\omega) = \{\omega' | P_i(\omega) = P_i(\omega'),\ i = n,\ldots,m\}$. For convenience we will introduce some notation. On $\mathcal{F} \times \mathbb{Z}^+$ we define a multiplication by $(B_1,n_1) \circ (B_2,n_2) = (B_1 \cap T^{n_1}(B_2), n_1+n_2)$. When we say $\omega \in (B,n)$ we mean $\omega \in B$, by the "block length" $\ell(B,n)$ we mean n, and if $a = (B,n)$ by (a,m) we mean $(B,n+m)$, and $T(a) = (T(B),m+1)$. This notation simply generalizes to arbitrary sets the idea of concatenation of names.

By an "almost open" set we will mean a set which is within measure zero of open, and similarly "almost closed" and if a set is both almost open and almost closed we will call it "almost clopen". Notice that this need not mean that the set is within measure zero of clopen. It is within the algebra of almost clopen sets where all our work will take place.

If the process (T,P), $P = \{R_1,R_2,\ldots\}$ on $(\Omega, \ ,\mu)$, $h(P) < \infty$, possesses the following structure we will call it "finitarily Bernoulli". For every $\varepsilon > 0$, $n \in \mathbb{N}$ we must have a "nice partition", $C(\varepsilon,n)$, of Ω into almost clopen sets where $C(\varepsilon',n')$ refines $C(\varepsilon,n)$ if $\varepsilon' \leq \varepsilon$ and $n' \geq n$. These should behave like partitions into sets of small diameter. We describe this as follows.

There is a countable partition $A(\varepsilon) = \{A_i(\varepsilon),\bar{A}_i(\varepsilon)\}$ of Ω into almost clopen sets where $\bigcup_i A_i(\varepsilon)$ nest down in ε a.s. to ϕ. There also is, for each i, a subset $\hat{A}_i(\varepsilon)$, $A_i(\varepsilon) \subset \hat{A}_i(\varepsilon) \subset A(\varepsilon)$. Now for any $\omega \in \Omega$ and $n \in \mathbb{N}$ we construct $t_0(\omega,\varepsilon,n) = \{\omega' | i \in \{0,\ldots,n-1\},\ T^{-i}(\omega)$ and $T^{-i}(\omega')$ are both in the same $A_j(\omega)$, or $P_i(\omega') = P_i(\omega)$ and $T^{-i}(\omega'),\ T^{-i}(\omega) \in \hat{A}_j(\varepsilon)\}$. This "thickening" of the orbit of ω from indices 0 to $(n-1)$ plays the role of an ε-neighborhood of this piece of orbit. What $t_0(\omega,\varepsilon,n)$ consists of is a restircted $\bar{d}$-neighborhood of the orbit of ω, the $\bar{d}$ changes forced to lie in the $A_i(\varepsilon)$. Notice the symmetry of the definition, if $\omega' \in t_0(\omega,\varepsilon,n)$, then

$\omega \in t_0(\omega',\varepsilon,n)$. Also t_0 has a "nice concatenation property", $t_0(\omega,\varepsilon,n) = \prod_{i=0}^{n-1} (t_0(T^{-i}(\omega),\varepsilon,1),1)$. We define, for $E \in \mathcal{F}$, $t_0(E,\varepsilon,n) = \bigcup_E t_0(\omega,\varepsilon,n)$, an almost open set. For any $c \in C(\varepsilon,n)$ if $\omega \in c$, we demand that $c \subseteq t_0(\omega,\varepsilon,n)$. This forces $C(\varepsilon,n)$ to be a "small" partition. Set $t_1(E,\varepsilon,n) = \bigcup\{c \in C(\varepsilon,n) \mid \mu(E \cap c) > 0\}$, a second kind of thickening.

We need a bit more than just this kind of smallness. We demand that for any $\varepsilon_0 > 0$, there are $\alpha_1(\varepsilon_0),\alpha_2(\varepsilon_0),\ldots$ so that if $\varepsilon_i < \alpha_i(\varepsilon_0)$ and we define

$$E_1 = t_0(\omega,\varepsilon_1,n)$$

and then

$$E_{i+1} = t_0(E_i,\varepsilon_i,n),$$

i.e., we thicken over the piece of orbit $0,\ldots,n-1$ by ever smaller amounts, then $E_i \subset t_0(\omega,\varepsilon_0,n)$ for all i, i.e., in so doing we remain in a neighborhood of the piece of orbit. The purpose of the elaborate definition of t_0 is to help it to satisfy this last condition.

The thickenings t_0 and t_1 are to be thought of as occurring across the indices $0,\ldots,n-1$ in an orbit. When we speak of a t_*,ε-thickening of the orbit of ω across the block of indices $k,\ldots,k+n-1$ we mean $T^k(t_*(T^{-k}(\omega),\varepsilon,n))$. The important differences between t_0 and t_1 to keep in mind is that t_0 has the nice concatenation property but only gives almost open sets, where t_1 is actually a lumping of elements of an almost clopen partition.

Next, for each $\varepsilon > 0$ we must have $N(\varepsilon) \overset{\varepsilon}{\nearrow} \infty$ and an almost clopen "spacer" set $S(\varepsilon)$, $\mu(S(\varepsilon)) > 0$, satisfying the strongest requirements of our definition.

(i) For all ε',n,m, $t_1(T^n(S(\varepsilon)),\varepsilon',n+N(\varepsilon)+m) = T^n(S(\varepsilon))$, i.e., thickening, no matter where or by how much, does nothing to a spacer.

(ii) There is an N_0 so that for any ε, if ε' is small enough, then for any $-N(\varepsilon')+N_0 \leqslant n \leqslant N(\theta)-N_0$, $T^n(S(\varepsilon')) \cap S(\varepsilon) = \phi$.

(iii) For any ε_0, there is an $\overline{N}(\varepsilon_0)$ so that for any $\varepsilon_1,\ldots,\varepsilon_{k_1},\ldots,\varepsilon_k$ and $\varepsilon'_1,\ldots,\varepsilon'_{k'_1},\ldots,\varepsilon'_{k'}$, if $n_i < -N(\varepsilon_i) - \overline{N}(\varepsilon_0)$ and $m_i > 0$ then

$$\mu\Big(\big(\bigcap_{i=1}^{k_1} T^{n_i}(S(\varepsilon_i)) \bigcap_{i=k_1+1}^{k} T^{n_i}(S(\varepsilon_i)^c)\big) \cap \big(\bigcap_{i=1}^{k'_1} T^{m_i}(S(\varepsilon'_i)) \bigcap_{i=k'_1+1}^{k'} T^{m_i}(S(\varepsilon'_i)^c)\big)\Big)$$

$$= \mu\big(\bigcap_{i=1}^{k_1} T^{n_i}(S(\varepsilon_i)) \bigcap_{i=k_1+1}^{k} T^{n_i}(S(\varepsilon_i)^c)\big)\mu\big(\bigcap_{i=1}^{k'_1} T^{m_i}(S(\varepsilon'_i)) \bigcap_{i=k'_1+1}^{k'} T^{m_i}(S(\varepsilon'_i)^c)\big)2^{\pm\varepsilon}$$

i.e., there is a "uniform rate of mixing" on the spacer process. We will abbreviate this USM (uniform spacer mixing).

(iv) There is a $\sigma(\varepsilon) > 0$ so that for any $\varepsilon' \leq \varepsilon$, $c' \subset C(\varepsilon,n')$, $c' \subset T^{n'-n(\varepsilon'')}(S(\varepsilon''))$, $c' \subset c \subset C(\varepsilon,n)$, $c \subset T^{n-N(\varepsilon''')}(S(\varepsilon'''))$ then if $\mu((S(\varepsilon'),N(\varepsilon')) \circ (c',n')) \neq 0$, $\mu((S(\varepsilon'),N(\varepsilon'))(c',n') \mid (S(\varepsilon'),N(\varepsilon'))(c,n)) > \sigma(\varepsilon)^{n-n'}$, i.e., the sizes of sets in $C(\varepsilon,n)$ which end in a spacer block decrease in n at a bounded exponential rate, the bound dependent on ε. We call this property 4.

(v) For any ε, $i = 1,\ldots,k_1,\ldots,k$, $\varepsilon_i \geq \varepsilon$, $0 \leq j(i) \leq J$, $\mu(\bigcap_{i=1}^{k_1} T^{j(i)}(S(\varepsilon_i)) \bigcap_{i=k_1}^{k} T^{j(i)}(S(\varepsilon_i)^c))$ is either 0 or $> \sigma(\varepsilon)^J$, i.e., spacer cylinders for ε_i which are not measure zero have sizes going to zero at a bounded exponential rate. We call this property 5.

This structure must now satisfy the following "conditional block independence". For any $\varepsilon_1,\ldots,\varepsilon_k$, $n_1,\ldots,n_k \in \mathbb{N}$ sets $E_1,\ldots,E_k$ and $\overline{\varepsilon}_i \geq \max(\varepsilon_i,\varepsilon_{i+1})$ setting

$$\overline{E}_i = (t_1(E_i,\overline{\varepsilon}_i,n_i),n_i)$$

and

$$\overline{S}_i = (S(\varepsilon_i),N(\varepsilon_i))$$

then

$$\mu(\overline{E}_1 \circ \overline{S}_1 \circ \overline{E}_2 \circ \overline{S}_2 \circ \ldots \circ \overline{S}_{k-1}\ \ \overline{E}_k \circ \overline{S}_k \mid \overline{E}_1 \circ \overline{S}_1 \circ \ldots \circ \overline{E}_{k-1} \circ \overline{S}_{k-1})$$

$$= \mu(\overline{S}_{k-1} \circ E_k \circ \overline{S}_k \mid \overline{S}_{k-1})2^{\pm\varepsilon_{k-1}}.$$ We will abbreviate this CBI.

CBI is easily necessary as we shall soon see. It is similar to Bowen's notion of specification (10) which Lind has used very effectively to show the Bernoulliness of ergodic group automorphisms (8). Lind has recently partially extended his work to the almost continuous category using a condition even more like CBI (9). This now completes our definition.

Notice in this definition that the $A_i(\varepsilon)$, could be empty, but we must have spacers. Also notice that it is sufficient for this condition that we only verify that the needed structures exist for a sequence of ε's tending to zero.

As we have said before different aspects of the definition can be strengthened still maintaining its necessity but losing its ease of applicability. The structure of thickening and spacers is simply not unique and what we have tried to do is give as weak a condition as possible so that thickenings which arise naturally within an example will work. We leave it to the interested reader to attempt further weakening of our condition.

To help set this definition in mind we will give two examples and then prove its necessity.

Let (T,P) be a finite state mixing Markov process. Set $A_i(\varepsilon) = \phi$, $\bar{A}_1(\varepsilon) = \hat{A}_1(\varepsilon) = \Omega$. Thus for any E, $t_o(E,\varepsilon,n) \subset P_0^{n-1}$ consists of all atoms of P_0^{n-1} which intersect E. Let $C(\varepsilon,n) = P_0^{n-1}$ so $t_0 = t_1$. It is an easy check that these thickenings satisfy the needed conditions.

Now choose N so that there are two words $a_0 \in P_0^{N-1}$, $a_1 \in P_0^N$ both of which begin with symbol R_1 and end with a symbol with positive transition probability into R_1 and $a_1 \cap a_0 = \phi$, $a_1 \cup a_0 \neq \Omega$. Thus any name of the form $(a_{i(1)},N+i(1))(a_{i(2)},N+i(2))\cdots(a_{i(s)},N+i(s))$ has positive probability. Choose $N_0 > N(2N+1)$ so large that the set B of all substrings of length N_0 in such a concatenation of $(a_i,N+i)$'s has less than full probability.

Let $S(\varepsilon) = (a_0,N)^{2N}((a_0,N)(a_1,N+1))^{n(\varepsilon)}$ where $n(\varepsilon) > N + \frac{1}{\varepsilon}$ and now $N(\varepsilon) = 2N^2 + n(\varepsilon)(2N+1)$. Clearly

$S(\varepsilon) \subset B_0^{N(\varepsilon)-1}$ and for any $\varepsilon, \varepsilon'$, if $-N(\varepsilon) + N_0 < t < N(\varepsilon') - N_0$, $T^t(S(\varepsilon)) \cap S(\varepsilon') = \phi$. As $S(\varepsilon) \subset P_0^{N(\varepsilon)-1}$, $t_0(T^{-n}(S(\varepsilon)), \varepsilon', n+N(\varepsilon)+n) = T^{-n}(S(\varepsilon))$.

CBI and properties 4,5 follow directly from the Markov property and USM from recurrence.

We will now describe a different example, the study of which led to the general case. Start with the two-shift $(\overline{T}, \overline{P})$ acting on $\overline{\Omega} = \{0,1\}^Z$ with $\overline{P} = \{0,1\}$ and independent 1/2,1/2 measure. Let R_α be rotation by α on $\mathbb{T}^1$, the circle. Let $A = (0, \pi]$, $B = (\pi, 2\pi]$, and set $\Omega = \overline{\Omega} \times \mathbb{T}^1$ with partition $P = \overline{P} \times \{A,B\}$. Now for some fixed irrational α, let $g; \overline{\Omega} \to \mathbb{T}^1$ be

$$g(\overline{\omega}) = \begin{cases} \alpha & \text{if } \overline{\omega} \in 0 \\ 0 & \text{if } \overline{\omega} \in 1 \end{cases} .$$

Now define $T(\overline{\omega}, \theta) = (\overline{T}(\overline{\omega}), g(\overline{\omega}) + \theta)$, the usual skew product. For $\omega \in \Omega$, let $\omega = (\overline{\omega}(\omega), \theta(\omega))$.

On $\overline{\Omega} \times \mathbb{T}^1$ we have a choice of two topologies, one arising from its representation as a sequence space on symbols $\overline{P} \times \{A,B\}$, and the other arising from $\{0,1\}^Z$ cross the usual topology. Call these τ and τ'. Now $\{A,B\}$ is a partition into almost open sets in τ' and in fact it is easy to see that an open set in the process $(R_\alpha, \{A,B\})$ differs from an open set in the usual circle topology by a subset of the countable set $C = \{(n\pi + m\alpha)\}$ and furthermore any set in τ differs by a subset of $\overline{\Omega} \times C$ from a subset in τ' and vice versa. Thus the actions of T on τ and on τ' are almost continuously isomorphic. Thus although we will concern ourselves with the process topology τ, in this example our work extends to the system's more natural topology τ'. Our hope is that this will prove to be a common phenomenon.

This map T first arose as an example of a very weak Bernoulli map that is not weakly Bernoulli (11). This adds interest to the fact that we now show it to be finitarily Bernoulli.

For the process (T,P), and for $\varepsilon > 0$, ε of the form

2^{-n}, set

$$\bar{A}_1(\varepsilon) = \{(\bar{\omega},\theta) \mid \bar{\omega} \in 0,\ \theta \in (-\tfrac{\varepsilon}{2}, \tfrac{\varepsilon}{2})\},$$

$$\bar{A}_2(\varepsilon) = \{(\bar{\omega},\theta) \mid \bar{\omega} \in 0,\ \theta \in (\pi - \tfrac{\varepsilon}{2}, \pi + \tfrac{\varepsilon}{2})\},$$

$$\bar{A}_3(\varepsilon) = \{(\bar{\omega},\theta) \mid \bar{\omega} \in 1,\ \theta \in (-\tfrac{\varepsilon}{2}, \tfrac{\varepsilon}{2})\},$$

$$\bar{A}_4(\varepsilon) = \{(\bar{\omega},\theta) \mid \bar{\omega} \in 1,\ \theta \in (\pi - \tfrac{\varepsilon}{2}, \pi + \tfrac{\varepsilon}{2})\},$$

$A_1(\varepsilon)$ the rest of Ω, $\hat{A}_1(\varepsilon) = \Omega$.

It is straightforward now that for any $(\bar{\omega},\theta)$ and ε,n the set $t_0(\omega,\varepsilon,n)$ is the $\bar{P}$,n-name of Ω cross an $\varepsilon/2$ neighborhood on the circle of the P,n-name of ω, hence contains $\bar{P}_0^{n-1}(\bar{\omega}) \times I$ where I is an interval on $\mathbb{T}^1$ containing $(\theta - \frac{\varepsilon}{2}, \theta + \frac{\varepsilon}{2})$.

Let $C(\varepsilon,n) = \bar{P}_0^{n-1} \times \{(\frac{k\varepsilon}{2}, \frac{(k+1)\varepsilon}{2}] \mid k = 1,2,\ldots,\frac{2}{\varepsilon}\}$ be the nice partition. It follows that $t_1(\omega,\varepsilon,n) \subset t_0(\omega,\varepsilon,n)$. It is an easy exercise to check that t_0, t_1 satisfy the needed conditions (hint: choose $\Sigma\,\alpha_i(\varepsilon) < \varepsilon$).

To get CBI choose $N(\varepsilon)$ so large that for any interval $I \subset \mathbb{T}^1$ of length $\geq \varepsilon/2$,

$$\frac{1}{(N(\varepsilon)-2)} \Sigma_{i=1}^{N(\varepsilon)-2} \chi_I(i\alpha) = \ell(I)2^{\pm\varepsilon},$$

and so that $N(\varepsilon) \overset{\varepsilon}{\nearrow} \infty$.

Now set $S(\varepsilon) = \{(\bar{\omega},\theta) \mid j,\ 2 \leq j \leq N(\varepsilon)-3,\ \bar{P}_0(\bar{\omega}) \in 1,$ $\bar{P}_1(\bar{\omega}),\ldots,\bar{P}_j(\bar{\omega}) \in 0,\ \bar{P}_{j+1}(\bar{\omega}),\ldots,\bar{P}_{N(\varepsilon)-2}(\bar{\omega}) \in 1,\ \bar{P}_{N(\varepsilon)-1}(\bar{\omega}) \in 0\}$. Notice that $S(\varepsilon) \subset \bar{P}_0^{N(\varepsilon)-1}$ so that for all ε',n,m $t_1(T^{-n}(S(\varepsilon)),\varepsilon',n+N(\varepsilon)+m) = T^{-n}(S(\varepsilon))$. We easily check that for any ε,ε', if $-N(\varepsilon)+3 \leq t \leq N(\varepsilon')-3$ then $T^t(S(\varepsilon)) \cap S(\varepsilon') = \phi$. USM, properties 4 and 5 follow as $S(\varepsilon)$ are all cylinders of the independent $(T,\bar{P})$ process of length $N(\varepsilon)$.

Once more let $f \in P_{-\infty}^{-1}$ and $\bar{\varepsilon} \leq \varepsilon$, $\bar{E} = (t_1(E,\varepsilon,n),n)$,

$\overline{S} = (S(\overline{\varepsilon}),N(\overline{\varepsilon}))$. We compute $\mu(\overline{S} \circ \overline{E}|f)$. Note that $P_{-\infty}^{-1} \perp \overline{S}$ as $\overline{S} \subset \overline{P}_0^{N(\overline{\varepsilon})}$ and further that $\overline{E}$ can be written as a disjoint union of sets $S(I,a)$ where $I \subset \mathbb{T}^1$ is an interval of length $\geq \varepsilon \geq \overline{\varepsilon}$ where $\theta(S(I,a)) \in I$ and the $\overline{P}_0^{n-1}$-name of any point in $S(I,a)$ is a.

Now

$$\mu(T^{-N(\overline{\varepsilon})}(\omega) \in I|\theta(T^{-1}(\omega)) = \theta_0, \omega \in \overline{S})$$

$$= \ell(I)2^{\pm\varepsilon} \quad \text{for all } \theta_0 .$$

Thus for any $f \in P_{-\infty}^{-1}$

$$\mu(T^{-N(\overline{\varepsilon})}(\omega) \in I|\omega \in f, \omega \in S(\varepsilon))$$

$$= \ell(I)\mu(a)2^{\pm\overline{\varepsilon}}$$

and hence

$$\mu(\overline{S} \circ \overline{E}|f,\overline{S}) = \mu(\overline{E})2^{\pm\overline{\varepsilon}} .$$

As $\overline{E} \perp \overline{S}$, $f \perp \overline{S}$, we now have CBI, and (T,P) is finitarily Bernoulli.

III. The Necessity of Finitarily Bernoulli

Suppose (T,P) is an independent process and $Q \subset P_{-\infty}^{\infty}$ generates under T. Furthermore, outside a set of measure zero, the topologies generated by cylinder sets of (T,P) and (T,Q) are identical. This is the same as saying any atom $q_i \in Q$ is almost surely a union of cylinder sets of (T,P) and vice versa any atom $p_i \in P$ is a.s. a union of cylinders of (T,Q). Any process (T',Q') almost continuously isomorphic to (T,P) can be embedded in (T,P) as such a (T,Q). What we wish to show is that (T,Q) is finitarily Bernoulli.

For each integer m, let $A_i(1/m)$ be the atoms of P_{-m}^{m} which are contained within a single Q atom, and $\overline{A}_i(1/m)$ those which are not. Clearly $A(1/m) = P_{-m}^{m}$ hence almost clopen and $\bigcup_i A_i(\varepsilon) \not\stackrel{\varepsilon}{\sim} \phi$. Set $\hat{A}_i(1/m) = A_i(1/m)$. This now defines $t_0(\omega, \frac{1}{m}, n)$ to be $P^{m}_{-m}(\omega)$.

Set $C(\varepsilon,n) = P^{1/\varepsilon + n}_{-1/\varepsilon}$ so that $t_1(E,\varepsilon,n) = t_0(E,\varepsilon,n)$. It is an easy check that these thickenings satisfy the needed conditions (hint: $\alpha_i(\varepsilon) = \varepsilon$).

To build the $S(\varepsilon)$ pick δ so small that a 10δ neighborhood in $\bar{d}$ of a.e. ω in (T,Q) has measure zero. Now choose ε_0 so small that $\mu(\bigcup_i \bar{A}_i(\varepsilon_0)) < \delta$ and as $\bigcup_i A_i(\varepsilon_0)$ is almost clopen, choose n_0 so large that for at least 1/2 the atoms $\alpha \subset Q_0^{n_0-1}$, if $\alpha' \in t_0(\alpha,\varepsilon_0,n_0)$, then $\bar{d}_{n_0}(\alpha,\alpha') < 5\delta$ and a 5δ neighborhood of α has measure less than 1/10. Let $a_0 = t_0(\alpha_0,\varepsilon_0,n_0)$, $a_1 = t_0(\alpha_1,\varepsilon_0,n_0)$ for two such α,α' with $\bar{d}_{n_0}(\alpha_0,\alpha_1) > 10\delta$. Set $\bar{a}_0 = (a_0,n_1)$, $\bar{a}_1 = (a_1,n_1+1)$ where $n_1 = n_0 + \frac{2}{\varepsilon_0} + 1$. It is clear $\bar{a}_0 \cap \bar{a}_1 = \phi$ and $\bar{a},\bar{a}_1 \subset P^{n_1-1-1/\varepsilon_0}_{-1/\varepsilon_0}$.

Thus for any string

$$\bar{a} = \bar{a}_{i(1)} \circ \bar{a}_{i(2)} \circ \cdots \circ \bar{a}_{i(s)}, \quad \mu(\bar{a}) = \prod_{i=1}^{s} \mu(\bar{a}_{i(s)}).$$

Choose $N_0 > 4n_1^2 + (2n_1+1)^2$ so long that the collection of all atoms of $Q_0^{N_0-1}$ whose names occur along a substring of N_0 consecutive digits in the name of some point in such an $\bar{a}$ has measure less than one.

Let $S(\varepsilon) = \bar{a}_0^{2n_1} \circ (\bar{a}_0 \circ \bar{a}_1)^{n_1(2/\varepsilon + 2)})$, $N(\varepsilon) = (S(\varepsilon)) + 2/\varepsilon_0$.

It follows that for any ε,ε', if $-N(\varepsilon) + N_0 \leq t \leq N(\varepsilon') - N_0$, $T^t(S(\varepsilon)) \cap S(\varepsilon') = \phi$. We also have $S(\varepsilon) \subset P^{N(\varepsilon)-1+1/\varepsilon_0}_{-1/\varepsilon_0}$ and so for all $\varepsilon' \leq \varepsilon_0, n, m$

$$t_1(T^n(S(\varepsilon)),\varepsilon',n+N(\varepsilon)+m) = T^{-n}(S(\varepsilon))$$

and this consists of all atoms of $P^{n+N(\varepsilon)+m+1/\varepsilon'}_{-1/\varepsilon'}$ intersecting

$T^n(S(\varepsilon))$. USM, Properties 4 and 5 follow from the fact that spacers are cylinders in an independent process. To check CBI, given $\varepsilon_1,\ldots,\varepsilon_k$, $n_1,\ldots,n_k$, $E_1,\ldots,E_k$, $\overline{\varepsilon}_i \geq \max(\varepsilon_i,\varepsilon_{i+1})$, let

$$\overline{E}_i = (t_1(E_i,\overline{\varepsilon}_i,n_i),n_i) \subset P_{-1/\overline{\varepsilon}}^{n_i-1+1/\overline{\varepsilon}_i}$$

and $\overline{S}_i = (S(\overline{\varepsilon}_i),N(\overline{\varepsilon}_i))$.

To compute $\mu(\overline{E}_1 \circ \overline{S}_1 \circ \ldots \circ \overline{S}_{k-1} \circ \overline{E}_k \circ \overline{S}_k)$, write $\overline{S}_i = \overline{S}_i \circ \overline{S}_i''$ where

$$\overline{S}_i' = \overline{a}_0^{2n_1} \circ (a_0 \circ a_1)^{n_1(1/\varepsilon + 1)}, \quad \overline{S}_i'' = (a_0 \circ a_1)^{n(1/\varepsilon + 1)}$$

and we see that

$$\mu(\overline{E}_1 \circ \overline{S}_1 \circ \ldots \circ \overline{S}_{k-1} \circ \overline{E}_k \circ \overline{S}_k)$$

$$= \mu(\overline{E}_1 \circ \overline{S}_1')\mu(\overline{S}_1'' \circ \overline{E}_2 \circ \overline{S}_2') \ldots \mu(\overline{S}_{k-2}' \circ \overline{E}_{k-1} \circ \overline{S}_{k-1}')\mu(\overline{S}_{k-1}' \circ \overline{E}_k \circ \overline{S}_k)$$

$$= \frac{\mu(\overline{E}_1 \circ \overline{S}_1)}{\mu(\overline{S}_1'')} \frac{\mu(\overline{S}_1 \circ \overline{E}_2 \circ \overline{S}_2)}{\mu(\overline{S}_1') \circ \mu(\overline{S}_2'')} \cdots \frac{\mu(\overline{S}_{k-1} \circ \overline{E}_k \circ \overline{S}_k)}{\mu(\overline{S}_{k-1}'')}$$

$$= \mu(\overline{E}_1 \circ \overline{S}_1)\mu(\overline{S}_1 \circ \overline{E}_2 \circ \overline{S}_2) \ldots \mu(\overline{S}_{k-1} \circ \overline{E}_k \circ \overline{S}_k) \prod_{i=1}^{k-1} \mu(\overline{S}_i)^{-1},$$

and CBI follows, again without error, and necessity is proven. We see that we could in fact ask that CBI have no errors. Our earlier example shows that this may be very difficult to achieve.

IV. The Isomorphism

(a) Introduction. We now show that any two finitarily Bernoulli processes of the same entropy are finitarily isomorphic. Let (T,P) and (T',P') be such processes with all their requisite machinery of thickenings and spacers. We will proceed, as in the Smorodinsky Kean arguments (2)(3)(4) by first building skeletal processes in (T,P) and (T',P'). Rather than build identical skeleta in both the processes, we will show that both are finitarily isomorphic to an intermediary process (T_1,P_1) which is the direct product of the skeleta process of (T,P) and (T',P'), respectively, and any independent process of appropriate entropy. We next show that in (T_1,P_1) we can again verify the finitarily Bernoulli condition, but for either set of spacers. These spacers are statistically identical to those in (T,P) and (T',P'), respectively. This now will allow us to construct identical skeletal processes in (T,P) and (T_1,P_1) and (T',P') and (T_1,P_1) and verify their finitary isomorphism.

In constructing both finitary isomorphisms, (T,P) to (T_1,P_1) and (T_1,P_1) to (T',P') we will work within the skeletal structure, again following Smorodinsky and Keane, by successive applications of the marriage lemma, at each level of the skeleta, building up the isomorphism. There is an obstacle to this in that the distributions across two consecutive blocks between markers of the skeleta will concatenate only ε-independently. If the distributions of names in the two processes across each block have been coupled to form societies we cannot be sure that the distributions across the concatenated blocks have been so coupled. We overcome this obstacle by "thickening" the coupling first by the thickenings built into the definition of finitarily Bernoulli, giving an ε-independence, then further by an explicit method to overcome the remaining ε. The marriage lemma and these notions of thickening play such a crucial role that we now turn to them.

(b) Societies, couplings, the marriage lemma, and thickenings. Let $S = \{s_1,\ldots,s_k\}$ and $S' = \{s'_1,\ldots,s'_{k'}\}$ be finite sets, with probability measures μ and μ'. Let us first spend a little time understanding the Smorodinsky-Keane Marriage lemma. We include a proof which is only a simple modification of theirs but which clarifies the issues at hand. We call $R \subset S \times S'$ a "society" if for any subset $A \subset S$, the set $R(A) = \{s' \in S' | (s,s') \in R$ for some $s \in A\}$ satisfies $\mu'(R(A)) \geq \mu(A)$. As it stands this appears to be an asymmetric definition, but it is a simple argument on compliments that $R' = \{(s',s) | (s,s') \in R\}$, the "dual" of R is also a society.

By a "coupling" of (S,μ) and (S',μ') we mean a probability measure ω on $S \times S'$ with marginals μ and μ' on the two coordinate subalgebras. A coupling induces a society $R_\omega = \{(s,s') | w(s,s') > 0\}$. We will call such a society a "coupling society". We leave it as an exercise for the reader to construct a society which is not a coupling society. Even so something close to this is true.

Lemma 4b.1. For any society R on the finite probability spaces (S,μ) and (S',μ') there is a coupling ω so that $R_\omega \subseteq R$.

Proof. We show this by induction on $\mathrm{card}(\{s_i \in S | \mu(s_i) > 0\}) + \mathrm{card}(\{s_i' \in S' | \mu'(s_i') > 0\})$. If this number is 2 or 3 the result follows, as either S or S' is a point mass.

Select a fixed element $(s_i, s_i') \in R$, $\mu(s_i) > 0$, $\mu'(s_j') > 0$. Let $U = \{A \subset S | s_i \notin A$ but $s_j' \in R(A)\}$, a finite collection of subsets.

Let $\alpha = \min_{A \in U} (\mu'(R(A)) - \mu(A)) \geqslant 0$.

Case I. If $\alpha = 0$, then for some set $A \subset S|_{\{s_i\}}$, $\mu(A) = \mu'(R(A))$, so $\mu(R(A)^c) = \mu(A^c)$, and so $R'(R(A)^c) = A^c$. In this case the society splits as the sum of two societies, R on $A \times R(A)$, and R' on $R(A)^c \times A^c$, and as both A and A^c have some measure, we can use induction on the two pieces.

Case II. If $\alpha \geqslant \min(\mu(s_i), \mu(s_j'))$, let $\bar{\alpha} = \min(\mu(s_i), \mu(s_j'))$ and define new measures

$$\hat{\mu}(s_\ell) = \begin{cases} \dfrac{\mu(s_\ell)}{1-\bar{\alpha}} & \text{if } \ell \neq i \\[2ex] \dfrac{\mu(s_i) - \bar{\alpha}}{1 - \bar{\alpha}} & \text{if } \ell = i \end{cases}$$

and

$$\hat{\mu}'(s_\ell') = \begin{cases} \dfrac{\mu'(s_\ell)}{1-\bar{\alpha}} & \text{if } \ell \neq j \\[2ex] \dfrac{\mu'(s_j) - \bar{\alpha}}{1 - \bar{\alpha}} & \text{if } \ell = j\,. \end{cases}$$

Now R is still a society on $(S,\hat{\mu})$ and $(S',\hat{\mu}')$, as for any set $A \subset S$, if $s_i \notin A$, $s_j' \in R(A)$, then $\hat{\mu}'(R(A)) - \hat{\mu}(A) = \frac{\mu'(R(A)) - \mu(A) - \bar{\alpha}}{1 - \bar{\alpha}} \geqslant \frac{\alpha - \bar{\alpha}}{1 - \bar{\alpha}} > 0$ and in the other possible cases the situation is better. But now either $\hat{\mu}(S_i) = 0$ or $\hat{\mu}'(S_j') = 0$ and in either case by induction we can get a coupling $\hat{\omega}$ with $R_{\hat{\omega}} \subseteq R$. Now define

$$\omega((s_\ell,s_t')) = \begin{cases} \hat{\omega}((s_\ell,s_t'))(1-\bar{\alpha}) & \text{if } s_\ell \neq s_i \text{ or } s_t' \neq s_j' \\ \bar{\alpha} & \text{if } s_\ell = s_i,\ s_t' = s_j'. \end{cases}$$

this couples (S,μ) and (S',μ') with $R_\omega \subseteq R$.

<u>Case III</u>. If $\alpha < \min(\mu(S_i),\mu'(S_j'))$, then define $\hat{\mu}$ and $\hat{\mu}'$ by

$$\mu(s_\ell) = \begin{cases} \dfrac{\mu(s_\ell)}{1-\alpha} & \text{if } s_\ell \neq s_i \\ \dfrac{\mu(s_i)-\alpha}{1-\alpha} & \text{if } s_\ell = s_i, \end{cases}$$

and

$$\hat{\mu}(s_\ell') = \begin{cases} \dfrac{\mu'(s_\ell')}{1-\alpha} & \text{if } s_\ell' \neq s_j' \\ \dfrac{\mu'(s_j')-\alpha}{1-\alpha} & \text{if } s_\ell' = s_j'. \end{cases}$$

As before, R is still a society on $(S,\hat{\mu})$ and $(S',\hat{\mu}')$. But now $\hat{\mu}(S_i) > 0$ and $\hat{\mu}'(S_j') > 0$. Looking at

$$\hat{\alpha} = \min_{A \subset U} \{\hat{\mu}'(R(A)) - \hat{\mu}(A)\} = 0,$$

though, we see that we can apply Case I to $(S,\hat{\mu})$ and $(S',\hat{\mu}')$ to get a coupling $\hat{\omega}$, and now define

$$\omega((s_\ell,s_t')) = \begin{cases} \hat{\omega}((s_\ell,s_t'))(1-\alpha) & \text{if either } s_\ell \neq s_i \text{ or } s_t' \neq s_j' \\ \hat{\omega}((s_i,s_j))(1-\alpha) + \alpha & \text{otherwise.} \end{cases}$$

This couples (S,μ) and (S',μ') with $R_\omega \subseteq R$. ∎

Smorodinsky and Keane prove a version of this lemma where both measures need not be probability measures, but we do not need this stronger fact. We complete the proof of the Marriage through an elegant argument due separately to Ornstein and Smorodinsky.

Lemma 4b.2. For any coupling society R_ω on (S,μ) and (S',μ'), there is a coupling society $R_{\hat{\omega}} \subseteq R_\omega$ so that $\mathrm{card}(\{s_i \in S \mid \mathrm{card}(R_{\hat{\omega}}(s_i)) > 1\}) \leqslant \mathrm{card}(S') - 1$.

Proof. We construct a graph G_ω whose nodes are the elements of S'. For each $s_i \in S$ and each pair s'_ℓ, $s'_t \in R_\omega(s_i)$ we have an edge $p(i,\ell,t)$ connecting s'_ℓ and s'_t.

Suppose in G_ω we have a loop

$$s'_{j_1} \overset{p(i,j_1,j_2)}{\longleftrightarrow} s'_{j_2} \overset{p(i,j_2,j_3)}{\longleftrightarrow} \cdots \overset{p(i_{\ell-1},j_{\ell-1},j_\ell)}{\longleftrightarrow} s'_{j_\ell} \overset{p(i_\ell,j_\ell,j_1)}{\longleftrightarrow} s'_{j_1},$$

where all s'_{j_i} are distinct.

Let $\alpha = \min\limits_{t=1,\ldots,\ell} (\omega(s_{i_t}, s'_{j_t}), \omega(s_{i_t}, s'_{j_{(t+1)\mathrm{mod}\ \ell}})) > 0$

w.l.o.g. assume $\alpha = \omega(s_{i_1}, s'_{j_1})$ and define a new measure by

$$\hat{\omega}((s_{i_t}, s'_{j_t})) = \omega((s_{i_t}, s'_{j_t})) - \alpha,$$

$$\hat{\omega}((s_{i_t}, s'_{j_{(t+1)\mathrm{mod}\ \ell}})) = \omega((s_{i_t}, s_{j_{(t+1)\mathrm{mod}\ \ell}})) + \alpha, \quad \text{and}$$

$$\hat{\omega}(s_\ell, s'_t) = \omega(s_\ell, s'_t) \quad \text{otherwise.}$$

It is an easy check that $\hat{\omega}$ is again a coupling of (S,μ) and (S',μ'); we have simply pushed mass around the cycle. But as $\hat{\omega}((s_{i_1}, s_{j_1})) = 0$, $G_{\hat{\omega}}$ no longer has this cycle.

Continuing in this manner we can finally reduce to a coupling $\hat{\omega}$ where $G_{\hat{\omega}}$ is acyclic. The number of edges in an acyclic graph of n nodes is at most $n-1$, so the number of s_i such that $R_{\hat{\omega}}(s_i)$ contains more than one point is at most $\operatorname{card}(s') - 1$. ∎

Corollary 4b.3. For any society R on (S,μ) and (S',μ'), there is a coupling society $R_\omega \subseteq R$ with $\operatorname{card}(\{s_i \in S \mid \operatorname{card}(R(s_i)) > 1\}) \leq \operatorname{card}(s') - 1$. ∎

This completes our discussion of the Marriage lemma per se. We would like to point out a number of simple facts concerning societies. We leave their proofs to the reader.

(i) If R is a society on (S,μ) and (S',μ') and also on $(S,\bar{\mu})$ and $(S',\bar{\mu}')$, then it also is on any $(S,\hat{\mu})$ and $(S,\hat{\mu}')$ where $\hat{\mu}$ and $\hat{\mu}'$ are the same convex combinations of μ, $\bar{\mu}$ and μ', μ', respectively.

(ii) If R_1 is a society on (S,μ) and (S',μ'), and R_2 is a society on (S',μ') and (S'',μ''), then $R_1 \circ R_2 = \{(s,s'') \mid \exists\, s',\ (s,s') \in R_1,\ (s',s'') \in R_2\}$ is a society on (S,μ) and (S'',μ'').

(iii) If we have measures $\bar{\mu}$ and $\bar{\mu}'$ on $S_1 \times S_2$ and $S_1' \times S_2'$ so that R_1 is a society on (S_1,μ) and (S_2,μ') where

$$\mu(A) = \bar{\mu}(A \times S_2),\ \mu'(A) = \mu'(A \times S_2'),$$

and so that for any $(s,s') \in R_1$, $R_2(s,s')$ is a society on (S_2,μ_s) and (S_2',μ_s) where

$$\mu_s(A) = \bar{\mu}(A|s) \quad \text{and} \quad \mu_{s'}'(A) = \bar{\mu}'(A|s'),$$

then

$$\bar{R} = \{((s_1,s_2),(s_1',s_2')) \mid (s_1,s_1') \in R_1,\ (s_2,s_2') \in R_2(s_1,s_1')$$

is a society on $(\bar{\mu},\ S_1 \times S_2)$ and $(\bar{\mu}',\ S_1' \times S_2')$.

Smorodinsky and Keane used this last fact on product measures only to extend societies from one level of skeleta to the next. It is easily proven in terms of couplings.

What we need is a strengthened version of (iii) to carry us over the lack of independence from one block to the next in our version of finitary isomorphism.

For any function $f;\cdot[0,1] \to \mathbb{R}^+$, we can define an f-fat society as one for which $\mu'(R(A)) \geq \mu(A) + f(\mu(A))$. Any society is 0-fat. The f which we will find most useful is $f(x) = \varepsilon(x-x^2)$. If R is f_ε-fat it is not true that R' need be, but it is true that R' is $\frac{f_\varepsilon}{1+\varepsilon}$-fat. We now prove a lemma which extends (iii) above and is the hinge around which much of our argument will turn.

Let $(S_1 \times S_2, \bar{\mu})$ and $(S_1' \times S_2', \bar{\mu}')$ be two finite probability spaces. Let $\mu_1(A) = \bar{\mu}(A \times S_2)$, $\mu_1'(A) = \bar{\mu}'(A \times S_2')$, $\mu_s(A) = \bar{\mu}(S_1 \times A|s)$ for $s \in S_1$, $\mu_{s'}'(A) = \bar{\mu}'(S_1' \times A|s')$ for $s' \in S_1'$. Let R_1 be a society on (S_1, μ_1) and (S_1', μ_1'), and for every $(s,s') \in R_1$ let $R_2(s,s')$ be a society on (S_2, μ_s) and $(S_2', \mu_{s'}')$.

Theorem 4b.4. If for some $\varepsilon \leq 1$, R_1 is an f_ε-fat society and for each $(s,s') \in R_1$, $R_2(s,s')$ is an f_ε-fat society, then

$$\bar{R} = ((s_1, s_2), (s_1', s_2')) | (s_1, s_1') \in R_1,\ (s_2, s_2') \in R_2(s_1, s_1')\}$$

is an f_ε-fat society on $(S_1 \times S_2, \bar{\mu})$ and $(S_1' \times S_2', \bar{\mu}')$.

Proof. Let A be a fixed subset of $S_1 \times S_2$. We want to show $\bar{\mu}'(\bar{R}(A)) \geq \bar{\mu}(A) + f_\varepsilon(\bar{\mu}(A))$. What we will do is replace $\bar{\mu}$ by a new measure $\hat{\mu}$ so that $\hat{\mu}(A) \geq \bar{\mu}(A) + f_\varepsilon(\bar{\mu}(A))$ and $\bar{R}$ is still a society on $(S_1 \times S_2, \hat{\mu})$ and $(S_1' \times S_2', \bar{\mu}')$.

Let $g_\varepsilon(x) = x + f_\varepsilon(x)$ and now for $B \subset S_2$, $s \in S_1$, set

$$\hat{\mu}_s(B) = g_\varepsilon(\bar{\mu}_s(B \cap A)) + \left(\frac{1 - g_\varepsilon(\bar{\mu}_s(B \cap A))}{1 - \bar{\mu}_s(B \cap A)}\right) \bar{\mu}_s(B \cap A^c) .$$

Thus, for any $B \subset A$, $s \in S_1$, $\hat{\mu}_s(B) = g_\varepsilon(\bar{\mu}_s(B))$ and for any other B,

$$\hat{\mu}_s(B) \leq g_\varepsilon(\bar{\mu}_s(B)) .$$

It follows that for any $(s,s') \in R_1$, $R_2(s,s')$ is a society on $(S_2,\hat{\mu}_s)$ and $(S_2',\overline{\mu}_s')$.

Now to define $\hat{\mu}_1$ on S_1. We write $\hat{\mu}_1(s) = f(s)\mu_1(s)$, where we need

(i) $\int_{S_1} fd\overline{\mu}_1 = 1$,

(ii) $\int_{S_1} \hat{\mu}_s(A)fd\mu_1(s) \geq g_\varepsilon(\overline{\mu}(A))$, and

(iii) for any $B \subset S_1$ $\int_B fd\overline{\mu} \leq g_\varepsilon(A)$.

From (i) we get $\hat{\mu}_1$ a probability measure, from (ii) $\hat{\mu}(A) \geq g_\varepsilon(\overline{\mu}(A))$, and from (iii) R_1 is still a society on $(S_1,\hat{\mu}_1)$ and $(S_1',\overline{\mu}_1')$.

Write $S_1 = \{S_1,S_2,\ldots,S_\ell\}$ where if $\overline{\mu}_{S_i}(A) = x_i$, $x_1 \geq x_2 \geq \ldots \geq x_\ell$. Let $f(S_1) = \dfrac{g_\varepsilon(\overline{\mu}_1(S_1))}{\overline{\mu}(S_1)}$ and for $i > 1$,

$$f(S_i) = \frac{g_\varepsilon(\overline{\mu}_1(\{S_1,\ldots,S_i\})) - g_\varepsilon(\overline{\mu}_1(\{S_1,\ldots,S_{i-1}\}))}{\overline{\mu}_1(S_i)} .$$

It is clear that $f > 0$ and (i) $\int fd\overline{\mu} = 1$.

To see (iii), let $B \subset S_1$ and now

$$\begin{aligned}\hat{\mu}(B) = \int_B fd\overline{\mu} &= \sum_{i=1}^{\ell-1} g_\varepsilon(\overline{\mu}(\{S_1,\ldots,S_{i+1}\}))-g_\varepsilon(\overline{\mu}(\{S_1,\ldots,S_i\})))1_B(S_{i+1} \\ &+ g_\varepsilon(\overline{\mu}(S_1))1_B(S_1) \\ &\leq \sum_{i=1}^{\ell-1} g_\varepsilon(\overline{\mu}(B\cap\{S_1,\ldots,S_{i+1}\})-g_\varepsilon(\overline{\mu}(B\cap\{S_1,\ldots,S_i\})) \\ &+ g_\varepsilon(\overline{\mu}(B\cap\{S_1\}))\end{aligned}$$

as g_ε is monotone increasing and strictly concave downward. This last collapses to $g_\varepsilon(B)$.

To show (ii), let $\overline{\mu}_1(S_i) = y_i$ and now

$$\hat{\mu}(A) = \int f(s)\hat{\mu}_s(A)d\bar{\mu}(s)$$

$$= \sum_{i=1}^{\ell-1} (g_\varepsilon(y_1 + \ldots + y_{i+1}) - g_\varepsilon(y_1 + \ldots + y_i))g_\varepsilon(x_{i+1})$$

$$+ g_\varepsilon(y_1)y_\varepsilon(x_1) \quad \text{where}$$

$$\bar{\mu}(A) = \sum_{i=1}^{\ell} x_i y_i .$$

Block the x_i into subsets of equal value and let $x_{\ell_1}, x_{\ell_2}, \ldots, x_{\ell_k}$ be the first element of each block. Let $\bar{x}_i = \bar{x}_{\ell_i}$ and $\bar{y}_i = \sum_{j=\ell_i}^{\ell_{i+1}-1} y_i$, $\bar{y}_0 = 0$ and now

$$\hat{\mu}(A) = \sum_{i=1}^{k} (g_\varepsilon(\bar{y}_0 + \ldots + \bar{y}_i) - g_\varepsilon(\bar{y}_0 + \ldots + \bar{y}_{i-1})g_\varepsilon(\bar{x}_i)$$

where $\bar{x}_1 > \bar{x}_2 > \ldots > \bar{x}_k$, $\bar{y}_i > 0$ for $i \neq 0$, $\sum_{i=1}^{k} \bar{y}_i \leq 1$, $\sum_{i=1}^{k} \bar{x}_i \bar{y}_i = \bar{\mu}(A)$.

For any vectors $\vec{x},\vec{y}$ with $x_1 > x_2 > \ldots > x_k > 0$, $y_i > 0$ with $\sum_{i=1}^{k} y_i < 1$, $\sum_{i=1}^{k} x_i y_i = \bar{\mu}(A)$, set $y_0 = 0$ and let

$$U_k(\bar{x},\bar{y}) = \sum_{i=1}^{k} (g_\varepsilon(\sum_{j=0}^{i} y_j) - g_\varepsilon(\sum_{j=0}^{i-1} y_j))\, g_\varepsilon(x_j)$$

The theorem will follow from:

<u>Lemma 4b.5</u>. For any k for any $\vec{x}$ and $\vec{y}$ as above,

$$U_k(\vec{x},\vec{y}) \geq g_\varepsilon(\bar{\mu}(A)) .$$

<u>Proof</u>. We argue this by modifying $\vec{x}$ and $\vec{y}$ only decreasing the value of U_k to a point where it is easily calculated. Fix $\vec{x},\vec{y}$ and for some fixed $s \in 1,\ldots,k-1$ and

$$t \in \left[\frac{-y_s(x_s - x_{s-1})}{x_{s-1}}, \; y_s + y_{s+1}\right], \quad (x_0 = 1, \; y_{k+1} = 0),$$

let

$$\vec{x}(t) = (x_1, x_2, \ldots, x_{s-1}, x_{s-1} + \frac{y_s(x_s - x_{s-1})}{t}, \ldots, x_k),$$

$$\vec{y}(t) = (y_1, y_2, \ldots, y_{s-1}, t, y_s + y_{s+1} - t, \ldots, y_k) .$$

The new vectors $\vec{x}(t)$, $\vec{y}(t)$ still satisfy the needed conditions except at the endpoints and $(\vec{x}(y_s), \vec{y}(y_s)) = (\overline{x}, \overline{y})$. It is a computation that

$$\left(\frac{d(U(\vec{x}(t), \vec{y}(t))}{dt}\right)_{t=y_s}$$

$$= \left(g_\varepsilon\left(\sum_{j=1}^{s} y_j\right) - g_\varepsilon\left(\sum_{j=1}^{s-1} y_j\right)\right) g_\varepsilon'(x_s) \left(\frac{x_s - x_{s-1}}{y_s}\right)$$

which is < 0 unless $y_s = 0$ or $x_s = 0$ or $x_{s-1} = x_s$, all of which will be false if $s \neq k-1$. Thus, $U_k(\vec{x}, \vec{y}) \geqslant$

$$U(\vec{x}(y_s + y_{s+1}), \vec{y}(y_s + y_{s+1}))$$

$$= U_k(\vec{x}', (y_1, \ldots, y_{s-1}, y_s + y_{s+1}, 0, y_{s+s}, \ldots, y_k)) .$$

These vectors are not allowed as $y'_{s+1} = 0$, but we can replace these with vectors one dimension less. Working inductively from $s = k-2$ down to $s = 1$ we get

$$\begin{aligned} &U_k(\vec{x}, \vec{y}) \geqslant \\ &= U_2((1, x_2'), (y_1', 1 - y_1')) = g_\varepsilon(y_1') \\ &\quad + (1 - g_\varepsilon(y_1')) g_\varepsilon(x_2') = g_\varepsilon(y_1') + g_\varepsilon(x_2') \\ &\quad - g_\varepsilon(y_1') g_\varepsilon(x_2') = g_\varepsilon(y_1' + x_2' - y_1' x_2') \\ &\quad + (\varepsilon - \varepsilon^2)(x_2' - x_2'^2)(y_1' - y_1'^2) \\ &\geqslant g_\varepsilon(\overline{\mu}(A)) \end{aligned}$$

and we are done.

We now know that f_ε-fatness goes over to products. We need to know how to build f_ε-fatness. We will do this by describing a procedure for taking an "average" of two societies, one of which is ε-fat, and getting back a society which is still $\varepsilon/_4$-fat. Let R_1 be an ε-fat society on (S_1,μ) and (S_1',μ') and let $R_2 \subseteq R_1$ be just a society on these two. Let $(S_1 \times S_2,\bar{\mu})$, be a probability space whose marginal on the first coordinate is μ. Let $B_s \subset (s \times S_2)$, $\bar{\mu}_s(B_s) \geq 1/2$. Define R on $(S_1 \times S_2,\bar{\mu})$ and (S_1',μ') to be

$$\{((S_1,S_2),S_1') \mid (S_1,S_1') \in R_1 \quad \text{if} \quad S_2 \in B_{S_1},$$
$$(S_1,S_1') \in R_2 \qquad \text{if} \quad S_2 \in B_{S_1}^c\} .$$

Lemma 4b.6. If R is defined as above, then R is an $f_{\varepsilon/_4}$-fat society.

Proof. Let $g(\alpha) = \inf_{\substack{A \subset S_1 \times S_2 \\ \bar{\mu}(A) = \alpha}} (\bar{\mu}(R(A))$. We need to see that $g(\alpha) \geq g_{\varepsilon/_4}(\alpha)$. Let $\bar{\mu}(A) = \alpha$. Write $S_1 = A_1 \cup A_2 \cup A_3$ where

$$s \in A_1 \quad \text{if} \quad (A \cap \{s \times S_2\}) \cap B_s \neq \phi$$

$$s \in A_2 \quad \text{if} \quad (A \cap \{s \times S_2\}) \cap B_s = \phi, \quad \text{but}$$
$$A \cap \{s \times S_2\} \neq \phi, \quad \text{and}$$

$$s \in A_3 \quad \text{if} \quad A \cap \{s \times S_3\} = \phi .$$

Now

$$\mu'(R(A)) = \mu'(R_1(A_1) \cup R_2(A_2))$$

$$\geq \begin{cases} \mu(A_1 \cup A_2) & \text{if} \quad f_\varepsilon(\mu(A_1)) \leq \mu(A_2) \\ \mu(A_1 \cup A_2) + (f_\varepsilon(\mu(A_1)) - \mu(A_2)) & \text{if} \quad f_\varepsilon(\mu(A_1)) > \mu(A_2) . \end{cases}$$

As $\bar{\mu}(A) \leqslant \mu(A_1) + \frac{1}{2}\mu(A_2)$, if $f_\varepsilon(\mu(A_1)) \leqslant \mu(A_2)$, then

$$\mu'(R(A)) \geqslant \mu(A_1) + \mu(A_2) \geqslant \bar{\mu}(A) + \frac{1}{2}\mu(A_2)$$

$$\geqslant \bar{\mu}(A) + \frac{1}{4} f_\varepsilon(\mu(A_1)) + \frac{1}{4}\mu(A_2) \geqslant g_{\varepsilon/4}(\bar{\mu}(A)),$$

and if $f_\varepsilon(\mu(A_1)) \geqslant \mu(A_2)$, then

$$\mu'(R(A)) \geqslant g_\varepsilon(\mu(A_1)) \geqslant \mu(A_1) + \frac{1}{2}\mu(A_2)$$

$$+ \frac{1}{4} f_\varepsilon(\mu(A_1)) + \frac{1}{4}\mu(A_2) \geqslant g_{\varepsilon/4}(\bar{\mu}(A)).$$

In our application of this lemma, S_1 and S_1' will consist of the fillers between markers, S_2 a short cuing name within the marker, and R_1 and R_2 the societies on the fillers before and after an application of the marriage lemma. In these terms we interpret the lemma as saying that by using a cue at the beginning of a block to decide whether to use the new refined society, or simply keep the old approximation, the society becomes once more fat.

(c) Markers. We now begin to treat (T,P) explicitly, constructing the marker sets $M(i)$ which break up the orbit of a point into a skeletal structure. Precisely the same procedure is carried out in (T',P') but there need be no connection with that on (T,P).

Choose $\varepsilon_0 < 10^{-4}$ with $\mu(S(\varepsilon_0)) < 10^{-4}$. Let $\bar{M}_0 = \{S(\varepsilon_0), S(\varepsilon_0)^c\}$ and in $(T,\bar{M}_0)$ we construct some basic blocks. Choose n so large that

$$\mu((S(\varepsilon_0), N(\varepsilon_0)+n)^3) = \mu(S(\varepsilon_0))^3\, 2^{\pm\varepsilon_0},$$

$$\mu((S(\varepsilon_0), N(\varepsilon_0)+n)(S(\varepsilon_0)^c, N(\varepsilon_0)+n)(S(\varepsilon_0), N(\varepsilon_0)+n))$$

$$= \mu(S(\varepsilon_0))^2(1-\mu(S(\varepsilon_0)))2^{\pm\varepsilon_0}$$

using USM. Make sure $N(\varepsilon_0) > 10n$. Set

$m_0 = (S(\varepsilon_0),N(\varepsilon_0)+n)^3$, and

$m_1 = (S(\varepsilon_0),N(\varepsilon_0)+n)(S(\varepsilon_0)^c,N(\varepsilon_0)+n)(S(\varepsilon_0),N(\varepsilon_0)+n)$.

From USM we have for any sequence $i(1),\ldots,i(k) \in \{0,1\}$ and numbers $n(s) \geq 0$,

$$\mu\Big(\prod_{s=1}^{t-1}(m_{i(s)},n(s))\Big)\mu(m_{i(t)})\,\mu\Big(\prod_{s=t+1}^{k}(m_{i(s)},n(s))\Big)$$

$$=\mu\Big(\prod_{=1}^{k}(m_{i(s)},n(s))\Big)2^{\pm 2\varepsilon_0}.$$

Now choose a cylinder set $\overline{S} \in \overline{M}_{0,1}^{n_2-1}$ so that $\mu(\overline{S}) < \dfrac{1}{3(N(\varepsilon_0)+n)(3(N(\varepsilon_0)+n)+1)} = \dfrac{1}{n_3}$. Thus $\mu((\overline{S},1)(\overline{S}^c,1)^{n_3}) > 0$. Setting $S = ((\overline{S},1)(\overline{S}^c,1)^{n_3},N(\varepsilon_0))$ we have, for $0 < j \leq n_3$, $T^j(S) \cap S = \phi$, and for any n, m, ε', $m > \ell(S)+n$, that $t_1(T^n(S),\varepsilon',m) = T^n(S)$ as S is in $\overline{M}_{0,1}^{\ell(S)-N(\varepsilon_0)-1}$. Choose n_4 large enough that $\mu((S(\varepsilon_0),n_4)(S,n_4)(S(\varepsilon_0)) > 0$.

A "premarker" now will be a set of the form

$$M_k = (m_0)^{2\ell(m_0)}(\Omega,n_4)(S,n_4)(m_0)^{2\ell(m_0)}(m_1,1)^k$$

where $k > 10^4(\ell(S)+2n_4+4\ell(m_0)^2)$. From our choice of parameters and CBI, $\mu(M_k) > 0$.

Let $n_5 = \ell(m_0)(\ell(m_0)+1)$ and we see that if $0 < j \leq \ell(M_k)-n_5$, then for any ℓ, $M_k \cap T^j(M_\ell) = 0$ (small shifts are handled by S, larger by the m_0 and m_1 blocks), and if $j \geq \ell(M_k)$, then by USM $\mu(M_k \cap T^j(M_\ell)) = \mu(M_k)\mu(M_\ell)2^{\pm\varepsilon_0}$. For $\ell(M_k)-n_5 < j < \ell(M_k)$, letting

$$C_t = \frac{\mu(T^t((m_0)^{2\ell(m_0)}) \cap (m_1,1)^{2\ell(m_0)})}{\mu((m_0)^{2\ell(m_0)})\mu((m_1,1)^{2\ell(m_0)})}$$

we check that, again by USM,

$\mu(M_k \cap T^j(M_\ell)) = \mu(M_k)\mu(M\ )C_{j-\ell(M_k)+n_5}\ 2^{\pm 4\varepsilon_0}$. It is this explicit control on the sizes of such intersections independent of k which was the purpose of constructing premarkers (and next markers), rather than simply using an $S(\varepsilon)$. Thus there are constants $\sigma_1, \sigma_2 > 0$ independent of j,k and ℓ with $\mu(M_k)\mu(M_\ell)\sigma_1 \leqslant \mu(M_k \cap T^j(M_\ell)) \leqslant \mu(M_k)\mu(M_\ell)\sigma_2$ whenever $\mu(M_k \cap T^j(M_\ell)) \neq 0$.

To construct markers we will select inductively $\varepsilon_i \overset{i}{\searrow} 0$, choose n_6 and a set $B \subset \overline{M}_{0,1}^{n_6-1}$ with $2^{-3\varepsilon_0} \cdot \frac{2}{3} > \mu(B|S(\varepsilon_0), T^{n_6}(S(\varepsilon_0))) > \frac{1}{2} \cdot 2^{3\varepsilon_0}$, and now $S(\varepsilon_0) \cap T^n(S(\varepsilon_i)) \cap T^{n+N(\varepsilon_i)+n_8(i)} > 0$ (n chosen from USM earlier with error $2^{\pm\varepsilon_0}$) and now choose $k(i) > 10(N(\varepsilon_i) + 2n + \overline{N}(\varepsilon_i))$, $k(i+1) > 2k(i)$ and set

$$M_0(i) = (M_{k(i)}, n-N(\varepsilon_0))(S(\varepsilon_0) \cap B, n_6)(S(\varepsilon_0), n)(S(\varepsilon_i), N(\varepsilon_i)+n)M(k(i)),$$

$$M_1(i) = (M_{k(i)}, n-N(\varepsilon_0))(S(\varepsilon_0) \cap B^c, n_6)(S(\varepsilon_0), n)(S(\varepsilon_i), N(\varepsilon_i)+n)M(k(i)),$$

and the marker $M(i) = M_0(i) \cup M_1(i)$. The splitting of $M(i)$ into two pieces according to the block preceding $S(\varepsilon_i)$ is to provide the cue for fattening the societies ala Lemma 3b.6.

Notice that the vast majority of the length of the marker is in the $M_{k(i)}$.

We still have at our disposal the values ε_i and the magnitudes of $k(i)$. Notice, though, that $\mu(M(i)) > 0$, $2^{-\varepsilon_0}\frac{2}{3} > \mu(M_0(i)|M(i)) > \frac{1}{2}2^{\varepsilon_0}$, and independent of our choices of $k(i)$ and ε_i, if $0 < j \leqslant \ell(M(i)) - n_5$, then $\mu(M(i) \cap T^j(M(s)) = 0$, if $i \neq s$, $\mu(M(i) \cap M(s)) = 0$, and if $j > \ell(M(i))$, then $\mu(M(i) \cap T^j(M(s))) = \mu(M_i))\mu(M(s))2^{\pm\varepsilon_0}$, and for $\ell(M(i)) - n_5 < j < \ell(M(s))$, $\mu(M(i) \cap T^j(M(s))) = \mu(M(i))\mu(M(s))C_{j-\ell(M(i))+n_5}\ 2^{\pm 4\varepsilon_0}$ as for premarkers all by USM. The same bounds hold on $M_0(i)$ and $M_1(i)$ independent of how small we choose ε_i or large $k(i)$. It is also clear from USM that once ε_i is fixed $\mu(M(i))$ decreases exponentially to zero as a function of $k(i)$. In fact, we can write $\mu(M(i)) \leq a(i)a^{k(i)}$ where $a(i) < 1$ depends only on ε_i, and $a = \mu(m_1)2^{\varepsilon_0}$. As $\ell(M(i)) = \ell(i) + (\ell(m_1)+1)2k(i)$, we get $\mu(M(i)) = \bar{a}(i)\bar{a}^{\ell(M(i))}$ where $\bar{a}(i) < 1$ and $\bar{a} < 1$ does not depend on ε_i. This now allows the following lemma which is critical in constructing the finitary Bernoulli structure on (T_1,P_1).

Lemma 4c.1. If the values $k(i)$ increase rapidly enough, then for any i_1, i_2, $I \subset \mathbb{Z}^+$ if $\ell(M(i_1))-n_5 < j < \ell(M(i_1))$ and $\mu(M(i_1) \cap T^j(M(i_2))) \neq 0$ then

$$\frac{\sigma_1}{2}\,\mu(M(i_2)) < \mu(T^j(M(i_2))\,|\,M(i_1) \cap \bigcap_{j'=1}^{j-1} \bigcap_{i\in I} T^j(M(i)^c)) < 2\sigma_2\mu(M(i_2))$$

and if $j \geq \ell(M(i_1))$ then

$$\frac{1}{2}\,\mu(M(i_2)) < \mu(T^j(M(i_2))\,|\,M(i_1) \cap \bigcap_{j'=1}^{j-1} \bigcap_{i\in I} T^j(M(i)^c)) < 2\mu(M(i_2)).$$

(Remember $\sigma_1,\sigma_2 > 0$.)

Proof. We work by induction on j. Let $E(j) = \bigcap_{j'=1}^{j-1} \bigcap_{i\in I} T^j(M(i)^c) \cap M(i_1)$. If $j < \ell(M(i_1))$, then

$$\frac{\mu(M(i_2))\mu(M(i_1))\sigma_1}{\mu(M(i_1)) - \mu(\bigcup_{j'=1}^{j-1} \bigcup_{i\in I} M(i_1) \cap T^{j'}(M(i))}$$

$$= \frac{\mu(M(i_2))\mu(M(i_1))\sigma_1}{\mu(M(i_1))} < \mu(T^j(M(i_2))\,|\,E(j))$$

$$< \frac{\mu(M(i_2))\mu(M(i_1))\sigma_2}{\mu(M(i_1))(1 - n_5\sigma_2 \sum_{i=1}^{\infty} \mu(M(i)))}.$$

Choosing $\sum_{i=1}^{\infty} \mu(M(i)) < \frac{10^4}{n_5\sigma_2}$ we get the result for $j<\ell(M(i_1))$. This also applies for $j = \ell(M(i))$ with σ_1 and σ_2 replaced by 1. Assume the result up to $j-1$ for all i_1, i_2, I. First notice

$$\mu(T^j(M(i_2))|M(i_1) \bigcap_{i\in I} (\bigcap_{j'=1}^{j-\ell(M(i))-\overline{N}(\varepsilon_0)} T^{j'}(M(i)^c))$$

$$= \mu(M(i_2))2^{\pm\varepsilon_0}$$

by USM. Thus,

$$\mu(T^j(M(i_2)) \cap E(j))$$

$$= \mu(M(i_2))\mu(M(i_1) \bigcap_{i\in I} (\bigcap_{j=1}^{j-\ell(M(i))-\overline{N}(\varepsilon_0)} T^{j'}(M(i)^c)))2^{\pm\varepsilon_0}$$

$$- \sum_{i\in I} \sum_{j'=j-\ell(M(i))-\overline{N}(\varepsilon_0)+1}^{j-1} \mu(T^j)(M(i_2))\cap E(j') \cap T^{j'}(M(i)).$$

By ABI this is

$$\mu(M(i_2)(\mu(M(i_1) \bigcap_{i\in I} (\bigcap_{j'=1}^{j-\ell(M(i))-\overline{N}(\varepsilon_0)} T^{j'}(M(i)^c)))2^{\pm\varepsilon_0}$$

$$\pm \sigma_2 \sum_{i\in I} \sum_{j'=j-\ell(M(i))-N(\varepsilon_0)-1}^{j-1} \mu(T^{j'}(M(i)) \cap E(j')))$$

$$= \mu(M(i_2))\mu(E(j))(1-(\sigma_2-1) \frac{\sum_{i\in I} \sum_{j'=j-\ell(M(i))-N(\varepsilon_0)-1}^{j-1} \mu(T^{j'}(M(i))\cap E(j'))}{\mu(E(j))}$$

By our induction hypothesis the last term of this is less than

$$2(\sigma_2-1)\sigma_2 \sum_{i\in I} \mu(M(i)) \sum_{j'=j-\ell(M(i))-\overline{N}(\varepsilon_0)-1}^{j-1} (\frac{\mu(E(j'))}{\mu(E(j))}).$$

Again by our induction hypothesis

$$\frac{\mu(E(j'))}{\mu(E(j))} < (1 - 2\sigma_2 \sum_{i\in I} \mu(M(i)))^{j'-j} .$$

Make $k(i)$ so large that $1 - 2\sigma_2 \sum_{i=1}^{\infty} \mu(M(i)) > \sqrt{\bar{a}}$ and we conclude

$$\mu(M(i)) \sum_{j'=j-\ell(M(i))-\bar{N}(\varepsilon_0)-1}^{j-1} \left(\frac{\mu(E(j'))}{\mu(E(j))}\right)$$

$$\leq \bar{a}(i)\bar{a}^{\ell(M(i))}(\ell(M(i))+\bar{N}(\varepsilon_0) + 1)(\sqrt{\bar{a}})^{-\ell(M(i))-\bar{N}(\varepsilon_0)-1}$$

$$\leq C_i\ell(M(i))\bar{a}^{\ell(M(i))/2}$$

where C_i does not depend on $k(i)$. This clearly goes to zero in $k(i)$. Choose $k(i)$ so large that now

$$\sum_{i=1}^{\infty} C_i\ell(M(i))\bar{a}^{\ell(M(i)/2} < \frac{1}{4(\sigma_2-1)\sigma_2}$$

and we are done. ∎

Notice that this used critically the explicit control we have on $\mu(M(i))$ as a function of $\ell(M(i))$, control we need not have on $\mu(S(\varepsilon))$ as a function of $N(\varepsilon)$, and is the reason for explicitly constructing markers.

What Lemma 3c.1 tells us is that we have exponentially decaying return times between markers. It is interesting to note at this point that Smorodinsky (13) has constructed a countable state Markov chain which is not finitarily Bernoulli simply by forcing polynomial return times.

We now can conclude by this and CBI, letting $\bar{M}(i) = \bigvee_{k=1}^{i} \{M(k),M(k)^c\}$, that for any $c \in \bar{M}(i)_0^{n+1}$, $c \subset c' \in \bar{M}(i)_0^n$, $\mu(c) > 0$, then $\mu(c|c') \geq \bar{\sigma} > 0$ where $\bar{\sigma}$ is independent of c, c', i and n.

We still can choose ε_i smaller and $k(i)$ larger and maintain our earlier work.

Let $\overline{M} = \bigvee_{i=1}^{\infty} \overline{M}(i)$, and $(T,\overline{M})$ be the "marker process". For every $\omega \in \Omega$ let $m_i(\omega) \geq 0$ be the least integer with $T^{-m_i(\omega)}(\omega) \in M(i')$ for some $i' \geq i$, and $m_i'(\omega) > 0$ the least integer with $T^{-m_i'(\omega)+1}(\omega) \in M(i')$ for some $i' \geq i$. Now $\mathcal{S}_i(\omega)$, the i-skeleton of ω is the set of points whose T, $\overline{M}$ name from 0 to $m_i(\omega) + m_i'(\omega)$ is the $T,\overline{M}$ name of ω from $-m_i(\omega)$ to $m_i'(\omega)$. We will also identify $\mathcal{S}_i(\omega)$ with this $\overline{M}$-name. The length $\ell(\mathcal{S}_i(\omega))$ is $m_i(\omega) + m_i'(\omega) - 1$. As this is in keeping with Smorodinsky and Keane, we will not discuss subskeleta and the nesting of skeleta except to introduce the notation that any i-skeleton $\mathcal{S}_i$ decomposes, from left to right, into a unique sequence of $(i-1)$-skeleta, $\mathcal{S}_{i-1}(1,\mathcal{S}_i),\ldots,\mathcal{S}_{i-1}(k(\mathcal{S}_i),\mathcal{S}_i)$. The length of $\mathcal{S}_i$ is the sum of the lengths of these subskeleta. Let $S_i = \cup\mathcal{S}_i$, $S_{i,\ell} = \bigcup_{\ell(\mathcal{S}_i)=\ell} (\mathcal{S}_i)$ and $m(k,\mathcal{S}_i) = \sum_{k'=1}^{k} \ell(\mathcal{S}_{i-1}(k',\mathcal{S}_i))$.

Lemma 4c.2. If the values $k(i)$ increase rapidly enough, there are constants $\alpha_1,\alpha_2 < 1$, independent of i and ℓ, so that

$$\mu(S_{i,\ell}) < \alpha_1\alpha_2^{\ell} .$$

Proof. For $i',i'' \geq i$, $\ell > \ell(M(i'))$,

$$\mu(S_{i,\ell} \cap M(i') \cap T^{\ell+1}(M(i'')))$$

$$< \mu(M(i')\mu(M(i''))\left(\frac{\sigma_1}{2}\right)^{-\ell(M(i'))}$$

by Lemma 4c.1, which in turn is

$$< \bar{a}(i)\bar{a}^{(M(i))}\left(\frac{\sigma_1}{2}\right)^{\ell-\ell(M(i'))} < a_0^{\ell} ,$$

where $a_0 < 1$ does not depend on i' or i''. For any ℓ, as surely $\ell(M(i')) > i'$, there are at most $\ell + n_5$ markers $M(i')$ which initiate a block in $S_{i,\ell}$. Hence,

$\mu(S_{i,\ell}) < \ell a_0^\ell < \alpha_1\alpha_2^\ell$ where $\alpha_2 < 1$, α_i are independent of i and ℓ.

(d) Entropy and Fillers. We now inductively describe how the "filler partitions" $F(\mathcal{S}_i)$ of each i-skeleton are built. These are the pieces on which our coding will take place. To begin, $F(\mathcal{S}_1) = \{\mathcal{S}_1, \phi\}$ for 1-skeleta.

Now to define $F(\mathcal{S}_i)$. First let $M(i')$ initiate $\mathcal{S}_i$. Notice that $M(i')$ can be written as $(M^-(i'))(S(\varepsilon_{i''}),$ $N(\varepsilon_{i'}))(M^+(i'))$ where, as $(M^-(i'))$ and $(M^+(i'))$ are cylinders in $(T,\overline{M}_0)$, they are unaffected by any t_1,ε' thickening over a block of indices containing them in its length. Define $S_{i,\ell}^{i',m,m'} \subset S_{i,\ell}$ to be that subset beginning with an i'-marker with two consecutive (i-1)-markers at indices $m+1 < m'+1 < \ell$.

For $\overline{\varepsilon}_{i-1} \geqslant \varepsilon_{i-1}$, we define an "$\overline{\varepsilon}_{i-1}$, i-skeleta stopping time" ($\overline{\varepsilon}_{i-1}$,i-s.t.) to be a map $\overline{k}; S_i \to \mathbb{N}$ satisfying

(i) $1 \leqslant \overline{k}(\omega) < k(\mathcal{S}_i)$ if $\omega \in \mathcal{S}_i$, and

(ii) if $\omega,\omega' \in S_{i,\ell}^{i',m(\overline{k}(\omega),\mathcal{S}_i),m(\overline{k}(\omega)+1,\mathcal{S}_i)} = Z(\overline{k},\omega)$,

and

$$\omega' \in Z(\overline{k},\omega) \cap (M^-(i')) \circ (S(\varepsilon_{i'}),N(\mathcal{S}_{i'}))$$

$$\circ\ (t_1(T_{(\omega)}^{-\ell(M^-(i'))-N(\varepsilon_{i'})},\overline{\varepsilon}_{i-1},m(\overline{k}(\omega)\mathcal{S}_i)-\ell(M^-(i')-N(\varepsilon_i'))$$

$$= \overline{F}(\overline{\varepsilon}_{i'-1},\overline{k},\omega),$$

then $\overline{k}(\omega') = \overline{k}(\omega)$. Notice that $\overline{F}(\overline{\varepsilon}_{i-1},\overline{k},\omega)$ is not defined to lie in $\mathcal{S}_i$, but rather in $Z(\overline{k},\omega)$, and that if $\overline{k}$ is an $\overline{\varepsilon}_{i-1}$,i-s.t. it is also an $\overline{\varepsilon}'_{i-1}$,i-s.t. for $\varepsilon_{i-1} \leqslant \overline{\varepsilon}'_{i-1} \leqslant \overline{\varepsilon}_{i-1}$ as $\overline{F}(\overline{\varepsilon}'_{i-1},\overline{k},\omega) \subseteq \overline{F}(\overline{\varepsilon}_{i-1},\overline{k},\omega)$. Furthermore, if $0 < \overline{k} < k(\mathcal{S}_i)$, split the $T,\overline{M},\ell(\mathcal{S}_i)$-name of $\mathcal{S}_i$ into two $T,\overline{M}$-names, one, $\mathcal{S}_i^-(\overline{k})$, the initial $m(\overline{k},\mathcal{S}_i)+1$ digits, and the other, $\mathcal{S}_i^+(\overline{k})$ the final $\ell(\mathcal{S}_i)-m(\overline{k},\mathcal{S}_i)-1$ digits, and if $\overline{k}$ is an $\overline{\varepsilon}_{i-1}$,i-s.t. then $\overline{F}(\overline{\varepsilon}_{i-1},\overline{k},\omega) \subset \mathcal{S}_i^-(\overline{k}(\omega))$. Thus, if

$\omega' \in \overline{F}(\overline{\varepsilon}_{i-1},\overline{k},\omega)$, we must have $m(\overline{k}(\omega),\mathcal{S}_i(\omega)) = m(\overline{k}(\omega'),\mathcal{S}_i(\omega'))$ and $\overline{F}(\overline{\varepsilon}_{i-1},\overline{k},\omega) = \overline{F}(\overline{\varepsilon}_{i-1},\overline{k},\omega')$ and so the sets $\overline{F}(\overline{\varepsilon}_{i-1},\overline{k}) = \{\overline{F}(\overline{\varepsilon}_{i-1},\overline{k},\omega)\}_{\omega\in S_i}$ partition S_i. One easily computes with CBI that $\mu(\overline{F}(\overline{\varepsilon}_{i-1},\overline{k},\omega)|\mathcal{S}_i) = \mu(\overline{F}(\overline{\varepsilon}_{i-1},\overline{k},\omega)|\mathcal{S}_i^-(k(\omega))2^{\pm 16\varepsilon_{i-1}}$.

We will build $F(\mathcal{S}_i)$ by constructing, for ε_{i-1} small enough, $\mu(M(i'))$, $k(i')$, $i' \geq i$ small and large enough, an appropriately chosen $\overline{\varepsilon}_{i-1}, i$ -s.t. With it we make $F(\mathcal{S}_i)$ as follows. Let $\hat{Z} = S_{i,\ell}^{i',m,m'}$. Partition $\hat{Z} = T^{-m-\ell(M^-(\varepsilon_{i-1}))-N(\varepsilon_{i-1})-1}(Z)$ in two ways; the first,

$$F_1(Z) = \bigvee_{k'=\overline{k}}^{k(\mathcal{S}_i)} T^{-m(k',\mathcal{S}_i)-m-\ell(M^-(\varepsilon_{i-1}))-N(\varepsilon_{i-1})-1}(F(\mathcal{S}_{i-1}(k',\mathcal{S}_i))/\hat{Z}$$

inherited from the earlier fillers across the remaining block after $\overline{k}(\omega)$, and the second

$$F_2(Z) = C(\overline{\varepsilon}_{i-1},\ell-m-\ell(M^-(i-1))-N(\varepsilon_{i-1}))/\hat{Z}$$

consisting of $t_1(\cdot,\overline{\varepsilon}_{i-1},\cdot)$ thickened orbits over this remaining length.

Construct a third partition, $F_3(Z)$ by lumping sets in $F_2(Z)$ so that (i) $\mathrm{card}(F_3(Z)) \leq \mathrm{card}(F_1(Z))$, and (ii) to each element $f_3 \in F_3(Z)$ there is assigned an element $f' \in F_1(Z)$ so that every set $f'' \in F_2(Z)$, $f'' \subset F'$ intersects f'.

This is easily done by building $F_3(Z)$ up sequentially. Now any set $f \in F_3(Z)$ satisfies $t_1(f,\overline{\varepsilon}_{i-1},\ell-m-\ell(M^-(i-1))-N(\varepsilon_{i-1})) \cap Z = f$. Now $F(\mathcal{S}_i)$ will consist of all sets of the form $f = \overline{F}(\overline{\varepsilon}_{i-1},\overline{k},\omega) \circ f_3$ where $\overline{k}(\overline{F}) = \overline{k}$, $\overline{F} \subset S_{i,\ell}^{i-1,m,n'}$, $f_3 \in F_3(S_{i,\ell}^{i-1,m,n'})$ and hence f is of the form

$$(M^-(i')) \circ (S(\varepsilon_{i'}),N(\varepsilon_{i'}))\overline{E}_2 \circ (S(\varepsilon_{i-1}),N(\varepsilon_{i-1}))\overline{E}_1$$

where both $\overline{E}_1$ and $\overline{E}_2$ are $t_1(\cdot,\varepsilon_{i-1},\cdot)$ thickened.

Thus, after appropriate applications of CBI

$$\mu(f) = \mu(\overline{F})\mu(f_3|\overline{F}) = \mu(\overline{F})\mu(f_3|S^{i',m,m'}_{i,\ell})2^{\pm 8\varepsilon_{i-1}}.$$

Arguing inductively in the same manner, for any cylinder $\overline{f} = f_1 \circ f_2 \circ \ldots \circ f_n$, $f_j = \overline{F}_j \circ f_{3,j} \in F(\mathcal{S}_i(j))$, where $\overline{f} \subset Z(\overline{f}) = \prod_{j=1}^{n} Z(\overline{F}_j)$, the concatenation of the various $S^{i',m,m'}_{i,\ell}$'s containing the f_j, we can compute

$$\mu(\overline{f}|Z(\overline{f})) = \prod_{j=1}^{n} \mu(f_j|Z(\overline{F}_j))2^{\pm 8\varepsilon_{i-1}n}$$

$$= \prod_{j=1}^{n} \mu(\overline{F}_j|Z(\overline{F}_j))\mu(f_{3,j}|Z(\overline{F}_j))2^{\pm 8\varepsilon_{i-1}n},$$

and

$$\mu(\overline{f}) = \frac{\prod_{j=1}^{n} \mu(\overline{F}_j)\mu(f_{3,j})2^{\pm 8\varepsilon_{i-1}n}}{\mu(Z(\overline{f}))}.$$

These last computations remain correct if we replace $f_{3,j}$ by $f_{2,j} \in F_2(Z(\overline{F}_j))$. We will use this fact later.

Notice that by condition (ii) we had to lump within each $\mathcal{S}^+_i(\overline{k}(\omega))$ in forming f_3, so if $\omega \in f \in F(\mathcal{S}_i)$, then $f \subseteq \mathcal{S}_i$ so $F(\mathcal{S}_i)$ indeed partitions $\mathcal{S}_i$.

It is important to understand the sense in which $F_3(Z)$ differs from $F_1(Z)$, as in this modification across this latter bit of the skeleton we might possibly discard a lot of our coding up to this point by replacing the concatenation of fillers on which we have been coding with a new partition of thickened orbits, destroying the finitariness of our work. Consider two points $\omega_0, \omega \in \mathcal{S}_i$, $\omega_0, \omega \in Z = S^{i',m,m'}_{i,\ell}$. Let $\omega_0' = T^{-m}(\omega_0)$, $\omega' = T^{-m}(\omega)$. Suppose $\omega_0', \omega' \in f_3 \in F_3(Z)$, i.e., both ω_0 and ω are in the same atom of $F(\mathcal{S}_i)$. Suppose ω_0' is in that atom $f_1^0 \in F_1(z)$ which, by condition (ii) on $F_3(Z)$ corresponds to f_3, and $\omega \in f_1 \in F_1(Z)$. By the nice concatenation property of t_0 and the fact that t_1 sets are contained in t_0 sets, we can conclude $f_3 \subset t_0(f_1^0, \overline{\varepsilon}_{i-1}, \ell - m)$. Now let $k' > \overline{k}(\omega)$ and suppose $T^{-m(k',\mathcal{S}_i)}(\omega) = \omega'' \in \overline{f} \in F(\mathcal{S}_{i-1}(k',\mathcal{S}_i))$, and $T^{-m(k',\mathcal{S}_i)}(\omega_0) = \omega_0'' \in \overline{f}^0 \in F(\mathcal{S}_{i-1}(k',\mathcal{S}_i))$. It is how close $\overline{f}$ and $\overline{f}^0$ are which concerns us. Assume inductively that $\overline{f}$, $\overline{f}^0$ have the form we just put on sets in $F(\mathcal{S}_i)$. As f_3 intersects both $T^{-m(k',\mathcal{S}_i)}(\overline{f})$ and $T^{-m(k',\mathcal{S}_i)}(\overline{f}^0)$, both $\overline{f}$ and $\overline{f}^0$ are in

$$T^{m(k',\mathcal{S}_i)}(t_0(T^{-m(k',\mathcal{S}_i)}(f_3), \overline{\varepsilon}_{i-2}, \ell(\mathcal{S}_{i-1}(k',\mathcal{S}_i)))\ .$$

From the symmetry of t_0 we now conclude that $\overline{f}$ lies inside a set obtained from $\overline{f}^0$ by three successive t_0-thickenings, all across blocks of digits containing $0,\ldots,\ell(\mathcal{S}_{i-1}(k',\mathcal{S}_i))$, first by $\overline{\varepsilon}_{i-1}$, then by $\overline{\varepsilon}_{i-2}$ and finally by $\overline{\varepsilon}_{i-1}$. It is not obvious but this is all the control we will need. We leave this point in its present stage, returning after the construction of $\overline{k}$

The purpose of $\overline{k}$ is, as in the Smorodinsky, Keane arguments, to control $\text{card}(F(\mathcal{S}_i))$. To overcome the problem described in the previous paragraph and to provide some needed bounds on the sizes of sets in $F(\mathcal{S}_i)$, we also need $\overline{k}$ quite large. We will now write down what we need of $\overline{k}$, we will define $\overline{k}$, and then proceed to show that upper bounds for $i' \geq i-1$, $\overline{\varepsilon}_{i'}$, $\varepsilon_{i'}$ and lower bounds for $k(i'), i' > i$ will ensure the properties for $\overline{k}$.

What we want of $\overline{k}$ is:

(i) <u>For i-odd</u> and <u>all</u> $\mathcal{S}_i$,

$$2^{(h(T,P)-h(T,\overline{M}))(1-2^{-16^i})\ell(\mathcal{S}_i)} > \text{card } F(\mathcal{S}_i) \text{ and}$$

for all but a set $B_1 \subset S_i$, $\int_{B_i} \ell(\mathcal{S}_i) d\mu < 2^{-16^i}$, for all but a set $B_2(\mathcal{S}_i)$, $\mathcal{S}_i \subset B_1^c$, $\int_{\bigcup_{\mathcal{S}_i}(B_2(\mathcal{S}_i))} \ell(\mathcal{S}_i) d\mu < 2^{-16^i}$,

$$2^{-(h(T,P)-h(T,\overline{M}))(1-2^{-8\cdot 16^{i-1}})\ell(\mathcal{S}_i)} > \mu(f|\mathcal{S}_i) \text{ and } \underline{\text{for}} \ i$$

<u>even</u> and <u>all</u> $\mathcal{S}_i$

$$2^{(h(T,P)-h(T,\overline{M}))(1-2^{-4\cdot 16^i})\ell(\mathcal{S}_i)} > \text{card } F(\mathcal{S}_i)$$

and sets B_1, $B_2(\mathcal{S}_i)$ bounded as above, for $f \notin B_2(\mathcal{S}_i)$, $\mathcal{S}_i \notin B_1$, $2^{-(h(T,P-h(T,M))(1-2^{-2\cdot 16^i})\ell(\mathcal{S}_i)} > \mu(f|\mathcal{S}_i)$; and

(ii) for $\omega \in f \notin B_2(\mathcal{S}_i)$, $\mathcal{S}_i \notin B_1$,

$$m(\overline{k}(\omega),\mathcal{S}_i) > \begin{cases} (1-2\cdot 2^{-16^i})\ell(\mathcal{S}_i) & \text{for } \underline{i\text{-odd}} \\ (1-2\cdot 2^{-4\cdot 16^i})\ell(\mathcal{S}_i) & \text{for } \underline{i\text{-even}}\,. \end{cases}$$

The parity differences here will be reversed on the other process giving the needed differences in cardinality and size needed to effectively apply the Marriage lemma.

We now define $\overline{k}(\omega)$, once all the markers are built, $\overline{\varepsilon}_{i-1}$ chosen, to be, for i-odd (i-even) the smallest value $\overline{k}$ so that there is an $\mathcal{S}'_i \subset S_{i,\ell}$, $\overline{F}(\overline{\varepsilon}_{i-1},\overline{k},\omega) \cap \mathcal{S}'_i \neq \phi$ and for <u>some</u> $\omega' \in \overline{F}(\overline{\varepsilon}_{i-1},\overline{k},\omega) \cap \mathcal{S}'_i$,

$$\mu(\overline{F}(\overline{\varepsilon}_{i-1},\overline{k}+1,\omega')|\mathcal{S}^-_i(\overline{k}+1)$$

$$< 2^{-(h(T,P)-h(T,\overline{M}))(m(\overline{k}+1,\mathcal{S}_i)(1-2^{-16^i})+(\ell-m(\overline{k}+1,\mathcal{S}_i))2^{-16^i})}$$

$$\left(< 2^{-(h(T,P)-h(T,\overline{M}))(m(\overline{k}+1,\mathcal{S}_i)(1-2^{-4\cdot 16^{i-1}})+(\ell-m(\overline{k}+1,\mathcal{S}_i))2^{-16^i})}\right).$$

Note as $\overline{F}(\overline{\varepsilon}_{i-i},\overline{k},\omega) \subset S^{i',m,m'}_{i,\ell}$, $m(\overline{k}+1,\omega) = m(\overline{k}+1,\omega')$.

Our task now is to show that the parameters can be chosen to achieve (i) and (ii). If the reader is not in need of the details of such work, we recommend he skip ahead to the statement of Lemma 4d.11 which expresses the order in which the parameters must be chosen, and continues reading after its proof.

Our first step in getting these conditions is to develop

some explicit control of the sizes of sets $\overline{F}(\overline{\varepsilon}_{i-1},\overline{k},\omega)$ obtained from an $\overline{\varepsilon}_{i-1}$,i-s.t. Our first preliminaries are standard applications of the pointwise Birkhoff ergodic theorem and the pointwise Shannon-McMillan-Breiman Theorem, so our proofs are sketchy.

Lemma 4d.1. For any ergodic finite measure preserving transformation T of a Lebesgue space $(\Omega,\mathcal{F},\mu)$, for any set $A \subset \Omega$, $\mu(A) > 0$, and any $\varepsilon > 0$, there is a $\delta > 0$ so that for any set S, $0 < \mu(S) < \delta$, r_S the return time to S, and any $n; S \to \mathbb{Z}^+$, with $r_S(\omega) \geq n(\omega)$, but $\int_S n(\omega)d\mu > \varepsilon$, there is a set $B \subset S$, $\int_B n(\omega)d\mu < \varepsilon^2$ so that if $\omega \in S \cap B^c$, then $\frac{1}{n(\omega)} \sum_{i=0}^{n(\omega)-1} 1_A(T^i(\omega)) = \mu(A)(1\pm\varepsilon)$.

Proof. With $\overline{\varepsilon} = \frac{\varepsilon^4}{36}$, apply the Birkhoff Theorem to A for both T and T^{-1} to get an N, and a bad set B_0, $\mu(B_0) < \frac{\varepsilon^4}{36}$, and ergodic averages for both T and T^{-1} are within ε for $n > N$ for all $\omega \in B_0^c$. Choose δ so small that if $\mu(S) < \delta$ then $E_N = \{T^i(\omega) | \omega \in S, [\frac{n(\omega)}{3}] < i \leq [\frac{2n(\omega)}{3}], [\frac{n(\omega)}{3}] > N\}$ has $\mu(E_N) > \frac{\varepsilon}{6}$. Now $B = \{\omega \in S | \text{for all } i, [\frac{n(\omega)}{3}] < i < [\frac{2n(\omega)}{3}], \omega \in B_0\}$ certainly has $\int_B n(\omega)d\mu < 3\mu(B) + 2\delta < \varepsilon^2$, and if $\omega \in B^c \cap S$, as for some i, $[\frac{n(\omega)}{3}] + 1 < i \leq [\frac{2n(\omega)}{3}]$, $T^i(\omega)$ satisfies the pointwise theorem to within $\frac{\varepsilon^2}{36}$ hence $\frac{\varepsilon}{2}$ for both i on T^{-1} and $n(\omega)-i$ for T, the result follows. ∎

Lemma 4d.2. For any T as above, Q a countable partition, $h(Q) < \infty$, for any ε, there is a δ so that for any $S \subset \Omega$, $\mu(S) < \delta$, $n(\omega) \leq r_S(\omega)$, $\int_S n(\omega)d\mu < \varepsilon$, there is a $B \subset S$, $\int_B n(\omega)d\mu < \varepsilon^2$ so that for all $\omega \in B^c \cap S$, the cylinder set $Q_1^{n(\omega)}(\omega)$ satifies $\mu(a(\omega)) = 2^{-n(\omega)(h(T,Q)\pm\varepsilon)}$.

Proof. Argue in a similar manner to Lemma 4d.1, but use the Shannon-MacMillan-Breiman Theorem, only choose for E the set $\{T^i(\omega) | \omega \in S, 0 \leq i \leq [\frac{\varepsilon^2 n(\omega)}{h(Q)}]\}$ for the upper bound and

$$\left\{T^i(\omega)\,|\,\omega \in S,\ n(\omega) \leq i \leq \left[\frac{\varepsilon^2(r_s(\omega) + n(T^{r_s(\omega)}(\omega))}{h(Q)}\right]\right\}$$

under the action of T^{-1} for the lower bound. ■

<u>Lemma 4d.3</u>. For (T,Q) as above, $h(T,Q) > 0$, for any $\varepsilon_1, \varepsilon_2 > 0$, there are $\delta_1(\varepsilon_1)$, $\delta_2(\varepsilon_1,\varepsilon_2)$ so that if $\mu(S) < \delta_2$, $n(\omega) \leq r_s(\omega)$ but $\int_S n(\omega) > \varepsilon_2$, and F is any partition of S satisfying

(i) if $\omega,\omega' \in f \in F$, then $n(\omega) = n(\omega')$, and

(ii) if $\omega,\omega' \in f \in F$ then $\bar{d}_{n(\omega)}(Q_0^{n(\omega)}(\omega), Q_0^{n(\omega)}(\omega')) < \delta_1$,

then for all but a set $B \subset F$, $\int_B n(\omega)d\mu < \varepsilon_2^2$, any $f \in F \cap B^c$ satisfies

$$\mu(f) < 2^{-(h-\varepsilon_1)n(f)}.$$

<u>Proof</u>. Lump together the small states of Q to give a finite partition $\bar{Q}$, $h(T,\bar{Q}) > h(T,Q) - \varepsilon_1/4$. Using $\bar{Q}$ in Lemma 3d.2 set $\varepsilon = \min(\frac{\varepsilon_1^2}{16}, \frac{\varepsilon_0^2}{16})$ to get a δ. Be sure δ_2 is this small. We conclude that for all but a set of atoms $B \subset F$, $\int_{B_1} n(\omega) < \frac{\varepsilon_2^2}{16}$, $\bar{f} \in F \cap B^c$ has at most a fraction $\frac{\varepsilon_1^2}{16}$ of its measure in the set B_1 of Lemma 4d.2. Be sure δ_1 is so small that for all N, the number of names in a $\bar{d}_N,\delta_1$-ball in $\bar{Q}_0^N$ is at most $2^{h(T,Q)N(\varepsilon_1/2)}$ (this is a little Stirling's formula computation using $\bar{Q}$ finite). It follows that for $\bar{f} \in \bar{Q}_0^n$ $\mu(\bar{f} \cap B^c) < 2^{-(h(T,\bar{Q}-\varepsilon_1/4)n(f)}$, so for $f \in Q_n^n$, $f \subset f$ hence $\mu(f) < (1 + (\frac{\varepsilon_1}{4})^2)\,(\bar{f} \cap B^c) < 2^{-h(T,Q) - \varepsilon_1)n(f)}$. ■

<u>Lemma 4d.4</u>. Assume the markers $M(i),\ldots,M(i-2)$ are

constructed. Given any $\varepsilon_1, \varepsilon_2 > 0$ there is a $\delta_1(\varepsilon_1)$ so that if $\varepsilon_{i-1} \leq \bar{\varepsilon}_{i-1} < \delta_1(\varepsilon_1)$ and now $k(i-1)$ are fixed, then there is a δ_2 so that if $\sum_{i' \geq i} \mu(M(i'))\ell(M(i')) < \delta_2$ then for any $\bar{\varepsilon}_{i-1}, i$-s.t. $\bar{k}$, with $\int_{S_i} m(\bar{k}(\omega), \mathcal{S}_i) d\mu > \varepsilon_2$, there is a set $B \subset \bar{F}(\bar{\varepsilon}_{i-1}, \bar{k})$ with $\int_B m(\bar{k}(\omega), \mathcal{S}_i) d\mu < \varepsilon_2^2$ and if $\bar{F} \notin B$, then $\mu(\bar{F}) < 2^{-h(T,P)-\varepsilon_1) m(\bar{k}(\omega), \mathcal{S}_i)}$.

Proof. Choose $\delta_1(\varepsilon_1)$ so small that for $\bar{\varepsilon}_{i-1} < \delta_1$, $\mu(A) = \mu(\bigcup_i A_i(\bar{\varepsilon}_{i-1})) < \frac{\bar{\delta}_1(\frac{\varepsilon_1}{4})}{4}$, $\bar{\delta}_1(\frac{\varepsilon_1}{4})$ taken from Lemma 3d.3. Let $\bar{\delta}_2(\frac{\varepsilon_1}{4}, \frac{\varepsilon_2}{4})$ be from Lemma 4d.3.

Using this A in Lemma 4d.1 (if $\mu(A) = 0$, the following is unnecessary), and $\varepsilon^2/4$, we get a $\bar{\delta}_3$. Make sure δ_2 is so small that

$$\sum_{i' \geq i} \mu(M(i')) = \mu(S) < \min(\bar{\delta}_2, \bar{\delta}_3)$$

and

$$\sum_{i' \geq i} \mu(M(i'))\ell(M(i')) < \left(\frac{\bar{\delta}_1}{4}\right)^2 \left(\frac{\varepsilon_2}{4}\right)^2 < \frac{\bar{\delta}_2^2}{4} \int_{S_i} m(k(\omega) \mathcal{S}_i).$$

This sets our parameters. By the last inequality, for all but a set $B_0 \subset F = \bar{F}(\bar{\varepsilon}_{i-1}, k)$, $\int_{B_0} m(\bar{k}(\omega), \mathcal{S}_i) < \frac{\bar{\delta}_2 \varepsilon_2}{16} < \frac{\varepsilon_2}{16}$, we know $\frac{\ell(M(i'))}{m(\bar{k}(\omega), \mathcal{S}_i)} < \frac{\bar{\delta}_1 \varepsilon_2}{16} < \frac{\bar{\delta}_1}{16}$.

Using $S' = \{T^{m(i')+1}(\omega)\}_{\omega \in M(i') \cap S_i}$ in our application of Lemma 4d.1, $n(\omega) = m(\bar{k}(\omega), \mathcal{S}_i) - \ell(M(i')) - 1$, as $\int_{S'} n(\omega) = \int_S m(\bar{k}(\omega), \mathcal{S}_i) - \sum_{i' \geq i} \ell(M(i')\mu(M(i')) > \frac{\varepsilon}{2}$ we conclude that for all but a set $B_1 \subset S_i$, $\int_{B_1} m(\bar{k}(\omega), \mathcal{S}_i) < \frac{\varepsilon_2}{2}$, we have $\sum_{i=\ell(M(i'))}^{m(\bar{k}(\omega), \mathcal{S}_i)} 1_A(T^i(\omega)) < \frac{\delta_1}{4}(1 + \frac{\varepsilon_1}{4}) < \frac{1}{2}$.

From the form of t_0 we can conclude that for any

$\omega \notin B_1$, $\omega \in \overline{F} \notin B_0$, for any $\omega' \in \overline{F}$

$$\overline{d}_{m(\overline{k}(\omega),\mathcal{S}_i)}\left(P_0^{m(\overline{k}(\omega),\mathcal{S}_i)}(\omega),\ P_0^{m(\overline{k}(\omega),\mathcal{S}_i)}\right) < \frac{a}{16}\,\delta_1 .$$

Let $B_2 = \{\omega \in \overline{F} \in \overline{F}(\overline{\varepsilon}_{i-1},\overline{k}) \mid$ for all $\omega' \in \overline{F}, \omega' \in B_1\}$. Now $B_2 \subset F$ and clearly

$$\int_{B_2} m(\overline{k}(\omega),\omega)d\mu < \int_{B_1} m(\overline{k}(\omega),\omega)d\mu < \frac{\varepsilon_2}{2}.$$

Define $n(\omega) = \begin{cases} m(\overline{k}(\omega),\mathcal{S}_i), & \omega \notin B_2 \\ 0, & \omega \in B_2 \end{cases}$. It follows that $\int_{S_i} n(\omega)d\mu > \frac{\varepsilon_2}{2}$ and clearly

(i) if $\omega,\omega' \in \overline{F}$, $n(\omega) = n(\omega')$, and from our above work

(ii) if $\omega,\omega' \in \overline{F}$, $\overline{d}_{n(\omega)}(P_0^{n(\omega)}(\omega),P_0^{n(\omega)}(\omega')) < \frac{a}{16}\,\delta_1 < \delta_1$.

Thus Lemma 4d.3 applies, and outside a set B_3, $\int_{B_3} n'(\omega)d\mu < (\frac{\varepsilon_2}{4})^2$, $\mu(F) < 2^{-(h(T,P)-\varepsilon_1/4)n(\overline{F})}$. Let $B = B_2 \cup B_3$, and as

$$\int_{B_2 \cup B_3} m(\overline{k}(\omega),\mathcal{S}_i)d\mu = \int_{B_2} m(\overline{k}(\omega),\mathcal{S}_i)d\mu + \int_{B_3} n(\omega)d\mu < \frac{\varepsilon_2}{2}$$

we are done. ∎

We need a lower bound on $\mu(\overline{F})$. This comes via the usual technique, having shown most are small, as they concatenate with almost independence, few can be too small. First we need to control the sizes of cylinders beyond $m(\overline{k}(\omega),\mathcal{S}_i)$.

Lemma 4d.5. Given $\varepsilon_1,\varepsilon_2 > 0$ there is a $\delta(\varepsilon_1)$ so that if $\varepsilon_{i-1} \leq \overline{\varepsilon}_{i-1} < \delta_1(\varepsilon_1)$ and $k(i)$ are chosen, then there is a δ_2 so that if $\sum_{i' \geq i} \mu(M(i'))\ell(M(i')) < \delta_2$, then for any $\overline{\varepsilon}_{i-1}$, i-s.t. $\overline{k}$ with $\int_{S_i}((\ell(\mathcal{S}_i(\omega)) - m(\overline{k}(\omega),\mathcal{S}_i))d\mu > \varepsilon_2$, there is a set $B \subset \bigcup_{Z=S^{i',m,m'}_{i,\ell}} F_2(Z) = F$, $\int_B (\ell(\mathcal{S}_i(\omega)) - m(\bar{k}(\omega),\mathcal{S}_i) d\mu < \varepsilon_2^2$ so that if $f_2 \in F_2(S^{i',m,m'}_{i,\ell})$, $f_2 \notin B$ then $\mu(f) < 2^{-(H(T,P)-\varepsilon_1)(\ell-m)}$.

Proof. This is done exactly as Lemma 4d.4, only the partition F is across the latter part of i-skeleta rather than the initial. ∎

To get the lower bound, we now employ the partition $F'(\mathcal{S}_i)$ of all sets of the form $f = \bar{F}(\bar{\varepsilon}_{i-1},\bar{k},\omega) \cdot f_2$, where $\bar{F} \subset S^{i',m,m'}_{i,} = Z$, $f_2 \in F_2(Z) = C(\bar{\varepsilon}_{i-1}, \ell-m-\ell(M^-(i-1))-S(\varepsilon_{i-1}))/Z$.

We saw earlier that for any sequence $\bar{f} = \bar{F}_1 \circ f_{2,1} \circ \bar{F}_2 \circ \bar{f}_{2,2} \circ \dots \circ \bar{F}_n \circ f_{2,n}$ of positive probability

$$\mu(\bar{f}) = \frac{\prod_{j=1}^{n} \mu(\bar{F}_j)\mu(F_{2,j})2^{\pm 8\varepsilon_{i-1}}}{\mu(Z(\bar{f}))}.$$

This will get us lower bounds on $\mu(\bar{F})$ if we can show that $F' = \bigcup_i F'(\mathcal{S}_i)$ and Z, as partitions of S_i, have finite and small entropy, respectively, as then the Shannon-MacMillan Theorem will give us bounds on $\mu(\bar{f})$.

Lemma 4d.6. The partition Z on S_i has finite entropy tending to zero as a function of $\ell(M(i))$, and F' has finite entropy.

Proof. By Lemma 4c.2 there are constants $\alpha_1, \alpha_2 < 1$ with $\bar{\mu}(S_{i,\ell}) < \alpha_1 \alpha_2{}^{\ell}$, and as the pair $(m(\bar{k}(\omega),\mathcal{S}_i), m(\bar{k}(\omega)+1,\mathcal{S}_i))$ takes on at most $(\ell - \ell(M(i'))-n_5)^2$ values on $\bigcup_m S^{i',m,m'}_{i,\ell}$,

$$h(Z) < -\sum_{i' \geq i} \sum_{\ell=\ell(M(i'))-n_5} \alpha_1^{\ell}\alpha_2^{\ell}(\ell\, \ell n(\alpha_2) + \ell n(\alpha_1) + 2\ell n(\ell))$$

which is not only finite but, as $\ell(M(i')) > i'$, tends to zero in $\ell(M(i))$.

A set $f \in F'/S_{i,\ell}^{i',m,m'}$ can be viewed as given by a choice of two sets, one in $C(\overline{\varepsilon}_{i-1}, m-\ell(M(i')) + \ell(m^-(i-1)) + N(\varepsilon_{i-1}))$, and the other in $C(\overline{\varepsilon}_{i-1}, \ell-m-\ell(M^-(i-1)) - N(\varepsilon_{i-1}))$. As any $c \in C(\varepsilon,n)$, $c \subset c' \in C(\varepsilon,n-1)$ has $\mu(c|c') > \sigma(\varepsilon) > 0$, there are at most $\sigma(\overline{\varepsilon}_{i-1})^{-(m-\ell(M(i'))-\ell(m^-(i-1)) + N(\varepsilon_{i-1}))}$ sets in the first collection, $\sigma(\overline{\varepsilon}_{i-1})^{-(\ell-m-\ell(M^-(i-1))-N(\varepsilon_{i-1}))}$ in the second. Hence,

$$h(F'|Z) = \sum_{S_{i,\ell}^m} \mu(S_{i,\ell}^{i',m,m'})h(F/S_{i,\ell}^{i',m,m'})$$

$$< \sum_{S_{i,\ell}^{i',m,m'}} \mu(S_{i,\ell}^{i',m,m'})\ell n(\sigma(\overline{\varepsilon}_{i-1})^{-(\ell+1)} \cdot 2) < \infty$$

from our bounds on $\mu(S_{i,\ell}^{i',m})$. ∎

Lemma 4d.7. Given $\varepsilon_1 > 0$, there is a δ_1 so that if $\overline{\varepsilon}_{i-1} < \delta_1$ is chosen, then for any ε_2, there is a δ_2 so that if $\varepsilon_{i-1} < \delta_2$ and $k(i-1)$ are chosen, then there is a δ_3 so that if $\sum_{i' \geq i} \ell(M(i'))\mu(M(i')) < \delta_3$, then for any $\overline{\varepsilon}_{i-1}$,i-s.t. $\overline{k}$ with $\int_{S_i} m(\overline{k}(\omega),\mathcal{S}_i)d\mu < \varepsilon_2$, for all but a set $B \subset \overline{F}(\overline{\varepsilon}_{i-1},\overline{k})$, $\int_B m(\overline{k}(\omega),\mathcal{S}_i)d\mu < \varepsilon_2$, if $f \in \overline{F}(\overline{\varepsilon}_{i-1},\overline{k}) \cap B^c$, then $\mu(f) = 2^{-(h(T,P)\pm\varepsilon_1)m(\overline{k}(f),\mathcal{S}_i)}$.

Proof. We need only check > by Lemma 4d.4. Assume the result is false, i.e., there is an $\overline{\varepsilon}_1 > 0$ so that for any δ_1, we can find $\overline{\varepsilon}_{i-1} < \delta_1$ and a value $\overline{\varepsilon}_2$ so that for any δ_2, we can find $\varepsilon_{i-1} < \delta_2$ and $k(i-1)$ so that for any δ_3 we can now find $M(i')$, $i' \geq i$ with $\sum_{i' \geq i} \ell(M(i'))\mu(M(i')) < \delta_3$ and an $\overline{\varepsilon}_{i-1}$,i-s.t. with $\int_{S_i} m(\overline{k}(\omega),\mathcal{S}_i)d\mu > \varepsilon_2$, but the set B of those $f \in \overline{F}(\overline{\varepsilon}_{i-1},k)$ with $\mu(f) < 2^{-(h(T,P)-\varepsilon_1)m(\overline{k}(f),\mathcal{S}_i)}$ has $\int_B m(\overline{k}(\omega),\mathcal{S}_i)d\mu > \overline{\varepsilon}_2^{-2}$. Call such a $\delta_1,\delta_2,\delta_3$-counterexample.

Fix any δ_1, and we get $\overline{\varepsilon}'_{i-1} < \delta$ from above, and $\overline{\varepsilon}_2$. Now choose $\varepsilon_1,\varepsilon_2 < \dfrac{\overline{\varepsilon}_1\overline{\varepsilon}_2^{2}}{30h(T,P)}$ and from Lemma 4d.4 get $\delta_2 = \delta_1(\varepsilon_1)$. Select $\varepsilon_{i-1} = \overline{\varepsilon}_{i-1} < \min(\delta_2,\overline{\varepsilon}_{i-1}^{-1},\varepsilon_1)$ and setting $k(i-1)$, we get from Lemma 4d.4 a δ_2 which we call δ_3 and from above, $\delta_1,\delta_2,\delta_3$-counterexamples satisfying Lemma 4d.4 with the stated parameters. As $\overline{F}(\overline{\varepsilon}_{i-1},\overline{k}) \subset \overline{F}(\overline{\varepsilon}'_{i-1},\overline{k})$, we get for $f' \in B \cap \overline{F}(\overline{\varepsilon}'_{i-1},\overline{k})$, $\mu(f') < 2^{-(h(T,P)-\overline{\varepsilon}_1)m(\overline{k}(f'),\mathcal{S}_i)}$, for any $f \in \overline{F}(\overline{\varepsilon}_{i-1},\overline{k}) \cap B$, $\mu(f) < 2^{-(h(T,P)-\overline{\varepsilon}_1)m(\overline{k}(f),\mathcal{S}_i)}$, i.e., although the set B of our $\delta_1,\delta_2,\delta_3$-counterexample ostensibly only applies to $\overline{F}(\overline{\varepsilon}'_{i-1},k)$ as stated, as $\overline{\varepsilon}_{i-1} \leqslant \overline{\varepsilon}'_{i-1}$, the bounds hold just as well on B as a subset of $\overline{F}(\overline{\varepsilon}_{i-1},k)$.

Similarly, working through the parameters with Lemma 4d.5, if we find $\int(\ell(\mathcal{S}_i(\omega)) - m(\overline{k}(\omega),\mathcal{S}_i))d\mu > \varepsilon_2$, Lemma 4d.5 holds with $\varepsilon_1,\varepsilon_2$. Further, if δ_3 is small enough, then $\ell(M(i))$ will be large enough so that from Lemma 4d.6, $h(Z/S_i) < \varepsilon_2$. Call such a $\delta_1,\delta_2,\delta_3,\varepsilon_1,\varepsilon_2$-counterexample. We show the existence of such leads to a conflict.

Let $\overline{T}$ be the induced transformation on S_i, with $\overline{\mu} = \dfrac{\mu}{\mu(S_i)}$. We know $h(\overline{T},Z) < \varepsilon_2$ on a $\delta_1,\delta_2,\delta_3,\varepsilon_1,\varepsilon_2$-counterexample.

Writing $\overline{F} \in F_1^n$, we have seen $\overline{F} = \overline{F}_1 \circ f_{2,1} \circ \overline{F}_2 \circ f_{2,1} \circ \ldots \circ \overline{F}_n \circ f_{2,n}$ and
$$\mu(\overline{\mathbf{F}}) = \frac{\prod_{j=1}^{n} \mu(\overline{F}_j)\mu(f_{2,j})2^{\pm 8\varepsilon_{i-1}}}{\mu(Z(\overline{F}))}.$$

Applying the Shannon-MacMillan Theorem to $(\overline{T},F)$, for any ε', if n is large enough, for all but ε' of the $\overline{F} \in F_1^n$, $\mu(\overline{F}) = 2^{-(h(\overline{T},F)\pm\varepsilon')n}$ and $\mu(Z(\overline{F})) > 2^{-(\varepsilon_2+\varepsilon')n}$.

Look now at a $\delta_1,\delta_2,\delta_3,\varepsilon_1,\varepsilon_2$-counterexample. First assume $\int_{S_i} (\ell(\mathcal{S}_i(\omega)) - m(\bar{k}(\omega),\mathcal{S}_i))d\mu > \varepsilon_2$. Break the sets $\{\bar{F}(\mathcal{S}_i)\}$ into three classes A, B, C. On A, $2^{-(h(T,P)+\bar{\varepsilon}_1)m(\bar{k}(\omega),\mathcal{S}_i)} \leq \mu(\bar{F}) \leq 2^{-(h(T,P)-\varepsilon_1)m(\bar{k}(k(\omega),\mathcal{S}_i)}$, on B, $\mu(\bar{F}) < 2^{-(h(T,P)+\bar{\varepsilon}_1)m(\bar{k}(\omega),\mathcal{S}_i)}$ and C the rest. As we are in $\delta_1,\delta_2,\delta_3,\varepsilon_1,\varepsilon_2$-counterexample we know $\int_{A\cup B} m(\bar{k}(\omega),\mathcal{S}_i)d\bar{\mu} > (1-\varepsilon_2)\int_{S_i} m(\bar{k}(\omega),\mathcal{S}_i)d\bar{\mu}$ and $\int_B m(\bar{k}(\omega),\mathcal{S}_i)d\mu > \bar{\varepsilon}_2^2 \int_{S_i} \ell(\mathcal{S}_i)d\bar{\mu}$. Similarly split $F_2(S_{i,\ell}^{i',m,m'}) = F_2(z)$ into two sets $A'(z)$, $B'(z)$ so that for $f_2 \in A'(z)$, $\mu(f_2) < 2^{-(h(T,P)-\varepsilon_1)(\ell-m)}$ and we know $\int_{\underset{z}{\cup} A'(z)} (\ell(\mathcal{S}_i)-m(\bar{k}(\omega),\mathcal{S}_i))d\bar{\mu} > (1-\varepsilon_2)\int_{S_i} (\ell(\mathcal{S}_i)-m(\bar{k}(\omega)))d\mu$.

Fix $\varepsilon' < \frac{\bar{\varepsilon}_1\bar{\varepsilon}_2^2}{h(T,P)\cdot 30}$ and choose n so large, that by the ergodic and Shannon MacMillan Theorems on $(\bar{T},F)$ and $(\bar{T},Z)$, for all but ε' of the $\bar{F} \in F_1^n$, first we have exponential bounds to within ε' on both the sizes of $\bar{F}$ and $Z(\bar{F})$, second $\sum_{j=1}^{n} \ell(\mathcal{S}_i(f_j)) = (\frac{1}{\mu(S_i)} \pm\varepsilon')n$, and third, the density of elements $\bar{F}_j$ in A, B, and C is within ε' of their measure, as is the density of $f_{2,j}$ in $\cup A'(Z)$. We now compute

$$\mu(\bar{F}) < \prod_{\bar{F}_j\in A\cup B} 2^{-(h(T,P)-\varepsilon_1)m(\bar{k}(\omega),\mathcal{S}_i)} \prod_{\bar{F}_j\in B} 2^{-(\bar{\varepsilon}_1+\varepsilon_1))m(\bar{k}(\omega),\mathcal{S}_i)}$$

$$\prod_{f_{2,j}\in\cup A'(Z)} 2^{-(h(T,P)-\varepsilon_1)(\ell(\mathcal{S}_i)-m(\bar{k}(\omega),\mathcal{S}_i))}\, 2^{8\varepsilon_{i-1}n+\varepsilon_2 n}$$

$$< 2^{-(h(T,P)-\varepsilon_1)(1-\varepsilon_2-\varepsilon')\Sigma_{f_j}\ell(\mathcal{S}_i(f_j)}$$

$$\cdot 2^{(8\varepsilon_{i-1}+\varepsilon_2)n} \cdot 2^{-\bar{\varepsilon}_1\bar{\varepsilon}_2^2\,\Sigma_{f_j}\ell(\mathcal{S}_i(f_j))}$$

$$< 2^{-(h(T,P)(1-\varepsilon_2-\varepsilon')-\varepsilon_1-8\varepsilon_{i-1}-\varepsilon_2+\bar{\varepsilon}_1\bar{\varepsilon}_2^2}\left(\frac{1}{\mu(S_i)}-\varepsilon'\right)$$

$$< 2^{-(h(T,P)(1+\frac{\bar{\varepsilon}_1\bar{\varepsilon}_2^2}{2})(\frac{1}{\mu(S_i)}-\varepsilon')} < 2^{-h(\bar{T},F)(1+2\varepsilon')}$$

a conflict, as $\mu(\bar{F}) = 2^{-h(\bar{T},F)(1\pm\varepsilon')}$.

For the case when $\int_{S_i}((\ell(\mathcal{S}_i(\omega))-m(\bar{k}(\omega),\mathcal{S}_i)d\mu < \varepsilon_2$, argue to a contradiction in the same way, using just the $\bar{F} \in A \cup B$ which now cover all but $2\varepsilon_2$ of the indices of the name. ■

Lemma 4d.8. If δ_1 and then δ_2 are chosen as in Lemma d.7, then δ_3, and now if for $i' \geq i$, $k(i') \nearrow \infty$ rapidly enough, we can also conclude that for $f \in \bar{F}(\bar{\varepsilon}_{i-1},\bar{k}) \cap B^c$, that $\mu(f|\mathcal{S}_i^-(\bar{k}(\omega))) = 2^{-((h(T,P)-h(T,\bar{M}))\pm\varepsilon_1)m(\bar{k}(f),\mathcal{S}_i))}$.

Proof. Use $\frac{\varepsilon_1}{3}, \frac{\varepsilon_2}{3}$ in Lemma 4d.7. Make sure $k(i) \nearrow \infty$ fast enough that $h(T, \bigvee_{i'<i} \bar{M}(i')) > h(T,\bar{M}) - \frac{\varepsilon_1}{3}$. Now once $\bar{M}(i-1)$ is defined, in Lemma 4d.2, with $Q = \bigvee_{i'<i} \bar{M}(i)$, $\varepsilon = \min(\frac{\varepsilon_1}{3}, \frac{\varepsilon_2}{3})$ we get a δ, $n(\omega) = m(\bar{k}(\omega),\mathcal{S}_i)$ Be sure $\mu(S_i) < \delta_3 < \underline{\delta}$ and we conclude that outside a set B_1, $\int_{B_1} m(\bar{k}(\omega),\mathcal{S}_i)d\mu < \frac{\varepsilon_2}{4}$, $\mu(\mathcal{S}_i^-(\bar{k}(\omega))) = 2^{-(h(T,\bar{M})\pm\frac{2\varepsilon_1}{3})m(\bar{k}(\omega),\mathcal{S}_i)}$. Let B be the union of B_1 and the error set of Lemma 4d.7 and we are done. ■

We now return to $\bar{k}$ as defined earlier to complete our proof that selecting parameters appropriately it will satisfy the needed conditions.

Lemma 4d.9. If $M(i-1)$ has been defined, then for any $\alpha > 0$ if $k(i')$, $i' \geq i$ increase rapidly enough, then $\mu(\{\omega \in S_i | \exists\, \bar{k} < k(\mathcal{S}_i),\ \omega \in \mathcal{S}_i,$ and $m(\bar{k}+1,\omega)-m(\bar{k},\omega) < \alpha\ell(\mathcal{S}_i)\}) < \alpha\mu(S_i)$ (whenever we say $k(i')$, $i' \geq i$ increases rapidly enough we are simply setting lower bounds for each $k(i')$, $i' \geq i$).

Proof. Arguing as in Lemma 4c.2, using Lemma 4c.1, for any $i', i'' \geq i$, ℓ, $\ell' \geq \alpha\ell$, $k < \ell$,

$$\mu(S_{i,\ell} \cap M(i') \cap T^{\ell+1}(M(i'')) \cap T^{\overline{k}}(S_{i-1,\ell'}))$$

$$< \frac{(2\sigma_1^3)(1 - 2 \sum_{i' \geq i-1} \mu(M(i'))^{\alpha\ell - 2\ell(M(i-1))}}{(\frac{\sigma_2}{2})(1 - \frac{\sigma_1}{2}(\sum_{i' \geq i} \mu(M(i'))))^{\ell-1}}$$

$$< cd^{\ell}$$

where $d = \dfrac{(1 - 2 \sum_{i' \geq i-1} \mu(M(i')))^{\alpha}}{(1 - \frac{\sigma_1}{2} \sum_{i' \geq i} \mu(M(i')))}$. If $k(i') \nearrow$ rapidly enough, $i' \geq i$, as $M(i-1)$ is fixed, we can make $d < 1$. Now

$$\frac{\mu(\{\omega \in S_{i,\ell} \mid \exists\, \overline{k}, \overline{k} < k(\mathcal{S}_i), \omega \in \mathcal{S}_i \subset S_{i,\ell}, m(\overline{k}+1,\omega) - m(\overline{k},\omega) < \alpha\ell\})}{\mu(S_{i,\ell})}$$

$$< \ell^2 cd^{\ell},$$

and hence,

$$\frac{\mu(\{\omega \in S_i \mid \exists\, \overline{k} < k(\mathcal{S}_i), \omega \in {}_i, m(\overline{k}+1,\omega) - m(\overline{k},\omega) < \alpha\ell(\mathcal{S}_i)\})}{\mu(S_i)}$$

$$< \sum_{\ell = \ell(M(i))}^{\infty} \ell^2 cd^{\ell}.$$

As $d < 1$, if $\ell(M(i))$, i.e., $k(i)$, is large enough we have the result. ■

This lemma will allow us first, to get some a priori control on $\int_{S_i} m(\overline{k}(\omega), \mathcal{S}_i)\,d\mu$, to get into the hypothesis of Lemma 3d.8, which then will push $m(\overline{k}(\omega), \mathcal{S}_i)$ to nearly $\ell(\mathcal{S}_i)$. We assume $k(i) \nearrow$ rapidly enough that $h(T,\overline{M}) < \dfrac{h(T,P)}{2}$.

Lemma 4d.10. If $\bar{\varepsilon}_{i-1}$, ε_{i-1}, $k(i-1)$ have been chosen small and large enough, respectively, then there is an $\varepsilon_0(\bar{\varepsilon}_{i-1})$ so that for any $\varepsilon > 0$ if $k(i')$, $i' \geq i$ increase rapidly enough, then for $\bar{k}$ as defined, for all but a set B of ω, $\int_B \ell(\mathcal{S}_i) < \varepsilon$, $\dfrac{m(\bar{k}(\omega),\mathcal{S}_i)}{\ell(\mathcal{S}_i)} > \varepsilon_0$.

Proof. As $M(i-1)$ is fixed we can compute, for any $\bar{k}$, from Proposition 4 and Lemma 4c.1,

$$\mu(\bar{F}(\bar{\varepsilon}_{i-1},\bar{k},\omega))) > \mu((M^-(i'))(S(\varepsilon_{i'})N(\varepsilon_{i'})))$$

$$\times \sigma(\bar{\varepsilon}_{i-1})^{m(\bar{k},\mathcal{S}_i)} 2^{\pm 4\varepsilon_{i-1}} (1 - \frac{\sigma_2}{2}\mu(M(i-1)))^{m(\bar{k}+1,\mathcal{S}_i)-m(\bar{k},\mathcal{S}_i)}$$

$$\times (\frac{\sigma_1}{2}\mu(M(i-1)))2^{\pm 4\varepsilon_{i-1}}.$$

If, fixing ε_{i-1}, $k(i-1)$ is so large the $\mu(M(i-1))$ is sufficiently small, this is larger than

$$\sigma(\bar{\varepsilon}_{i-1})^{m(\bar{k},\mathcal{S}_i)} 2^{-(\frac{h(T,P)}{2})(\frac{2^{-16^i}}{100})(\ell(\mathcal{S}_i)-m(\bar{k},\mathcal{S}_i))}.$$

If $m(\bar{k},\omega)$ is sufficiently small that

$$-\ell n(\sigma(\bar{\varepsilon}_{i-1})) < \frac{(\ell(\mathcal{S}_i)-m(\bar{k},\mathcal{S}_i))}{m(\bar{k},\ _i)} 2^{-\frac{16^i}{100}} (\frac{h(T,P)}{2}),$$

then this is larger than

$$2^{-(h(T,P)-h(T,\bar{M}))2^{\frac{-16^i}{30}}(\ell(\mathcal{S}_i)-m(\bar{k},\mathcal{S}_i))}.$$

This will be satisfied if

$$2\varepsilon_0 = 2\left[\frac{-\ell n(\sigma(\bar{\varepsilon}_{i-1}))2^{\frac{-16^i}{100}}}{(\frac{h(T,P)}{2})} + 1\right] > \frac{m(\bar{k},\mathcal{S}_i)}{\ell(\mathcal{S}_i)} > \left[\frac{-\ell n(\sigma(\bar{\varepsilon}_{i-1}))^{2^{\frac{-16^i}{100}}}}{(\frac{h(T,P)}{2})} + 1\right] = \varepsilon_0.$$

Notice ε_0 depends only on $\bar{\varepsilon}_{i-1}$ through $\sigma(\bar{\varepsilon}_{i-1})$.

Using $\min(\frac{\varepsilon_0}{4}, \frac{\varepsilon}{4})$ in Lemma 4d.10 we know that if $k(i')$, $i' \geq i$, increases rapidly enough, then there is a set B, $\int_B \ell(\mathcal{S}_i) < \frac{\varepsilon}{2}$, and for $\omega \notin B$, there is a value $\bar{\bar{k}}(\omega)$ with $2\varepsilon_0 \ell(\mathcal{S}_i) > m(\bar{\bar{k}}+1, \mathcal{S}_i) > m(\bar{\bar{k}}, \mathcal{S}_i) > \varepsilon_0 \ell(\mathcal{S}_i)$.

It follows that for $\omega \notin B$, $\bar{k}(\omega) \geq \bar{\bar{k}}(\omega)$ and so $m(\bar{k}(\omega), \mathcal{S}_i) > \varepsilon_0 \ (\mathcal{S}_i)$. ∎

Our purpose in making $\bar{\varepsilon}_{i-1}$ a separate parameter from ε_{i-1} was to make this ε_0 dependent only $\bar{\varepsilon}_{i-1}$. Hence in Corollary 4d.8 we can choose ε_1, get $\delta_1(\varepsilon_1)$, make sure $\bar{\varepsilon}_{i-1} < \delta_1(\varepsilon_1)$ and <u>then</u> choose $\varepsilon_2 < \varepsilon_0(\bar{\varepsilon}_{i-1})$ to get δ_2, set the values $\varepsilon_{i-1} < \delta_2$, $k(i-1)$ and get δ_3 so that if $k(i') \nearrow$ rapidly enough, $i' \geq i$, we get the conclusion of Corollary 4d.8. Such epsil-antics could have been avoided by assuming $\sigma(\varepsilon)$ did not depend on ε, but this would exclude, for example, an easy treatment of β-transformations (to appear separately) and is an integral part of our proof (see sec. f). Furthermore, although it is clear some control on the rate of decrease of t_1 cylinders in n is needed, such as $\sigma(\varepsilon)$ provides, this is a serious restriction and should be kept as mild as possible.

Now we finally reach our goal.

<u>Lemma 4d.11</u>. For any $\varepsilon > 0$ there is a δ_1 so that if $\bar{\varepsilon}_{i-1} < \delta_1$ is chosen, there is a δ_2 so that if $\varepsilon_{i-1} < \delta_2$ and $k(i-1)$ are chosen, if $k(i') \nearrow$ rapidly enough, $i' \geq i$, then for $\bar{k}$ as defined, for all $\omega \notin B$, $m(\bar{k}(\omega), \mathcal{S}_i) >$ $(1-\varepsilon-2^{-16^i})\ell(\mathcal{S}_i)$ if i-odd ($> (1-\varepsilon-2^{-4\cdot 16^{i-1}}$ if i-even),

$\mu(f|\mathcal{S}_i^-(k(\omega))) = 2^{-(h(T,P-h(T,\bar{M}))(1\pm\varepsilon)m(\bar{k}(\omega),\mathcal{S}_i)}$ and

$\int_B \ell(\mathcal{S}_i) d\mu < \varepsilon$.

<u>Proof</u>. We prove only the odd case, the even being precisely analogous. From what was said above, for any ε', if our parameters are chosen appropriately in Lemma 3d.8, using Lemma 4d.9, then outside a set B_1, $\int_{B_1} m(\bar{k}, \omega) d\mu < \varepsilon'^2$, if $f \subset B^c$ then $\mu(f|\mathcal{S}_i^-(\bar{k}(\omega))) = 2^{-(h(T,P)-h(T,\bar{M})\pm\varepsilon')m(\bar{k}(f),\mathcal{S}_i)}$. We also know, as $m(\bar{k}(\omega), \mathcal{S}_i) > \varepsilon_0 \ell(\mathcal{S}_i)$ outside a set B_2,

$\int_{B_2} \ell(\mathcal{S}_i) < \varepsilon'$ that $\int_{B_1 \cup B_2} \ell(\mathcal{S}_i) < \int_{B_2} \ell(\mathcal{S}_i) + \int_{B_1 \cap B_2^c} \dfrac{m(\bar{k}(\omega),\mathcal{S}_i)}{\varepsilon_0}$ $< (\varepsilon' + \frac{\varepsilon'}{\varepsilon_0})$. Be sure $\varepsilon' < \frac{\varepsilon}{3\varepsilon_0}$.

Suppose for some $\omega \in S_{i,\ell}, \omega \in f \notin B_1 \cup B_2$, $m(\bar{k}(\omega),\mathcal{S}_i) < (1 - \frac{\varepsilon}{2} - 2^{-16^i})\ell$. We compute

$$\mu(f\,|\mathcal{S}_i^-(\bar{k}(\omega))) > 2^{-(h(T,P)-h(T,\bar{M}))((1-2^{-16^i})m(\bar{k}(\omega),\mathcal{S}_i)-\frac{2\varepsilon'}{\varepsilon}(\ell-m(\bar{k}(\omega),\mathcal{S}_i))}.$$

Select $\varepsilon' < \frac{\varepsilon 2^{-16^i}}{4}$ and this is

$$> 2^{-(h(T,P)-h(T,\bar{M}))((m(\bar{k}(\omega),\mathcal{S}_i)(1-2^{-16^i})+(\ell-m(\bar{k}(\omega),\mathcal{S}_i))(\frac{2^{-16^i}}{2}))}$$

We know from this and Proposition 4 that

$$\mu(\bar{F}(\bar{\varepsilon}_{i-1},\bar{k}(\omega)+1,\mathcal{S}_i)\,|\mathcal{S}_i^-(\bar{k}(\omega)+1,\mathcal{S}_i)) \; 2^{-(h(T,P)-h(T,\bar{M}))(m(\bar{k}(\omega),\mathcal{S}_i)(1-2^{-16^i})+(\ell-m(\bar{k}(\omega),\mathcal{S}_i))(\frac{2^{-16^i}}{2})+e)}$$

where

$$e = -\ln(\sigma(\bar{\varepsilon}_{i-1}))\left(\frac{m(\bar{k}(\omega)+1,\mathcal{S}_i)-m(k(\omega),\mathcal{S}_i)}{(\ell-m(\bar{k}(\omega),\mathcal{S}_i))\,(\frac{h(T,P)}{2})}\right).$$

This will conflict with the definition of $\bar{k}$ if $e < \frac{2^{-16^i}}{2}$, i.e., if

$$m(\bar{k}(\omega)+1,\omega)-m(\bar{k}(\omega),\omega) < \left(\frac{2^{-16^i}(\frac{h(T,P)}{2})}{-2\ \ln(\sigma(\bar{\varepsilon}_{i-1}))}\right)(\ell-m(\bar{k}(\omega),\mathcal{S}_i)).$$

This will be true if

$$m(\bar{k}(\omega)+1,\omega)-m(\bar{k}(\omega),\omega) < \left(\frac{2^{-16^i}(\frac{h(T,P)}{2})}{-\ln(\sigma(\bar{\varepsilon}_{i-1}))\varepsilon}\right)\ell \quad = \alpha\ell .$$

By Lemma 4d.9, as α is determined solely by ε, $\bar{\varepsilon}_{i-1}$, if $k(i')$, $i' \geq i$ increase rapidly enough, then outside a set B_3, $\int_{B_3} \ell(\mathcal{S}_i)d\mu < \frac{\varepsilon}{3}$, the above conflict must occur as no two $m(\bar{k}+1,\mathcal{S}_i)-m(\bar{k},\mathcal{S}_i) > \alpha\ell$, and hence for $\omega \notin B_1 \cup B_2 \cup B_3 = B$ we get $m(\bar{k}(\omega),\mathcal{S}_i) > (1-2^{-16^i}-\varepsilon)$ and further $\int_B \ell(\mathcal{S}_i)d\mu < (\varepsilon' + \frac{\varepsilon'}{\varepsilon_0} + \frac{\varepsilon}{3}) < \varepsilon$. ∎

Notice in Lemma 4d.11 in the order of choice of parameters we must satisfy upper bounds for $\bar{\varepsilon}_{i-1}$ and ε_{i-1}, but no bounds for $k(i-1)$. At the previous stage of construction lower bounds will have been set for all $k(i')$, $i' \geq i-1$ (as they are set at this stage for $i' \geq i$). Hence, we choose $k(i-1)$ to satisfy this bound, guaranteeing the success of our previous work. We are left only to satisfy the bounds on $k(i')$, $i' \geq i$ which we do at stage i' of the construction.

Lemma 4d.12. Setting $\varepsilon = 2^{-4\cdot 16^i}$ in Lemma 4d.11, and when ε_{i-1} is chosen, making $\varepsilon_{i-1} < \frac{\varepsilon}{16(h(T,P)-h(T,\bar{M}))}$, if the parameters $\bar{\varepsilon}_{i-1}$, ε_{i-1}, and $k(i')$, $i' \geq i$ are as in Lemma 4d.11, then $\bar{k}$ satisfies the two desired conditions.

Proof. We consider only i-odd. Condition (ii) follows automatically. As $\int_B \ell(\mathcal{S}_i)d\mu < \varepsilon$, for all but a set B_1 of the $\mathcal{S}_i$, $\int_{B_1} \ell(\mathcal{S}_i) < 2^{-2\cdot 16^i} < 2^{-16^i}$, for all but a set $B_2(\mathcal{S}_i)$, $\int_{B_2(\mathcal{S}_i)} \ell(\mathcal{S}_i) < 2^{-2\cdot 16^i} < 2^{-16^i}$, if $f \in F(\mathcal{S}_i) \cap B_2^c(\mathcal{S}_i)$, $f = \bar{F} \circ f_3$, then

$$\mu(f|\mathcal{S}_i) < \mu(\bar{F}|\mathcal{S}_i) = \mu(\bar{F}|\mathcal{S}_i(\bar{k}(\omega)))2^{\pm 16\varepsilon_{i-1}}$$

$$< 2^{-(h(T,P)-h(T,\bar{M}))(1-2^{-4\cdot 16^i})m(\bar{k}(\omega),\mathcal{S}_i)}\, 2^{16\varepsilon_{i-1}}$$

$$< 2^{-(h(T,P)-h(T,\bar{M}))((1-2^{-4\cdot 16^i})(1-2^{-4\cdot 16^i}-2^{-16^i}) + 2^{-4\cdot 16^i})}$$

$$< 2^{-(h(T,P)-h(T,\bar{M}))(1-2\cdot 2^{-16^i})}$$

and all that remains to show is the upper bound on $\mathrm{card}(F(\mathcal{S}_i))$. From the definition of $\bar{k}$, for all $\bar{F} = \bar{F}(\bar{\varepsilon}_{i-1},\bar{k}(\omega),\mathcal{S}_i)$,

$\mathcal{S}_i \subset S_{i,\ell}$, $\mu(\overline{F}|\mathcal{S}_i^-(\overline{k}(\omega))$

$$> 2^{-(h(T,P)-h(T,\overline{M}))((m(\overline{k}(\omega),\mathcal{S}_i)(1-2^{-16^i})+(\ell-m(\overline{k}(\omega),\mathcal{S}_i))2^{-16^i})},$$

and hence for all $\overline{F}$, $\mu(\overline{F}|\mathcal{S}_i)$

$$> 2^{-(h(T,P)-h(T,\overline{M}))((m(\overline{k}(\omega),\mathcal{S}_i)(1-2^{-16^i})+(\ell-m(\overline{k}(\omega),\mathcal{S}_i)2\cdot 2^{-16^i}}.$$

For a fixed $\mathcal{S}_i$ and value k, $S_i \subset Z = S_{i,\ell}^{i'm(\overline{k},\mathcal{S}_i),m(\overline{k}+1,\mathcal{S}_i}$ assuming the cardinality bound holds on all $F(\mathcal{S}_{i-1}(k',\mathcal{S}_i))$, $\mathrm{card}(F_3(z)) \leqslant \mathrm{card}(F_1(z))$

$$\leqslant \prod_{k'=\overline{k}}^{k(\mathcal{S}_i)} 2^{(h(T,P)-h(T,\overline{M}))(1-2^{-4\cdot 16^{i-2}})(m(k'+1,\mathcal{S}_i)-m(k',\mathcal{S}_i))}$$

$$= 2^{(h(T,P)-h(T,\overline{M}))(1-2^{-4\cdot 16^{i-2}})(\ell-m(\overline{k},\mathcal{S}_i))}.$$

Hence if $\omega \in \overline{F}$, $\overline{k}(\omega) = \overline{k}$, then there are at most this many sets $\overline{F} \circ f_3$ in $F(\mathcal{S}_i)$.

Hence $\mathrm{card}(F(\mathcal{S}_i)) = \sum_{\overline{F}} \mathrm{card}(F_3(Z(\overline{F}))$

$$= \sum_{\overline{F}} \mu(\overline{F}|\mathcal{S}_i)\left(\frac{\mathrm{card}(F_3(Z(\overline{F})))}{\mu(\overline{F}|\mathcal{S}_i)}\right)$$

$$< 2^{(h(T,P)-h(T,\overline{M}))(m(\overline{k}(\omega),\mathcal{S}_i)(1-2^{-16^i})+(\ell-m(\overline{k}(\omega),\mathcal{S}_i)(1-2^{-4\cdot 16^{i-2}}+2\cdot 2^{-16^i}}$$

$$< 2^{(h(T,P)-h(T,\overline{M}))(1-2^{-16^i})\ell(\mathcal{S}_i)}. \quad \blacksquare$$

We now return to the problem discussed earlier of the difference between $F_3(Z)$ and $F_1(Z)$. As we saw then, if $\omega'' \in \overline{f} \in \overline{F}(\mathcal{S}_{i-1}(k',\mathcal{S}_i))$ and $\omega_0'' \in \overline{f}^\circ \in F(\mathcal{S}_{i-1}(k',\mathcal{S}_i))$, and both $T^{-m(k',\mathcal{S}_i)}(\omega'')$ and $T^{-m(k',\mathcal{S}_i)}(\omega_0'')$ lie in the same $\overline{F} \circ f_3 \in F(\mathcal{S}_i)$, then $\overline{f}$ is contained in a set obtained from $\overline{f}_0$ by three t_0 thickenings, one by $\overline{\varepsilon}_{i-1}$, then $\overline{\varepsilon}_{i-2}$ and finally $\overline{\varepsilon}_{i-1}$. We now set an inductive upper bound on the $\overline{\varepsilon}_{i-1}$ so that the effect of this is finitarilly controlled. Define $\hat{F}(\mathcal{S}_i)$ to be $F(\mathcal{S}_i)$ if $\mathcal{S}_i$ begins with $M(i')$, $i' > i$, and to be $F(\mathcal{S}_i) \vee \{M_0(i), M_1(i)\}$ otherwise, i.e., split the initial marker into the two cuing sets. As $\varepsilon_{i-1} < \frac{\varepsilon_0}{12}$, it follows from CBI that for $f \in \overline{F}(\mathcal{S}_i)$, $\frac{2}{3} > \mu(M_0(i)|f) > \frac{1}{2}$. At stage i, sets of the form $Q(j,\overline{F},\mathcal{S}_i) = T^{-j}(\overline{F})$ where $\overline{F} \in F(\mathcal{S}_i)$, $0 \leq j \leq \ell(\mathcal{S}_i)$, partition Ω into countably many almost clopen sets. Call this partition $Q(i)$. A set in $Q(i)$ consists of those points at index j in filler $\overline{F} \in \hat{F}(\mathcal{S}_i)$ of the skeleton $\mathcal{S}_i$. Hence there is a $J(i)$ so large that for all but a set B of Ω, $\mu(B) < 2^{-16^i}$, the $P_{-J(i)}^{J(i)}$ name of ω determines which set in $Q(i)$, ω lies in. Select $\hat{\varepsilon}_i$ so small that $\hat{G}(i) = \{\omega | T^j(\omega) \in \bigcup_i A_i(\varepsilon_i)$ for all j, $J(i) \geq j \geq -J(i)\}$ has $\mu(\hat{G}(i)) > 1 - 2^{-16^i}$. Now as $\bigcup_i A_i(\hat{\varepsilon}_i)$ is almost clopen select $J'(i)$ so large that $\mu(\{\omega | P_{-J'(\omega)}^{J'(\omega)}(\omega) \subset A(\hat{\varepsilon}_i)(\omega)\}) > 1 - 2^{-16^i}$ and now select $\hat{\hat{\varepsilon}}_i$ so small that $\hat{\hat{G}}(i) = \{\omega | T^j(\omega) \in A_0(\hat{\hat{\varepsilon}}_i)$, $J(i) + J'(i) \geq j \geq -J(i) - J'(i)\}$ has $\mu(\hat{\hat{G}}(i)) > 1 - 2^{-16^i}$.

We now set some further bounds on $\overline{\varepsilon}_{i'}$ and $k(i')$, $i' > i$. First just a needed basic bound,

$$2^{-(h(T,P)/2)(2^{-4\cdot 16^i} - 2^{-8\cdot 16^i})(\ell(M(i))} < 2^{-16^i}.$$

Be sure $k(i+1)$ is so large that $\mu(\{Q(j,\overline{F},\mathcal{S}_{i+1}) | \mathcal{S}_{i+1}$ begins with $M(i+1)$ and $\ell(\mathcal{S}_{i+1}) - M(i+1) > 4\cdot 2^{16^i}(J(i)+J'(i))\})$ $> 1 - 2^{-16^{i+1}}$, (this measure is termwise bounded in terms of $(1 - \sum_{i' \geq i+1} \mu(M(i'))\ell(M(i')))$ hence if $k(i') \nearrow \infty$ fast enough we get the bound). We also require that $\overline{\varepsilon}_{i+1} < \hat{\hat{\varepsilon}}_i$ and that the sequence $\overline{\varepsilon}_{i+1}, \overline{\varepsilon}_{i+2}, \overline{\varepsilon}_{i+1}, \overline{\varepsilon}_{i+2}, \overline{\varepsilon}_{i+3}, \overline{\varepsilon}_{i+2} \cdots \overline{\varepsilon}_{i'}, \overline{\varepsilon}_{i'+1}, \overline{\varepsilon}_{i'}, \ldots$ be termwise bounded by $\alpha_1(\hat{\varepsilon}_i), \alpha_2(\hat{\varepsilon}_i), \ldots, \alpha_{3i'-(i+1)2}(\hat{\varepsilon}_i), \alpha_{3i'-2i-1}(\hat{\varepsilon}_i), \alpha_{3i'-2i}(\hat{\varepsilon}_i)$.

Thus, by stage i' we will have acquired a further finite number of upper bounds on $\overline{\varepsilon}_i$ and lower bounds on $k(i')$.

We select parameters as we work through Lemma 4d.11 to also satisfy these. Call such a choice of markers and filler partitions a "codible filled marker scheme". The next lemma explains how our last restrictions on parameters control the error introduced by replacing F_1 with F_3.

Lemma 4d.12. Let $M(i)$, $Q(i)$ form a codible filled marker scheme on (T,P). There is, then, a set $\mathcal{G}(i) \subset Q(i+1)$, $\mu(\mathcal{G}(i)) > 1 - 9\cdot 2^{-16^{i-1}}$ so that for $\overline{Q}_{i+1} \in \mathcal{G}(i)$, there is a $Q_i \in Q(i)$, $\overline{Q}_{i+1} \subset Q_i$, and for any sequence $Q_i \in Q(i')$, $i' \geq i+1$, $Q_{i+1} = \overline{Q}_{i+1}$, $\mu(Q_{i'+1} \cap Q_{i'}) > 0$, we have $Q_i \subset Q_i$ for all i'.

Proof. Write $\overline{Q}_{i+1} = Q(j',\overline{F}',\mathcal{S}'_{i+1})$. First assume $\mathcal{S}_{i+1}$ begins with $M(i+1)$. This omits 2^{16^i} of Ω and is $Q(i+1)$ measurable. Next assume $\ell(\mathcal{S}_{i+1}) - \ell(M(i+1)) > 4\cdot 2^{16}(J(i)+J'(i))$ $m(\overline{k}(\overline{F}'),\mathcal{S}_{i+1}) > (1-2\cdot 2^{-16^i})\ell(\mathcal{S}_{i+1})$, and $\ell(M(i+1)) + J(i) + J'(i) < j' < m(\overline{k}(\overline{F}'),\mathcal{S}_{i+1}) - J(i) - J'(i)$. This omits a further $5\cdot 2^{-16^i}$ of Ω, and once more is $Q(i+1)$ measurable. Next assume there is an atom $a \in P^{J(i)+J'(i)}_{-J(i)-J'(i)}$, $Q_{i+1}(\omega) \subset a$. This omits only points in $\hat{\hat{G}}(i)$ hence at most 2^{-16^i} of Ω and what remains is by definition $Q(i+1)$ measurable. Next assume $Q_{i+1} \subset \hat{G}(i)$. As $Q_{i+1} \subset a$, and $\hat{G}(i)$ is 2^{-16^i} filled by atoms a, this omits 2^{-16^i} of Ω and what remains is still $Q(i+1)$ measurable. Lastly assume $\overline{a} \in P^{J(i)}_{-J(i)}$ containing Q_{i+1}, as $Q(i+1)$ is in $\hat{G}(i)$, determines the set in $Q(i)$ it lies in. This omits a further 2^{-16^i} of Ω and is $Q(i+1)$ measurable. We have omitted at most $9\cdot 2^{-16^i}$ of Ω. Call the remaining $Q(i+1)$ measurable set $\mathcal{G}(i)$.

If $\overline{Q}_{i+1} \in \mathcal{G}(i)$ then $\overline{Q}_{i+1}$ is contained in $\hat{G}(i) \cap \overline{a}$, $\overline{a} \in {}^{J(i)}_{-J(i)}$, so $Q_i \subset \overline{Q}_{i+1}$.

Now suppose $Q_{i+1} = \overline{Q}_{i+1}, Q_{i+2}, \ldots, Q_{i'}, \ldots$ have $\mu(Q_{i-1} \cap Q_{i'}) > 0$. We know Q_{i+2} is contained in a set obtained from Q_{i+1} by three t_0-thickenings, first by $\overline{\varepsilon}_{i+2}$, then $\overline{\varepsilon}_{i+1}$, and finally $\overline{\varepsilon}_{i+2}$, inductively $Q_{i'}$ is contained in a set obtained by three t_0 thickenings of $Q_{i'-1}$, an $\varepsilon_{i'}$, then $\varepsilon_{i'-1}$, and finally $\varepsilon_{i'}$ thickening. All these thickenings are across blocks containing the indices of $\overline{Q}_{i+1}$. Thus by our bounds $Q_{i'}$ is contained in the t_0,ε_0-thickening of $\overline{Q}_{i+1}$ across its length. But now $\overline{Q}_{i+1} \subset \{\omega | T^j(\omega) \in \bigcup_i A_i(\varepsilon_0), -J(i) \leq j \leq J(i)\}$ hence this thickening is contained in $\overline{a}$. We conclude $Q_{i'} \subset \overline{a} \subset Q_i$ and we are done. ∎

We end this section with a simple remark. Suppose we have two finitarily Bernoulli processes (T,P) and (T',P'). We say they are "spacer compatible" if for some thickenings and spacers satisfying the conditions of the definition, there are sequences $\varepsilon_j \searrow 0$, $\varepsilon'_j \searrow 0$ so that $N(\varepsilon_j) = N'(\varepsilon'_j)$, and for any set $I \subset \mathbb{N}$

$$\mu\Big(\bigcap_{j\in I} T^{p(j)}(S(\varepsilon_j))\Big) = \mu'\Big(\bigcap_{j\in I} T'^{p(j)}(S'(\varepsilon'_j))\Big),$$

i.e., the spacer processes on these subsequences are identical. We say (T,P) and (T',P') are "marker matched" if we can construct in (T,P) and (T',P') codible filled marker schemes $M(i)$, $Q(i)$ and $M'(i)$, $Q'(i)$, inverting in (T',P') the roles of even and odd in the conditions on $\overline{k}'$, so that $(T,\overline{M})$ and $(T',\overline{M}')$ are identical processes. The following lemma is now virtually obvious.

Lemma 4d.13. If (T,P) and (T',P') are spacer compatible then they can be marker matched.

Proof. As all our bounds on ε_i, $\overline{\varepsilon}_i$ in the construction of codible filled marker schemes are one sided, we can always select from the subsequence ε_j, ε'_j, respectively, and furthermore make the same choice in both processes. As the bounds on $k(i)$ are one sided, we can select $k(i) = k'(i)$ and now as the maps are spacer compatible it follows that $(T,\overline{M}) \equiv (T',M')$. ∎

To finish the finitary isomorphism we now construct a map (T_1,P_1) spacer compatible with both (T,P) and (T',P'), and then show that if two processes are marker matched and of the same entropy, then they are finitarily isomorphic.

(e) Constructing (T_1,P_1). We begin by replacing the spacer sets $S(\varepsilon)$ in the finitarily Bernoulli structure of (T,P) with new spacers satisfying a needed condition. Choose $n_0 > N_0$, $n_0 > \overline{N}(1)$ so that by USM, for any ε,ε', $\mu(T^{n_0}(S(\varepsilon)) \cap S(\varepsilon')) = \mu(S(\varepsilon))\mu(S(\varepsilon'))2^{\pm 1}$. Select $\varepsilon_0,\varepsilon_1,\ldots \searrow 0$ so that for $i \neq i'$ if $-N(\varepsilon_i/2) + N_0 < j < N(\varepsilon_{i'}/2) - N_0$ then $T^j(S(\varepsilon_i/2)) \cap S(\varepsilon_{i'}/2) = \phi$ and $N(\varepsilon_i) > 2n_0$ and $\varepsilon_{2i} < \varepsilon_i/2$. Now define $\overline{S}(\varepsilon_i) = (S(\varepsilon_{2i}),N(\varepsilon_{2i})+n_0)(S(\varepsilon_{2i+1}),N(\varepsilon_{2i+1}))$, and $\overline{N}(\varepsilon_i) = N(\varepsilon_{2i}) + n_0 + N(\varepsilon_{2i+1})$.

Lemma 4e.1. The sets $\overline{S}(\varepsilon_i)$ and $\overline{N}(\varepsilon_i)$, leaving all the other structure fixed, can play the role of a spacer set and its length in (T,P).

Proof. Clearly $\overline{S}(\varepsilon_i)$ are almost clopen and (ii) if $i' \neq i$, $-\overline{N}(\varepsilon_i)$, $-\overline{N}(\varepsilon_i) + N_0 < j < \overline{N}(\varepsilon_{i'}) - N_0$, then $T^j(\overline{S}(\varepsilon_i)) \cap \overline{S}(\varepsilon_{i'}) = \phi$. (iii) USM for $\overline{S}$ follow from USM for S. (i) $\overline{S}$ spacers are unchanged by t_1 thickenings across their length as this is true of S spacers.

Lastly, (iv),v)) Proposition 4 and 5 for $\overline{S}$ follow with $\overline{\sigma}(\varepsilon_i) = 2^{-2\varepsilon_i/2}\,\sigma(\varepsilon_i)$ from CBI and Proposition 4 and 5 on S.

CBI on $\overline{S}$ follows from two applications of CBI on the original spacer, first splitting between the $S(\varepsilon_{2i})$ and $S(\varepsilon_{2i+1})$, and then adding back on just the $S(\varepsilon_{2i})$. ∎

These new spacers have an extra property "conditional spacer independence" (CSI). Let $\overline{S}_i = \{\overline{S}(\varepsilon_1),\ldots,\overline{S}(\varepsilon_i), (\bigcup_{j=1}^{i} \overline{S}(\varepsilon_j))^c\}$, and set $\overline{C}(i,n) = \bigvee_{j=1}^{i-1} (\overline{S}_j)_1^n \vee \bigvee_{j=i}^{\infty} (\overline{S}_j)_1^{n-\overline{N}(\varepsilon_j)}$. Now for any $i > i',i''$ and sets $A \subset \overline{C}(i',n)$, $B \subset \overline{C}(i'',m)$, then

$$\mu((A,n)\circ(\overline{S}(\varepsilon_i)\circ\overline{N}(\varepsilon_i))\circ(B,m)\,|\,(A,n)\circ(\overline{S}(\varepsilon_i);\overline{N}(\varepsilon_i)))$$

$$=\mu((\overline{S}(\varepsilon_i)\circ\overline{N}(\varepsilon_i))\circ(B,m)\circ(\overline{S}(\varepsilon_i)\circ\overline{N}(\varepsilon_i)))2^{\pm\varepsilon_i} \quad \text{if}$$

$$\mu((A,n)\circ(\overline{S}(\varepsilon_i),\overline{N}(\overline{\varepsilon}_i))) \neq 0.$$

This follows from two applications of CBI on S, as for any ε_j, $\overline{C}(i,n) \subset C(\varepsilon_j, n + \overline{N}(\varepsilon_{i-1})) \subset C(\varepsilon_j, n + N(\varepsilon_{2i}))$, hence we can first split between the $S(\varepsilon_{2i})$ and $S(\varepsilon_{2i+1})$ in $\overline{S}(\varepsilon_i)$, then add back on the $S(\varepsilon_{2i})$.

Thus we can always assume for some sequence $\varepsilon_i \searrow 0$ our spacers have CSI, but need never prove it.

Let (T,P) and (T',P') be finitarily Bernoulli processes of the same entropy. We distinguish the structures of the definition by putting a "prime" on those for (T',P') and a subscript 1 for (T_1,P_1).

To build (T_1,P_1) select $\varepsilon_j \searrow 0$, $\varepsilon_j' \searrow 0$, subsequences of the sequence on which CSI holds, so that:

(i) $h(s) = h(\{S(\varepsilon_1),\ldots,(\bigcup_i S(\varepsilon_i))^c\})$, and $h(S") = h(\{S"(\varepsilon_1'),\ldots,(\bigcup_i S'(\varepsilon_i'))^c\}) < h/4$;

(ii) $\mu(\{\omega \mid \omega \in S(\varepsilon_i), T^j(\omega) \in S(\varepsilon_i)^c$, $j = 1,\ldots, \frac{\min(N(\varepsilon_{i+1}),N'(\varepsilon_{i+1}'))}{10}\}) < \frac{\varepsilon i}{10}\,\mu(S(\varepsilon_i))$ as is the corresponding set in (T',P'); and

(iii) from USM on both processes, $\min(\frac{N(\varepsilon_i)}{2}, \frac{N'(\varepsilon_i)}{2}) > \max(\overline{N}(\varepsilon_i), \overline{N}'(\varepsilon_i))$.

Let $(\overline{T},B)$ be a finite state independent process with $h(T,S) + h(T',S') + h(B) = h$. Define $(T_1,P_1) = (T,S) \times (T',S') \times (\overline{T},B)$, a countable state process with $h(P_1) < \infty$ and $h(T_1,P_1) = h$. In (T_1,P_1) we have copies of the spacer processes $S(\varepsilon_j)$, $S'(\varepsilon_j')$ sitting as the first and second coordinates We now define thickenings so that (T_1,P_1) is finitaril Bernoulli by either choice, hence spacer compatible with both (T,P) and (T',P'). We only consider by symmetry, the (T,P) case. We now construct the needed structure on (T_1,P_1). Let

$$S_i = \{S(\varepsilon_1),\ldots,S(\varepsilon_i),(\bigcup_{j=1}^{i} S(\varepsilon_j))^c\},$$

and
$$C_1(\varepsilon_i,n) = \bigvee_{j=1}^{i-1} (S_j)_0^{n-1} \vee \bigvee_{j=i}^{\infty} (S_j)_0^{n-\overline{N}(\varepsilon_j)-1} \vee \bigvee_{j=1}^{i-1} (S_j')_0^{n-1} \vee B_0^{n-1}.$$
This is certainly clopen as, for each n, it is a finite intersection of clopen sets. Also for $\varepsilon_i \leqslant \varepsilon_j$, $n' \geqslant n$, $C_1(\varepsilon_i,n')$ clearly refines $C_1(\varepsilon_j,n)$. Let

$$S_i = \{S(\varepsilon_1),\ldots,S(\varepsilon_{i-1}),(\bigcup_{i'<i} S(\varepsilon_i))^c\},$$

$$S_i' = \{S'(\varepsilon_1'),\ldots,S'(\varepsilon_{i-1}'),(\bigcup_{i'<i} S'(\varepsilon_i))^c\} ,$$

and define

$$\{A_j(\varepsilon_i),\overline{A}_j(\varepsilon_i)\} = (S_i \times S_i' \times B)_0^{\min(\frac{N(\varepsilon_i)}{3},\frac{N'(\varepsilon_i')}{3})}$$

where an atom is an $A_j(\varepsilon_i)$ if for <u>some</u> $i' \leqslant i$, $0 \leqslant j' \leqslant \min(\frac{N(\varepsilon_i)}{3},\frac{N'(\varepsilon_i')}{3})$, at index j', $A_j(\varepsilon_i)$ is in $S(\varepsilon_{i'})$, <u>and</u> for some $i'' < i$, $0 \leqslant j'' \leqslant \min(\frac{N(\varepsilon_i)}{3},\frac{N'(\varepsilon_i')}{3})$, at index j'', $A_j(\varepsilon_i)$ is in $S'(\varepsilon_{i''}')$. As $\varepsilon_i \downarrow$, $(S_i \times S_i' \times B)_0^{\min(\frac{N(\varepsilon_i)}{3},\frac{N'(\varepsilon_i')}{3})}$ refine, and from the growth rate of $N(\varepsilon_i), N'(\varepsilon_i'), \mu(\bigcup_j A_j(\varepsilon_i)) > 1 - \frac{\varepsilon_{i-1}}{5}$. Clearly $\bigcup_j \overline{A}_j(\varepsilon_i) \overset{i}{\downarrow}$.

To see that $t_1(\omega,\varepsilon_i,n) \subset t_0(\omega,\varepsilon_i,n)$ note that if $\omega_1' \in t_1(\omega_1,\varepsilon_i,n)$ then the P_1,n-names of ω_1 and ω_1' must agree on occurrences of $S(\varepsilon_1)\ldots S(\varepsilon_{i-1})$, $S'(\varepsilon_1'),\ldots,S'(\varepsilon_{i-1}')$ and B. A difference for an $S(\varepsilon_j)$, $j \geqslant i$ or $S'(\varepsilon_j)$, $j \geqslant i$ then can only occur at an index in some $\overline{A}_j(\varepsilon_i)$, in which case $t_1(\omega,\varepsilon_i,n)$ will be contained in $\overline{A}_j(\varepsilon_i)$ at this index.

If we take $t_0(\omega,\varepsilon_i,n)$ and further t_0 thicken it across blocks containing $0,\ldots,n-1$ by values $\leqslant \varepsilon_i$, we will not change the set, as t_0 arises from a refining sequence of partitions.

Our spacers are of course $S(\varepsilon_i)$, $N(\varepsilon_i)$. We check (i) through (v).

(i) t_1-thickening a spacer $S(\varepsilon_i)$, by any amount across a block containing $0,\ldots,N(\varepsilon_i)-1$ will just give back $S(\varepsilon_i)$ from the form of $C_1(\varepsilon_j, n+N(\varepsilon)+m)$.

(ii) This follows automatically as it depends only on the $S(\varepsilon_i)$.

(iii) The same is true of USM.

(iv) Proposition 4 with $\sigma_1(\varepsilon_i) = \sigma(\varepsilon_i)\sigma'(\varepsilon_i)\min_{b\in B}(\overline{\mu}(B))$ follows from Proposition 4 on (T,P), Proposition 5 on (T',P') and the independence of $(\overline{T},B)$.

(v) Proposition 5, as with USM, is only a condition on spacers, hence, follows automatically.

CBI follows from CSI on the first coordinate, USM on the second, and independence of the third. Hence, we conclude:

Lemma 4e.2. The process (T_1,P_1) is finitarily Bernoulli and spacer compatible with both (T,P) and (T',P').∎

(f) The Isomorphism. We have now reduced the problem to showing that two marker matched finitarily Bernoulli maps (T,P) and (T',P') are finitarily isomorphic. Let $M(i)$, $Q(i)$ and $M'(i)$, $Q'(i)$ be their respective codible marker schemes. As $(T,M) \equiv (T',M')$, any i-skeleton $\mathcal{S}_i$ can be regarded as being in both marker schemes. Let ν_i be the conditional measure on $\hat{F}(\mathcal{S}_i)$ given $\mathcal{S}_i$, ν_i', that of $\hat{F}'(\mathcal{S}_i)$ given $\mathcal{S}_i$.

We now inductively define a sequence of societies, R_i between $(\hat{F}(\mathcal{S}_i),\nu_i)$ and $(F'(\mathcal{S}_i),\nu_i')$ if i-odd and $(\hat{F}'(\mathcal{S}_i),\nu_i')$ and $(F(\mathcal{S}_i),\nu_i)$ if i-even, where R_i is $f_{2^{-6i}}$-fat (remember fatness is not quite symmetric). We then will verify that the R_i converge to a finitary isomorphism.

To get started, R_1 is the universal society. Assume, inductively for each $\mathcal{S}_{i-1}$ (i-1)-even, we have R_{i-1}, an $f_{2^{-6(i-1)}}$-fat society. To construct R_i on $(\hat{F}(\mathcal{S}_i),\nu_i)$,

$(F'(\mathcal{S}_i), \nu_i')$, first write $F_k = \bigvee_{k'=1}^{k} T^{m(k,\mathcal{S}_i)+1}(F(\mathcal{S}_{i-1}(k,\mathcal{S}_i))$, $1 \leq k \leq k(\mathcal{S}_i)$ and $F_k' = \bigvee_{k'=1}^{k} T'^{m(k,\mathcal{S}_i)+1}(\hat{F}'(\mathcal{S}_{i-1}(k,\mathcal{S}_i))$.

The F_k and $\hat{F}_k'$ form nested algebras of sets. An element of F_k can be written $f_1 \circ f_2 \circ \ldots \circ f_k$, $f_{k'} \in F(\mathcal{S}_{i-1}(k',\mathcal{S}_i))$. Let $f \in F(\mathcal{S}_{i-1}(k,\mathcal{S}_i))$ and $\bar{f} \in F_{k-1}$. From CBI we compute

$$\nu_{i,\bar{f}}(f) = \nu_i(\bar{f} \circ f | \bar{f}) = \nu_{i-1}(f) 2^{\pm 8\varepsilon_{i-1}} .$$

Symmetrically an element of F_k' has the form $f_1' \circ f_2' \circ \ldots \circ f_k'$ where $f_{k'}' \in \hat{F}(\mathcal{S}_{i-1}(k',\mathcal{S}_i))$, and for $f' \in \hat{F}_k(\mathcal{S}_{i-1}(k,\mathcal{S}_i))$ $\bar{f}' \in F_{k-1}$, $\nu_{i,\bar{f}'}'(f') = \nu_i'(\bar{f}' \circ f' | \bar{f}') = \nu_{i-1}(f') 2^{\pm 8\varepsilon_{i-1}}$.

<u>Lemma 4f.1</u>. R_{i-1} is an $f_{2^{-6i+3}}$-fat society on $(\hat{F}'(\mathcal{S}_{i-1}(k,\mathcal{S}_i), \nu_{i,\bar{f}'}')$ and $(F(\mathcal{S}_{i-1}(k,\mathcal{S}_i), \nu_{i,\bar{f}})$ for any $\bar{f} \in F_{k-1}$, $\bar{f}' \in \bar{F}_{k-1}'$.

Letting $g_\varepsilon(x) = \begin{cases} \varepsilon x, & x \leq 1/2 \\ \varepsilon(1-x), & x \geq 1/2 \end{cases}$ it is an easy check that $g_\varepsilon(x) \geq f_\varepsilon(x) \geq g_{\varepsilon/4}(x)$.

Let $A' \subset F'(\mathcal{S}_{i-1}(k,\mathcal{S}_i))$. First assume $\nu_{i,f'}'(A) \leq 1/2$ (in this computation we assume all exponents are so small that $(1-2^{-\varepsilon}) < 2\varepsilon$, $(2^\varepsilon - 1) < \varepsilon$).

$$(\nu_{i,\bar{f}}(A')) - \nu_{i,\bar{f}'}'(A))$$

$$> (\nu_{i-1}(R_{i-1}(A')) - \nu_{i-1}'(A') 2^{16\varepsilon_{i-1}}) 2^{-8\varepsilon_{i-1}}$$

$$> (f_{2^{-6(i-1)}}(\mathcal{S}_{i-1}'(A')) - (2^{24\varepsilon_{i-1}} - 2^{8\varepsilon_{i-1}}) \nu_{i,\bar{f}'}'(A')) 2^{-8\varepsilon_{i-1}}$$

$$\geq (g_{2^{-6i+4}}(\mathcal{S}_{i-1}'(A')) - (2^{24\varepsilon_{i-1}} - 2^{8\varepsilon_{i-1}}) \nu_{i,\bar{f}}'(A')) 2^{-8\varepsilon_{i-1}} .$$

Now from the form of g_ε, as $\nu_{i,f'}'(A') \leq 1/2$, this is

$$> (2^{-6i+4} \cdot 2^{-8\varepsilon_{i-1}} - (2^{24\varepsilon_{i-1}} - 2^{8\varepsilon_{i-1}}))2^{-8\varepsilon_{i-1}} \nu'_{i,\overline{f}'}(A')$$

$$> (2^{-6i+4-8\varepsilon_{i-1}} - 25 \cdot 2^{-16^i}) \nu'_{i,f'}(A')$$

$$> (2^{-6i+3}) \nu'_{i,f'}(A') = g_{2^{-6i+3}}(\nu'_{i,f'}(A'))$$

$$> f_{2^{-6i+3}}(\nu'_{i,f'}(A')) .$$

If $\nu'_{i,f'}(A') > 1/2$, we compute symmetrically

$$(\nu_{i,f}(R_{i-1}(A')) - \nu'_{i,f'}(A'))$$

$$= (\nu'_{i,f'}(A'^c) - \nu_{i,f}(R_{i-1}(A')^c))$$

$$> (\nu'_{i-1}(A'^c)2^{-16\varepsilon_{i-1}} - \nu_{i-1}(R_{i-1}(A')^c))2^{8\varepsilon_{i-1}}$$

$$> (f_{2^{-6(i-1)}}(\nu'_{i-1}(A')^c) - (2^{8\varepsilon_{i-1}} - 2^{-8\varepsilon_{i-1}})\nu'_{i,f'}(A'^c))2^{8\varepsilon_{i-1}} .$$

Now as $\nu_{i,f'}(A'^c) \leqslant 1/2$ we can continue as in the previous case.

<u>Corollary 4f.2</u>. The society R'_{i-1} on $(F'_{k(\mathcal{S}_i)}, \nu'_i)$ and $(F_{k(\mathcal{S}_i)}, \nu_i)$ given by $\{(f'_1 \circ \dots \circ f'_{k(\mathcal{S}_i)}, f_1 \circ \dots \circ f_{k(\mathcal{S}_i)})/ (f_j, f'_j) \in R_{i-1}\}$ is $f_{2^{-6i+3}}$-fat.

<u>Proof</u>. Using Lemma 4f.1, apply Theorem 4b.4 sequentially across the $k(\mathcal{S}_i)$ subskeleta of $\mathcal{S}_i$. ■

Define $R_{i,1}$ on $(F'(\mathcal{S}_i), \nu'_i)$ and $(F(\mathcal{S}_i), \nu_i)$ to consist of all pairs (f',f) such that for some $f'_1 \in F_{k(\mathcal{S}_k)}$, $f_1 \in F_{k(\mathcal{S}_i)}$, $(f'_1, f_1) \in R'_{i-1}$ and both $\nu_i(f \cap f_1) > 0$ and $\nu'_i(f' \cap f'_1) > 0$. This is a standard way of transfering a society on one pair of partitions to another and preserves fatness.

<u>Corollary 4f.3</u>. $R_{i,1}$ is $f_{2^{-6i+3}}$-fat.

<u>Proof</u>. For $A' \subset F'(\mathcal{S}_i)$, let $A_1' = \{f_1' \in F_{k(\mathcal{S}_i)} \mid \nu_i(A' \cap f_1') > 0\}$, and now $\nu_i(R_{i'}(A')) \geqslant \nu_i(R_{i-1}'(A_1'))$ $\geqslant \nu_i'(A_1') + f_{2^{-6i+3}}(\nu_i'(A_1)) \geqslant \nu_i(A') + f_{2^{-6i+3}}(\nu_i'(A'))$.

Let $R_{i,2}$ be the dual society to $R_{i,1}$, between $(F(\mathcal{S}_i),\nu_i)$, $(F'(\mathcal{S}_i),\nu_i')$. It is $f_{\frac{2^{-6i+3}}{1+2^{-6i+3}}}$-fat, hence $f_{2^{-6i+2}}$-fat. If $\mathcal{S}_i$ does not begin with $M(i)$, let $R_i = R_{i,2}$. If $\mathcal{S}_i$ begins with $M(i)$, apply the Marriage lemma to get a society $R_{i,3} \subseteq R_{i,2}$ satisfying $\mathrm{card}(\{f \in F(\ _i) \mid$ two f_1', f_2' with $(f,f_1'),(f,f_2') \in R_{i,3}\})$ $\mathrm{card}\ F(\ _i)$.

To get R_i we use Lemma 4b.6 to "average" $R_{i,2}$ and $R_{i,3}$ as follows. Any set $f \in F(\mathcal{S}_i)$ has the form $(M(i)^-)$ $(S(\varepsilon_i),N(\varepsilon_i))\overline{E}_1(S(\varepsilon_{i-1}),N(\varepsilon_{i-1}))\overline{E}_2$ where E_1, E_2 are t_1, $\overline{\varepsilon}_{i-1}$ thickened. Divide f into two sets

$$B_f = M_0(i) \cap f, \quad B_f' = M_1(i) \cap f .$$

We compute $\frac{2}{3} \geqslant \nu_i(B_f | f) \geqslant \frac{1}{2}$ from the size of $M_0(i)$ and CBI.

Let $R_i = \{(f,f') \mid f \in \hat{F}(\mathcal{S}_i),\ f' \in F(\mathcal{S}_i)$ where either $f = M_0(i) \cap f_1$ and $(f_1,f') \in R_{i,2}$ or $f \in M_1(i) \cap f_1$ and $(f_1,f') \in R_{i,3}\}$. By Lemma 4b.6 R_i is $f_{2^{-6i+2}/4}$ hence $f_{2^{-6i}}$-fat. The construction of R_i for i even is precisely symmetric.

What now remains is to see that the R_i converge to a finitary isomorphism. In (T,P) we define, for $a' \in P_{-n}'^{n}$, a sequence of almost clopen sets $\varphi_i(a')$ which increase to the preimage under the isomorphism of a'. Let $\omega \in \varphi_i(a')$ if for some i'-odd, $i' \leqslant i$, $\omega \in Q(j,f,\mathcal{S}_{i'})$, $\omega \in \mathcal{G}(i')$, $f = f_0 \cap M_0(i')$, where $f_0 \in F(\mathcal{S}_{i'})$, and there is a unique f' $\in F'(\mathcal{S}_{i'})$ with $(f_0,f') \in R_{i,3}$, furthermore $Q'(j,f',\mathcal{S}_{i'})$ $\subset \mathcal{G}'(i'-1)$, which by Lemma 4d.12 is $Q'(i')$ measurable. Thus for some $Q'(j',f'',\mathcal{S}_{i-1}(k,\mathcal{S}_i))$ with $j = j'+m(k,\mathcal{S}_{i'})$,

$Q'(j,f',\mathcal{S}_{i'}) \subset Q'(j',f'',\mathcal{S}_{i'-1}(k,\mathcal{S}_i))$ and lastly we ask that $Q'(j',f'',\mathcal{S}_{i'-1}(k,\mathcal{S}_{i'})) = a'$.

$\varphi_i(a')$ is a countable union of sets of the form $\mathcal{G}(i') \cap Q(j,f,\mathcal{S}_{i'})$, hence is almost open, and clearly $\varphi_{i+1}(a') \supset \varphi_i(a')$. Let $\varphi(a') = \bigcup_i \varphi_i(a')$. We will now see that $\mu(\bigcup_{a' \in P'^{m}_{-m}} \varphi(a')) = 1$ and that if $a_1' \neq a_2'$ then $\varphi(a_1') \cap \varphi(a_2') = \phi$. It will follow that $\varphi(a')$ is almost closed. We will then see how φ extends to an almost continuous homomorphism (T,P) to (T',P') and lastly that φ', the map constructed at even steps, is φ^{-1}. We take a first step.

<u>Lemma 4f.4</u>. For all m, $\mu(\bigcup_{a' \in P^{m}_{-m}} \varphi(a)) > \frac{1}{4}$.

<u>Proof</u>. For any ε, as we have seen before, there is an I so that if $i' \geqslant I$, then $\mu'(U_1') = \mu'(\{Q' \in Q'(i'-1)) \mid$ for some $a' \in P^{m}_{-m}$, $Q' \subset a\}) > 1-\varepsilon$.

Thus for $i' \geqslant I$, by Lemma 4d.12, $\mu'(U_2') =$
$\mu'(\{Q'(j,f',\mathcal{S}_{i'}) \in Q'(i') \mid Q'(j,f',\mathcal{S}_{i'}) \subset \mathcal{G}'(i'-1) \cap U_1\}) >$
$1 - \varepsilon - 9 \cdot 2^{-16^{i'-1}}$.

By property (i) of $\bar{k}$, $\bar{k}'$ the $\bar{\varepsilon}_i$,i-s.t.'s, $\mu(U_3) =$
$\mu(\{Q(j,f,\mathcal{S}_{i'}) \mid$ there are more than one f',

$$(f_0,f') \in R_{i,3}\}) \leqslant \int_{\mathcal{S}_{i'} \text{ does not begin with } M(i')} \ell(\mathcal{S}_{i'}) + \int_{B_1} \ell(\mathcal{S}_{i'}) + \int_{\bigcup_{\mathcal{S}_i} B_2(\mathcal{S}_{i'})} \ell(\mathcal{S}_{i'}) + \sum_i \mu(\mathcal{S}_{i'})\ell(\mathcal{S}_{i'})\operatorname{card}(F'(\mathcal{S}_{i'}))$$
$$\cdot\, 2^{-(h(T,P)-h(T,\overline{M}))(1-2^{-8\cdot 16^{i'}})\ell(\mathcal{S}_{i'})}$$
$$\leqslant 3 \cdot 2^{-16^{i}} + \sum_{i'} \mu(\mathcal{S}_{i'})\ell(\mathcal{S}_{i'})2^{-(h(T,P)-h(T,\overline{M}))(2^{-8\cdot 16^{i'}}-2^{-4\cdot 16^{i'}})\ell(\mathcal{S}_{i'})}$$
$$\leqslant 3 \cdot 2^{-16^{i'}} + \sum_{i'} \mu(\mathcal{S}_{i'})(\ell(\mathcal{S}_{i'})2^{-h(T,P)(2^{-4\cdot 16^{i'}}-2^{-8\cdot 16^{i'}})\ell(M(i'))/2}$$
$$\leqslant 4 \cdot 2^{-16^{i}}$$

by a bound we set on $\ell(M(i))$.

Thus $\mu(U_4) = \mu(\{Q(j,f,\mathcal{S}_{i'}) \mid f = M_0(i) \cap f_0 \cap \mathcal{Z}(i')$ and there is a unique $f' \in U_2'$, $(f_0,f') \in R_{i,3}\}) > \frac{1}{3} - 4 \cdot 2^{-16^{i'}} - 8 \cdot 2^{16^{i'-1}} > \frac{1}{4}$ if ε small enough, i.e., i' large enough. But now $U_4 = U_4(i') \subset \bigcup_{a' \in P'^{m}_{-m}} \varphi_{i'}(a')$. ■

<u>Corollary 4f.5</u>. For all m, $\mu(\bigcup_{a' \in P'^{m}_{-m}} \varphi(a')) = 1$.

<u>Proof</u>. Let $\bar{\varphi}(m) = \bigcup_{a' \in P'^{m}_{-m}} (a')$. It is clear

$$\bigcap_{j=-t}^{t} T^j(\bar{\varphi}(n)) \supset \bar{\varphi}(n+t)$$

Hence for all m,

$$\mu(\bigcap_{j=-\infty}^{\infty} T^j(\bar{\varphi}(m))) > \frac{1}{4}.$$

Ergodicity of (T,P) implies the result. ■

<u>Lemma 4f.6</u>. If $Q(j,f,{}_{i})$ intersects $\varphi_{i'}(a')$, $i \geq i'$, and (f,f') for i-odd, or (f',f) for i-even, is in R_i, then $Q'(j,f',\mathcal{S}_i) \subset a'$.

<u>Proof</u>. Let $\omega \in Q(j,f,\mathcal{S}_i) \cap \varphi_{i'}(a')$, $i \geq i'$. Hence for some $i'' \leq i'$, i'' odd, $\omega \in Q(j',f_1,\mathcal{S}_{i''}) \cap \mathcal{Z}(i'')$ and by Lemma 4d.12, $Q(j,f,\mathcal{S}_i) \subset Q(j',f_1,\mathcal{S}_{i''})$. There also is a unique $f_1' \in F'(\mathcal{S}_{i''})$, $(f_1,f_1') \in R_{i''}$ and $Q'(j',f_1',\mathcal{S}_{i''}) \subset \mathcal{Z}(i''-1)$. As (f,f') i odd or (f',f), i even is in R_i, there exists a sequence $Q'_{i''},Q'_{i''+1},\ldots,Q'_i$, each $Q'_j \in Q'(j)$, and each consecutive pair intersecting, with $Q'_{i''} = Q'(j',f_1',\mathcal{S}_{i''})$ and $Q'_i = Q'(j,f',\mathcal{S}_i)$. By Lemma 4d.12 and the definition of $\varphi_{i'}(a')$ we conclude that $Q'(j,f_1',\mathcal{S}_i) \subset Q'(j',f_2',\mathcal{S}_{i''-1}) \subset a'$. ■

<u>Lemma 4f.7</u>. For all i, if a_1' and a_2' are distinct elements of P'^{m}_{-m}, then $\varphi_i(a_1')$ and $\varphi_i(a_2')$ are disjoint.

<u>Proof</u>. From Lemma 4f.6, then if $\omega \in \varphi_i(a_1')$, for all

$i' \geq i$, if $\omega \in Q(j,f,\mathcal{S}_i)$, (f,f') i-odd or (f',f) i-even, is in R_i, then $Q'(j,f',\mathcal{S}_i) \subset a_1'$. If $\omega \in \varphi_i(a_2')$ then $Q'(j,f',\mathcal{S}_i) \subset a_2'$. But $a_1' \cap a_2' = \phi$. ■

Lemma 4f.8. For all i, a', $\mu(\varphi_{i'}(a')) \leq \mu'(a')$.

Proof. Let $A_i = \{Q(j,\bar{F},\mathcal{S}_i) \mid Q(j,\bar{F},\mathcal{S}_i) \cap \varphi_{i'}(a') \neq \phi\}$. In each $\mathcal{S}_i$, let $A(j,\mathcal{S}_i) = \{f \in F(\mathcal{S}_i) | Q(j,f,\mathcal{S}_i) \in A_i\}$. Now $\nu'(R_i(A(j)\mathcal{S}_i)) \geq (A(j,\mathcal{S}_i))$ as R_i is a society. As $R_i(A(j,\mathcal{S}_i)) \subset T'^{-j}(a') \cap \mathcal{S}_i$, $\nu'(R_i(A(j,\mathcal{S}_i))) \leq \mu'(T'^{-j}(a')|\mathcal{S}_i)$. Now

$$\mu(A_i) = \sum_{\mathcal{S}_i} \sum_{j=1}^{\ell(\mathcal{S}_i)} \nu_i(A(j,\mathcal{S}_i))\mu(\mathcal{S}_i)$$

$$\leq \sum_{\mathcal{S}_i} \sum_{j=1}^{\ell(\mathcal{S}_i)} \nu'(R_i(A(j,\mathcal{S}_i)))\mu(\mathcal{S}_i)$$

$$\leq \sum_{\mathcal{S}_i} \sum_{j=1}^{\ell(\mathcal{S}_i)} \mu'(T^{-j}(a') \cap \mathcal{S}_i) = \sum_{\mathcal{S}_i} \sum_{j=1}^{\ell(\mathcal{S}_i)} \mu'(a' \cap T'^{j}(\mathcal{S}_i)) = \mu'(a') . ■$$

Corollary 4f.9. The sets $\varphi(a')$ are almost clopen and $\mu(\varphi(a')) = \mu'(a')$.

Proof. The sets $\varphi_i(a') \nearrow i$, are disjoint as a' ranges over P'^{m}_{-m}, almost open, and their union increases to a set of full measure. Hence $\varphi(a')$ is almost clopen and $\varphi(a')$ partition Ω a.s. As $\mu(\varphi_i(a')) \leq \mu'(a')$, we must have $\mu(\varphi(a')) = \mu(a')$. ■

Corollary 4f.10. Letting $\bar{P}' = \{\varphi(a')\}_{a' \in P'}$, for any $a' = \{a'_{-m}, a'_{-m+1}, \ldots, a'_m\} \in P'^{m}_{-m}$, $\varphi(a') = \bigcap_{j=-m}^{m} T^{-j}(\varphi(a_j'))$, hence $(T,\bar{P}') \equiv (T,P)$.

Proof. If $\omega \in \varphi_i(a')$ clearly $T^j(\omega) \in \varphi_i(a_j')$, hence $\varphi_i(a') \subseteq \bigcap_{j=-m}^{m} T^{-j}(\varphi_i(a_j'))$. Hence $\varphi(a') \subseteq \bigcap_{j=-m}^{m} T^{-j}(\varphi(a_j'))$. As a' ranges over P^n_{-n} both sides of this expression partition Ω. Equality follows. ■

Corollary 4f.11. Using the corresponding φ' we construct in (T',P') an almost clopen partition $\overline{P}$ with $(T',\overline{P}) \equiv (T,P)$. ∎

For any $\omega \in \Omega$, let $\varphi(\omega)$ be that point in (T',P') whose (T',P') name is the $(T,\overline{P}')$ name of ω, and similarly $\overline{\varphi}'$, $\Omega' \to \Omega$. These are both almost continuous homomorphisms

All that remains to be seen is that $\overline{\varphi}' = \overline{\varphi}^{-1}$. Let $P_1 = \overline{\varphi}' \circ \overline{\varphi}(P)$, i.e., that partition in (T,P) with $(T,\overline{P}' \vee P_1)$ $(T',P' \vee \overline{P})$. All we need is the following.

Lemma 4f.12. The partitions P and P_1 of Ω are a.s. equal.

Proof. Given $\varepsilon > 0$, if i is sufficiently large, for all but ε of the $Q_i \in Q(i)$, Q_i is contained in a single atom $P(Q_i)$ of P. Furthermore for all but ε of the Q_i, Q_i intersects some $\varphi_{i'}(a')$, $i \geq i'$, $a' \in P^m_{-m}$, m so large that for all but ε of the $a' \in P^m_{-m}$, $a' \subset Q_{i''}(a')$, where i'' is so large that for all but ε of the $Q'_{i''}$, $Q'_{i''}$ intersects some $\varphi'_{i'''}(a)$, $i'' \geq i'''$, $a \in P$. It follows that once i is large enough, for all but 4ε of the Q_i, Q_i is in a single atom $P(Q_i)$ and for $Q_i = Q(j,f,\mathcal{S}_i)$, if (f,f') for i odd, (f',f) for i even is in R_i, then $Q'(j,f',\mathcal{S}_i) \subset a' \subset Q'_{i''}$, and as $Q'_{i''}$ intersects $\varphi'_{i'''}(a)$, $i \geq i'''$, by Lemma 3f.6 as (f,f') or (f',f) is in R_i, $Q(j,f,\mathcal{S}_i) = Q_i \subset a$. Hence we must have $a = P(Q_i)$ and for all but 4ε of the $Q(j,f,\mathcal{S}_i)$, if $\omega \in Q(j,f,\mathcal{S}_i)$ then $\overline{\varphi}(\omega) \in a'$ where for any $\omega' \in a'$, $\overline{\varphi}'(\omega') \in P(\omega)$. Hence $\overline{\varphi}'(\overline{\varphi}(\omega)) \in P(\omega)$. Letting $\varepsilon \to 0$ we conclude the result. ∎

This now proves:

Theorem 4f.13. If (T,P) and (T',P') are marker matched finitarily Bernoulli processes of the same finite entropy, they are finitarily isomorphic. ∎

Using the intermediary map (T_1,P_1) we conclude:

Theorem 4f.14. If (T,P) and (T',P') are finitarily Bernoulli processes of the same finite entropy, then they are finitarily isomorphic. ∎

This now completes our work. As the reader has surely become aware, the machinery of this argument is quite involved. The cause of this is, of course, a desire to state as mild and flexible a condition as possible. Because of the now readily apparent lack of uniqueness of the structures of the condition, it will most assuredly be more easy to verify of a map than refute. The already extant counterexamples to finitary isomorphism show that it is not so difficult to directly contradict the existence of a finitary isomorphism, so this is no real handicap. Hopefully the condition will prove useful in investigating natural maps known to be Bernoulli; ergodic automorphisms of tori with eigenvalues of modulus one, or more generally, ergodic group automorphisms, countable state Markov chains with exponential return times, discrete time maps of a Totoki flow, or the continued fraction transformation, are a few examples.

Our characterization has placed the finitarily Bernoulli maps on a similar theoretical footing to the general v.w.B maps. One can ask if more of the theory lifts. Are finitary factors of a Bernoulli shift finitarily Bernoulli? Are isometric extensions by an almost continuous cocycle of a Bernoulli shift finitarily Bernoulli if they are weakly mixing? Are the finitary isomorphisms between two finitarily Bernoulli processes dense in the weak topology on couplings?

We can also ask if this almost continuous theory has analogues for other notions of equivalence. Is there a finitary relative isomorphism theory analogous to that of Thouvenot? Is there a finitary isomorphism theory for actions of $\mathbb{R}$, $\mathbb{Z}^n$, $\mathbb{R}^n$ or more general groups ($\mathbb{R}$ perhaps, but what are spacers in $\mathbb{Z}^n$ or $\mathbb{R}^n$?)? Is there a finitary version of the Kakutani equivalence theory of Feldman, Katok, Ornstein and Weiss? Is there a finitary orbit equivalence theory analogous to Dye equivalence for measure preserving maps, or even a lifting to Krieger's work for nonsingular maps?

As we have said earlier, another direction of investigation which may be fruitful is to attempt to simplify or modify the condition as stated. Two approaches might be taken. One can remain within the basic format of the given condition and seek simpler variants equally as strong. On the other hand, one might seek an entirely different structure, for example, is there a metric on sequences of symbols which, for finitary isomorphism plays the role $\bar{d}$ plays for isomorphism, and can a suitable condition be built around this metric? The metric almost certainly exists, somehow embodied in our definition of t_0. Whether it gives rise in any simpler way to a characterization of the finitarily Bernoulli maps is doubtful.

Bibliography

(1) Akoglu, Rahe and Del Junco, "Finitary codes between Markov processes" Z. Wahrsch. 47 (1979), 305-314.

(2) Keane, M. and Smorodinsky, M. "A class of finitary codes" Israel J. Math. 26, 3-4 (1977), 352-377.

(3) ________ "Bernoulli schemes of the same entropy are finitarily isomorphic" Ann. of Math. 109 (1979), 397-406.

(4) ________ "Finitary isomorphism of irreducible Markov shifts" Israel J. Math. 34, 4 (1979), 281-286.

(5) Shields, P. "Almost block independence" Z. Wahrsch. 49 (1979), 119-123.

(6) Ornstein, D. "Bernoulli shifts of the same entropy are isomorphic" Adv. in Math. 4 (1970), 337-352.

(7) Ornstein, D. and Weiss, B. "Finitely determined implies very weak Bernoulli" Israel J. Math. 17 (1974), 94-104.

(8) Lind, D. "Skew products with group automorphisms" Israel J. Math. 28, 3 (1977), 205-247.

(9) ________ "Finitarily splitting skew products" (see this volume).

(10) Bowen, R. "Periodic points and measures for axiom A diffeomorphism" Trans. Amer. Math. Soc. 154 (1971), 377-397.

(11) Shields, P. "Weak and very weak Bernoulli partitions" Monatsh. Math. 84 (1977), 133-142.

(12) Denker, M. and Keane, M. "Almost topological dynamical systems" Israel J. Math. 34, 1-2 (1979), 139-159.

Daniel J. Rudolph
Department of Mathematics
Stanford University
Stanford, California, 94305

FINITARILY SPLITTING SKEW PRODUCTS

D. S. Lind*

§1. Introduction

Skew products with ergodic automorphisms of compact abelian groups arise naturally in several contexts. For example, suppose S is an automorphism of the compact group G, and H is an S-invariant closed subgroup. By taking a measurable cross section to the quotient map $G \to G/H$, the transformation S can be regarded as a skew product of the quotient automorphism $S_{G/H}$ with the restriction S_H of S to H. We can study S by studying the simpler components, $S_{G/H}$ and S_H, and how they are joined in a skew product. This method was used in proving that ergodic automorphisms of compact groups are measure theoretically isomorphic to Bernoulli shifts [3]. Crucial to this method is the result that if S_H is ergodic, then the skew product S measure theoretically splits into the direct product $S_{G/H} \times S_H$.

In this example, however, there is additional structure. The base map $S_{G/H}$ is continuous, and the cross section to the quotient map can be chosen to be almost continuous (or finitary), i.e., continuous off a meager null set (Theorem 2). It is natural to ask whether the isomophism of S with $S_{G/H} \times S_H$ can also be made almost continuous. We show that the answer is "yes," and prove a general almost continuous splitting theorem assuming a mild condition on the base map.

One of our motivations is the search for "natural" examples of transformations that are measure theoretically but not almost continuously isomorphic to Bernoulli shifts, i.e., that are Bernoulli but not "finitarily Bernoulli." Ergodic toral automorphisms with off-diagonal 1's in Jordan blocks of eigenvalues of modulus one (called central skew automorphisms) do not obey weak specification [5], and fine enough smooth

*Supported in part by NSF Grant MCS77-04915

partitions are never weak Bernoulli. Hence they are natural candidates for such examples. However, our work shows that a central skew automorphism is almost continuously isomorphic to a finite factor of a diagonalizable automorphism. Techniques developed by Rudolph [6] should be sufficient to show the latter to be finitarily Bernoulli, and thus prove that all ergodic toral automorphisms are finitarily Bernoulli.

Skew products are closely related to cocycles, and the splitting of skew products can be cast in the form of "straightening out" certain cocycles with values in the affine group of G. In §3 we formulate this precisely, and show how this suggests analogous results about skew product actions of groups more complicated than the integers.

In §4 we give the modifications in the specification argument of [4] needed to get finitary splitting, and in §5 are the necessary lemmas to go from automorphisms with specification to general ergodic automorphisms.

In §6 we show that if the group automorphism is "hyperbolic" in a certain general sense, then skew products with it finitarily split with no conditions on the base map. The proof uses a Neumann series argument shown to us in the toral case by W. Parry.

The author wishes to give his hearty thanks to the organizers of the Special Year in Ergodic Theory at the University of Maryland and to the Workshop on Ergodic Theory, Institute for Advanced Studies, Hebrew University, Jerusalem, July-August 1980, for their support while this paper was being written.

§2. Finitary splitting

We first set up the general framework to state the theorem. Let (X,d) be a compact metric space, and μ be a nonatonic Borel probability measure on X. By throwing out the largest open μ-null set, we can and will assume that μ is strictly positive on nonempty open sets. In particular, if $B(x,\varepsilon)$ denotes the ball of radius $\varepsilon > 0$ centered on $x \in X$, then $\mu(B(x,\varepsilon)) > 0$.

An almost <u>continuous</u> (or <u>finitary</u>) automorphism of (X,μ)

is given by a pair (X_1, U), where $X \setminus X_1$ is meager and null, $U\colon X_1 \to X_1$ is a homeomorphism in the induced topology on X_1, and U preserves the trace of μ on X_1. In this case we will call U a map of X, it being understood that U is defined only up to an invariant meager null set. Two such maps are finitarily isomorphic if by removing further meager null sets there is measure-preserving homeomorphism conjugating them. For further discussion of maps, see Denker and Keane [1]. There they show that there exists a totally bounded metric d_1 on X_1 equivalent to the original metric d such that U is uniformly continuous on (X_1, d_1). Hence U extends to the compact completion of (X_1, d_1). Therefore we can and will think of U on X_1 as the restriction of a homeomorphism of a compact metric space (X, d_1) to an invariant residual set X_1 of full measure.

From now on we will assume that U is ergodic on (X, μ).

Let G be a metrizable compact abelian group (hereafter abbreviated "compact group") and S be a continuous, algebraic automorphism of G. Then S preserves Haar measure m on G. If $\alpha\colon X \to G$ is measurable, the skew product transformation $U \times_\alpha S\colon X \times G \to X \times G$ defined by

$$(U \times_\alpha S)(x,g) = (Ux,\ Sg+\alpha(x))$$

preserves $\mu \times m$.

We showed in [4] that if S is ergodic, then $U \times_\alpha S$ is isomorphic to the direct product $U \times S$ via an isomorphism of the form $W(x,g) = (x, g + \beta(x))$, where $\beta\colon X \to G$ is measurable. This amounts to solving the functional equation

$$\alpha(x) = \beta(Ux) - S\beta(x) \tag{1}$$

for β, given α, U, and S.

Suppose now that $\alpha\colon X \to G$ is almost continuous, i.e. continuous after removing a meager null set from X. Then $U \times_\alpha S$ is a map, and it is finitarily isomorphic to $U \times X$ if there is an almost continuous solution β to (1). We will show that such a β exists if U satisfies the following condition.

Definition: A map U of (X,d) is compressible if for every positive integer n and every $\varepsilon > 0$, there is an aperiodic $x \in X$ such that $\operatorname{diam} U^j x:\{0 \leq j \leq n\} < \varepsilon$.

The purpose of compressibility is to guarantee that modifications on long pieces of U orbit during the construction of β can be made topologically as well as measure theoretically small.

Theorem 1: Suppose U is compressible and S is ergodic. Then for each almost continuous skewing function $\alpha: X \to G$ there exists an almost continuous solution β to the functional equation (1). Hence the finitary skew product $U \times_\alpha S$ is always finitarily isomorphic to the direct product $U \times S$.

Note that if U has a fixed point x_0, which is a limit of aperiodic points, then small perturbations of x_0 yield long pieces of orbit with small diameter, and so U is compressible. Since automorphisms fix the identity, Theorem 1 applies to the situation described in §1 to show that S is finitarily isomorphic to $S_{G/H} \times S_H$.

§3. Cohomological interpretation

The measurable splitting of skew products in [4] can be interpreted as "straightening out" certain kinds of cocycles. Recast in this form, the result is similar to Zimmer's rigidity theorem for ergodic actions of semi-simple Lie groups [7]. A beautiful exposition of Zimmer's work has recently been given by Furstenberg [2]. This interpretation suggests a suitable framework for questions involving skew product actions of groups more complicated than the integers $\mathbb{Z}$.

The transformation U gives a measure-preserving action of $\mathbb{Z}$ on (X,μ). The automorphism S induces a homomorphism π from $\mathbb{Z}$ to the automorphism group of G by $\pi(n) = S^n$. This defines the semi-direct product group $\mathbb{Z} \times_\pi G$ with multiplication $(n_1,g_1)\cdot(n_2,g_2) = (n_1+n_2, g_1+S^{n_1}g_2)$. This semi-direct product acts affinely on G by $(n,g)\cdot g' = S^n g' + g$. The skew product $U \times_\alpha S$ yields a cocycle $\sigma: \mathbb{Z} \times X \to \mathbb{Z} \times_\pi G$

defined by

$$(U \times_\alpha S)^n(x,g) = (U^n x, \sigma(n,x)\cdot g).$$

This σ clearly obeys the cocycle equation

$$\sigma(n_1+n_2,x) = \sigma(n_1,U^{n_2}x)\sigma(n_2,x).$$

More explicitly, $\sigma(n,x) = (n,\alpha_n(x))$, where

$$\alpha_n(x) = \alpha(U^{n-1}x) + S\alpha(U^{n-2}x) + \cdots + S^{n-1}\alpha(x)$$

for $n \geq 1$, and a similar formula for $n \leq -1$. Thus σ is "level-preserving" in the sense that $\sigma(\{n\} \times X) \subset \{n\} \times G$, and every level-preserving cocycle corresponds to a skew product $U \times_\alpha S$.

The direct product $U \times S$ corresponds to the trivial cocycle $\tau(n,x) = (n,0)$. If $\beta\colon X \to G$ solves (1), put $\psi(x) = (0,\beta(x)) \in \mathbb{Z} \times_\pi G$. An easy computation shows that $\alpha_n(x) = \beta(U^n x) - S^n\beta(x)$, and therefore that

$$\psi(U^n x)^{-1}\sigma(n,x)\psi(x) = \tau(n,x). \tag{2}$$

Thus σ is cohomologous to τ via the coboundary defined by ψ, i.e., σ can be "straightened out."

We remark that proving σ is cohomologous to the trivial cocycle is equivalent to solving (1). For suppose $\psi\colon X \to \mathbb{Z} \times_\pi G$ obeys (2). If $\psi(x) = (p(x),\beta(x))$, then for $n = 1$ the first coordinates of (2) yield

$$-p(Ux) + 1 + p(x) = 1,$$

so $p(x)$ is U-invariant. Since we assume U to be ergodic, $p(x) = n_0$ almost everywhere. Applying S^{-n_0} to the second coordinates shows that β solves (1).

The splitting theorem of [4] therefore says exactly that every level-preserving cocycle $\sigma\colon \mathbb{Z} \times X \to \mathbb{Z} \times_\pi G$ is cohomologous to one that is independent of X. Zimmer's work [7] shows that every cocycle $\sigma: H \times X \to K$, where H is a suitable Lie group acting ergodically on (X,μ) preserving μ and K is another suitable Lie group, is cohomologous to one that is independent of X. The results are similar, but the groups operating are quite different.

This suggests the following general question. Let Γ be a countable discrete group acting ergodically on (X,μ), and π be a homomorphism from Γ to the automorphism group of G. The semi-direct product $\Gamma \times_{\pi} G$ acts affinely on G. When is a level-preserving cocycle $\sigma: \Gamma \times X \to \Gamma \times_{\pi} G$ cohomologous to the trivial cocycle $\tau(\gamma,x) = (\gamma,0)$? Equivalently, when does the corresponding skew product action of Γ on $X \times G$ split into the direct product action? When $\Gamma = \mathbb{Z}^n$, G is a torus, and $\pi(\gamma)$ is hyperbolic for $\gamma \neq e$ we have shown all cocycles are trivial. However, results available now are quite fragmentary.

§4. Automorphisms with weak specification

If S obeys weak specification, the measurable solution β to (1) found in [4, §4] is not necessarily almost continuous for the following reason. Uncontrolled modifications produced by using weak specification at each stage of the construction of β occur in long gaps between Rohlin stacks in X. Although these gaps become measure theoretically negligible, so the approximations converge a.e., they can also become topologically dense. In fact, this attempted argument for almost continuity of β breaks down when U is an irrational rotation of the circle. For this U the almost continuous solvability of (1) is still in doubt. This is the reason to impose compressibility on U.

First we establish what we need of compressible transformations. We denote the closure of a subset E by $\mathrm{cl}(E)$. A subset F of X is almost open if F and $X \setminus F$ agree with open sets up to a meager null set. This means that the "essential boundary" of F has measure zero. It is easy to see that for fixed $x \in X$, the ball $B(x,\delta)$ is almost open for all but countably many values of δ.

<u>Lemma 1</u>: Suppose U is a compressible map of (X,d,μ). Let $\{n_j\}$ and $\{m_j\}$ be increasing sequences of natural numbers, and $\varepsilon_j \searrow 0$. Then there are almost open sets $F_j \subset X$ of positive measure such that

(i) $\{U^i F_j : -n_j \le i < m_j\}$ is a disjoint collection for each j,

(ii) if $L_k = \cup \{U^i F_k : -n_j \le i < 0\}$, then

$$\operatorname{diam}[\operatorname{cl}(\bigcup_{k \in j} L_k)] < \varepsilon_j .$$

<u>Proof</u>: By ergodicity and compressibility of U, there are $x_j \in X$ with infinite U-orbit such that

$$\operatorname{diam}\{U^i x_j : -n_j \le i < 0\} < \varepsilon_j/8.$$

Since (X,d) is assumed compact, by taking a convergent subsequence we can assume that there is an $x_0 \in X$ such that $d(x_j,x_0) < \varepsilon_j/8$. Continuity of U shows there are $\delta_j > 0$, $\delta_j < \varepsilon_j/8$ such that if $F_j = B(x_j,\delta_j)$, then F_j is almost open and $\{U^i F_j : -n_j \le i < m_j\}$ is disjoint. Also, $\mu(F_j) > 0$ since nonempty open sets have positive measure, which proves (i). Finally, $\cup\{L_k : k \ge j\} \subset B/x_0, 3\varepsilon_j/8)$, proving (ii).

Next we recall the weak specification property.

<u>Definition</u>: A homeomorphism f of a compact metric space (Y,d) satisfies <u>weak specification</u> if for every $\varepsilon > 0$ there is an integer $M(\varepsilon)$ such that for every $r \ge 2$ and r points $y_1,\dots,y_r$ in Y, and for every set of integers $a_1 \le b_1 < a_2 \le b_2 < \cdots < a_r \le b_r$ with $a_j - b_{j-1} \ge M(\varepsilon)$ $(2 \le j \le r)$, there is a $y \in Y$ with $d(f^i y, f^i y_j) \le \varepsilon$ for $a_j \le i \le b_j$, $1 \le j \le r$.

For further details about this property, see [4]. We show there that certain basic group automorphisms have weak specification. On the other hand, not all ergodic group automorphisms have this property [5], answering a question raised in [4].

For those that do, [4] gives a simple proof of skew splitness.

We indicate here the modifications in the proof of Theorem 4.2 of [4] needed to obtain an almost continuous solution β to (1), i.e., to show that the skew product finitarily splits.

Proposition: If U is compressible and S obeys weak specification, then $U \times_\alpha S$ finitarily splits for each almost continuous skewing function α.

Proof: Two changes in the proof in [4, §4] are needed. The first is to replace Rohlin towers with Kakutani skyscrapers to control gap size. The second is to use compressibility of U to force uncontrolled specification adjustments on pieces of orbit into a topologically small set.

Suppose $\alpha: X \to G$ is almost continuous. Choose $\varepsilon_j \searrow 0$ with $\sum \varepsilon_j < \infty$. Let $M(\varepsilon)$ be the number determined by weak specification of S. For $n_j = M(\varepsilon_j)$ and $m_j = M(\varepsilon_j)/\varepsilon_j$, choose almost open sets F_j in accordance with Lemma 1.

For $x \in F_j$ define

$$h_j(x) = \min\{n: n > 0,\ U^n x \in F_j\} - n_j.$$

Since F_j is almost open, h_j is almost continuous on F_j. Put $E_j = \cup\{U^i x: x \in F_j,\ 0 \le i < h_j(x)\} = X \backslash L_j$.

Now define $\beta_1: F_1 \to G$ arbitrarily but almost continuously. As in [4], β_1 extends to an almost continuous function on E_1 satisfying (1) where defined.

If ρ is a translation invariant metric on G, and $f: E_k \to G$, put $\|f\|_{E_k} = \sup\{\rho(0,f(x)): x \in E_k\}$.

We shall construct $\beta_2: E_2 \to G$, from which the inductive step for defining β_k will be clear. The decomposition of F_2 into subsets K consisting of those points in F_2 with the same return time to F_2 and the same entry times $a_1,\dots,a_r$ into F_1 before returning to F_2 is an almost open partition. Thus it is enough to define β_2 over such sets K.

Begin by defining β_2' to be constant on K, and using

(1) to extend β_2' to $\cup\{U^iK: 0 \le i < h_2(K)\}$, where $h_2(K)$ is the constant value of h_2 on K. Now β_1 is already defined on the blocks.

$$\cup\{U^iK: a_j \le i < b_j = a_j + h_1(U^{a_j}K)\}, \qquad 1 \le j \le r,$$

where $a_{j+1} - b_j = M(\varepsilon_1)$. Note the use of skyscrapers to control gap size. The calculation in [4] shows that the error $\beta_2' - \beta_1$ on $\{U^ix: a_j \le i < b_j\}$ is the orbit of a point. By weak specification, there is an adjustment $\beta_2(x)$ of $\beta_2'(x)$ such that this error is uniformly less than ε_1 for $1 \le j \le r$. Since only finite conditions are involved for each K, this adjustment can be made almost continuously. The new function β_2 has the properties that it is almost continuous on E_2, it solves (1) where defined, and $\|\beta_2-\beta_1\|_{E_1} < \varepsilon_1$.

Inductively we obtain almost continuous functions $\beta_k: E_k \to G$ satisfying (1) where defined, and with $\|\beta_{k+1}-\beta_k\|_{E_k} < \varepsilon_k$. Hence $\{\beta_k\}$ converges uniformly off $cl(\cup_{k\ge j}L_k)$. The latter sequence of sets nests down to a point. Hence $\beta = \lim \beta_k$ is almost continuous and satisfies (1) a.e.

§5. Extension to general group automorphisms

Not all ergodic group automorphisms obey weak specification (e.g. the central skew automorphisms mentioned in §1). However, each is built up from two basic kinds, namely irreducible solenoidal automorphisms and group shifts, by the processes of products, factors, inverse limits, and skew products with basic automorphisms. These basic automorphisms obey weak specifications, hence finitarily split over compressible base maps.

The proof in [4] shows that measurable splitting is preserved under products, factors, skew products, and the kinds of inverse limits encountered in constructing general automorphisms. We give here the facts needed to extend this proof to finitary splitting, and therefore to prove Theorem 1. The extra ingredient is finding an almost continuous cross section to a group quotient map. Bord cross sections are well-known, but we have been unable to find this result in the literature.

Theorem 2: Let G be a compact abelian group, H be a closed subgroup of G, and $\pi: G \to G/H$ be the quotient map. Then there is an almost continuous map $\theta: G/H \to G$ such that $\pi\theta$ is the identity.

Proof: Let G have countable discrete dual group Γ, and G/H have dual $\Delta \subset \Gamma$. Suppose $g \in G/H$, so $g: \Delta \to \mathbb{T}$ is a character. We will give a recipe to extend g to $\tilde{g}: \Gamma \to \mathbb{T}$, and then define $\theta(g) = \tilde{g}$. Since $\tilde{g}|_\Delta = g$, θ will be a cross section to π. We will show that there is a meager null set $E \subset G/H$ such that θ is continuous off E.

Choose $\{\gamma_k: k \geq 1\} \subset \Gamma$ that together with Δ generate Γ. Let Δ_n be a subgroup of Γ generated by Δ and $\gamma_1, \ldots, \gamma_n$, and put $\Delta_0 = \Delta$. We will extend g successively to each Δ_n, and hence yield an extension to $\Gamma = \cup\Delta_n$.

Define $\tilde{g}_0 = g$, and let $n \geq 1$. Suppose g has been extended to $\tilde{g}_{n-1}: \Delta_{n-1} \to \mathbb{T}$, consistently in the sense that $\tilde{g}_{n-1}|_{\Delta_k} = \tilde{g}_k$ $(0 \leq k \leq n-1)$. We will construct $\tilde{g}_n: \Delta_n \to \mathbb{T}$.

Consider γ_n. If there is an integer k such that $k\gamma_n \in \Delta_{n-1}$, defined k_n to be the least positive such integer. Otherwise define $k_n = \infty$. Thus k_n is the order of $\gamma_n + \Delta_{n-1}$ in Γ/Δ_{n-1}.

For $k = 1,2,\ldots,$ define a function $r_k: \mathbb{T} \to \mathbb{T}$ as follows. Let $t_0 = \exp(-2\pi i\xi_0)$ where ξ_0 is an irrational number within $\frac{1}{100}$ of $\frac{1}{2}$. For $t = \exp(2\pi i\theta)$ with $-\xi_0 \leq \theta < -\xi_0 + 1$, define $r_k(t) = \exp(2\pi i\theta/k)$. Thus $r_k(t)^k = t$, and r_k is continuous except at t_0, which is close to -1.

If $k_n < \infty$, define $\tilde{g}_n(\gamma_n) = r_{k_n}(\tilde{g}_{n-1}(k_n\gamma_n))$. Since $\gamma_n + \Delta_{n-1}$ generates Δ_n/Δ_{n-1}, this defines an extension $\tilde{g}_n$ of $\tilde{g}_{n-1}$ to Δ_n. Consistency follows because

$$\tilde{g}_n(k_n\gamma_n) = \tilde{g}_n(\gamma_n)^{k_n} = r_{k_n}(\tilde{g}_{n-1}(k_n\gamma_n))^{k_n} = \tilde{g}_{n-1}(k_n\gamma_n),$$

and $\mathbb{Z}\gamma_n \cap \Delta_{n-1}$ is generated by $k_n\gamma_n$.

If $k_n = \infty$, define $\tilde{g}_n(\gamma_n) = 1$. In this case $\mathbb{Z}\gamma_n \cap \Delta_{n-1} = 0$, so $\tilde{g}_n$ is automatically consistent.

This sequence $\{\tilde{g}_n\}$ of extensions converges to a limit $\tilde{g} = \theta(g): \Gamma \to \mathbb{T}$ such that $\tilde{g}|_{\Delta_n} = \tilde{g}_n$. Of course θ depends on our choice of γ_n and ξ_0.

A discontinuity of θ on G/H can be introduced at stage n only if $k_n < \infty$ and $\tilde{g}_{n-1}(k_n\gamma_n) = t_0$, the point of discontinuity of r_{k_n}. We will show that (a) the set E_n of $g \in G/H$ such that $\tilde{g}_{n-1}(k_n\gamma_n) = t_0$ is contained in a coset of a nowhere dense null subgroup of G, and (b) off $\cup_{n\geq 1}E_n = E$ the map θ is continuous.

(a) If $k_n < \infty$, let $a(n) = k_n$. If $k_n = \infty$, put $a(n) = 1$. Let $A(n) = a(n)a(n-1)\cdots a(1)$. If $k_n < \infty$, let $q_n: \Delta_n \to \Delta_{n-1}$ be multiplication by k_n. If $k_n = \infty$, then $\Delta_n = \Delta_{n-1} \oplus \mathbb{Z}\gamma_n$, and let $q_n: \Delta_n \to \Delta_{n-1}$ be projection to the first coordinate.

Suppose now $k_n < \infty$ and $\tilde{g}_{n-1}(k_n\gamma_n) = t_0$. Then

$$\begin{aligned} t_0^{A(n-1)} &= \tilde{g}_{n-1}(k_n\gamma_n)^{A(n-1)} \\ &= \tilde{g}_{n-1}(q_n\gamma_n)^{a(n-1)A(n-2)} \\ &= \tilde{g}_{n-2}(q_{n-1}q_n\gamma_n)^{A(n-2)} \\ &= \cdots \\ &= \tilde{g}_0(q_1q_2\cdots q_n\gamma_n), \end{aligned}$$

i.e. $g(q_1q_2\cdots q_n\gamma_n) = \exp(2\pi i A(n-1)\xi_0)$. Since $q_1q_2\cdots q_n\gamma_n \in \Delta$, we only need the following lemma, with $\xi_1 = A(n-1)\xi_0$, $\gamma = q_1q_2\cdots q_n\gamma_n$, and G replacing G/H.

<u>Lemma 2</u>: Let G be a compact abelian group with dual Γ, let $\gamma \in \Gamma$, and $t_1 = \exp(2\pi i\xi_1)$ where ξ_1 is irrational.

Then $E = \{g \in G: g(\gamma) = t_0\}$ is either empty or a coset of a closed nowhere dense null subgroup of G.

Proof: If γ has finite order, then $g(\gamma)$ is a root of unity, and hence $E = \emptyset$.

Suppose γ has infinite order. The mapping $g \to g(\gamma)$ is then a homomorphism from G onto $\mathbb{T}$, whose kernel K is therefore a closed nowhere dense null subgroup of G. Clearly E is a coset of K, completing the proof of the lemma.

(b) Suppose $g \notin E$. A small change in g produces a small change in $\tilde{g}_1$, hence in $\tilde{g}_2$, etc. If we define $\theta_n(g) = \tilde{g}_n$, then $\theta_n: G/H \to G/\Delta_n^{\perp}$ is continuous off E. Since G is the inverse limit of the $G/\Delta_n^{\perp}$, the limit θ of $\{\theta_n\}$ is also continuous off E.

Corollary: If $\alpha: X \to G/H$ is almost continuous, then there is an almost continuous lifting $\tilde{\alpha}: X \to G$ of α such that $\pi\tilde{\alpha} = \alpha$.

Proof: Let $\theta: G/H \to G$ be a cross section to π continuous off a meager null set E. Some care is needed, since $\alpha^{-1}(E)$ could have positive measure in X and then $\theta\alpha$ need not be almost continuous.

For fixed $z \in G/H$ let $\theta_z(y) = \theta(y-z) + \theta(z)$. Then θ_z is again a cross section to π, continuous off $E + z$. Define the measure $\alpha(\mu)$ on G/H by $\alpha(\mu)(F) = \mu(\alpha^{-1}F)$. By Fubini's theorem,

$$\int_{G/H} \alpha(\mu)(E+z)\,dm(z) = \int_{G/H} m(E-z)\,d\alpha(\mu)(z) = 0,$$

so $\alpha(\mu)(E+z) = 0$ for m-almost every $z \in G/H$. For such z, we have $\mu(\alpha^{-1}(E+z)) = 0$, and it follows that $\alpha^{-1}(E+z)$ is also meager since μ is positive on open sets. Thus $\tilde{\alpha} = \theta_z\alpha$ works, finishing the proof.

For the remainder of this section, say that S finitarily splits if for each compressible U and almost continuous α, there is an almost continuous solution β of (1). Using the Corollary to find almost continuous instead of measurable liftings of α, the proofs of the following lemmas are easy adaptations of those in [4, §5]. Together with the structure of general group automorphisms given in [4, §7], they are sufficient to show that every ergodic group automorphism finitarily splits, i.e. to prove Theorem 1.

Lemma 3: Let S be an automorphism of G, and H be an S-invariant closed subgroup of G. If S finitarily splits, so does $S_{G/H}$.

Lemma 4: If S_H and $S_{G/H}$ finitarily split, then so does S, a solution in G/H lift to solutions in G.

The meaning of the last statement is that if $\alpha: X \to G$ and $\beta_1: X \to G/H$ solves (1) for $\pi\alpha$, then there is a solution $\beta: X \to G$ for α such that $\pi\beta = \beta_1$.

Lemma 5: Suppose S is an automorphism of G, and that $H_k \searrow 0$ are S-invariant subgroups with $H_0 = G$. If S_{H_{k-1}/H_k} finitarily splits for $k \geq 1$, then S finitarily splits.

§6. Neumann series solution for hyperbolic automorphisms

Since a finitary solution of (1) for arbitrary base maps U has eluded us, it is interesting that for a rather general class of ergodic automorphisms (1) can be finitarily solved for every U. This method was pointed out to us by W. Parry.

Let G be a compact abelian group with translation invariant metric ρ. Say that an automorphism S of G is hyperbolic if there are almost continuous functions $\pi_s: G \to G$, $\pi_u: G \to G$ with ranges K_s and K_u, respectively, and constants $C > 0$ and $\lambda \in (0,1)$ such that

(i) $g = \pi_s(g) + \pi_u(g)$ for all $g \in G$,

(ii) $\rho(0,S^n g) < C\lambda^n$ if $g \in K_s$, $n \geq 0$,

(iii) $\rho(0,S^{-n} g) < C\lambda^n$ if $g \in K_u$, $n \geq 0$.

If S is a hyperbolic automorphism of $\mathbb{T}^n$ in the usual sense (no eigenvalues of modulus one), then it is hyperbolic in the above sense. For π_u can be obtained by embedding $\mathbb{T}^n$ into $\mathbb{R}^n$, projecting $\mathbb{R}^n$ along the stable eigenspace of S to the unstable eigenspace, and projecting back to $\mathbb{T}^n$. Similarly for π_s. Shifts on compact groups (e.g. the n-shift) are hyperbolic, where the maps π_s and π_u are to the future and to the past. Irreducible solenoidal automorphisms with no eigenvalues of modulus one are hyperbolic (see [3]). However, toral automorphisms with eigenvalues of modulus one and automorphisms of the full solenoid $\hat{\mathbb{R}}^n$ are not hyperbolic because they have isometric parts (as one can show).

We remark that hyperbolic group automorphisms are automatically ergodic. This follows because nonergodic automorphisms have a nontrivial isometric factor automorphism. By (ii) and (iii) K_s and K_u would map to the identity, violating (i).

<u>Theorem 3</u>: If S is a hyperbolic group automorphism of G and U is a map of X (not assumed compressible or even ergodic), then for each almost continuous function $\alpha: X \to G$ there is an almost continuous solution β to (1).

<u>Proof</u>: Let π_s, π_u, K_s, K_u, C, and λ be as in the definition of hyperbolicity for S.

We first claim that if we can solve (1) for α replaced by a translate $\alpha + g_0$, then we can solve it for α. For since S is ergodic, $(I-S)G = G$, so there is a $g_1 \in G$ such that $g_1 - Sg_1 = g_0$. If β is a solution for $\alpha + g_0$, then $\beta - g_1$ is a solution for α. Therefore an averaging argument as in the proof of the Corollary over translates of α shows that we may assume that $\alpha_s = \pi_s : X \to K_s$ and $\alpha_u = \pi_u\alpha : X \to K_u$ are almost continuous.

Now we just write down the solutions. Let

$$\beta_s(x) = \sum_{j=0}^{\infty} S^j \alpha_s(U^{-j-1}x),$$

$$\beta_u(x) = -\sum_{j=0}^{\infty} S^{-j-1}\alpha_u(U^j x).$$

Since $\alpha_s(U^{-j-1}x) \in K_s$, $\rho(0, S^j\alpha_s(U^{-j-1}x)) < C\lambda^j$, $j \geq 0$. Thus the series defining β_s converges uniformly where defined, and since $\alpha_s(U^{-j-1}x)$ is almost continuous, so is β_s. Similarly for β_u. An easy calculation shows that

$$\alpha_s(x) = \beta_s(Ux) - S\beta_x(x), \qquad \alpha_u(x) = \beta_u(Ux) - S\beta_s(x).$$

Since $\alpha(x) = \alpha_s(x) + \alpha_u(x)$, the function $\beta(x) = \beta_s(x) + \beta_u(x)$ is an almost continuous solution of (1).

The motivation for this solution and justification for the reference to Neumann series is as follows. Suppose S is a hyperbolic toral automorphism, and lift α to $\hat{\alpha}: X \to \mathbb{R}^n$. Form π_s projecting $\mathbb{R}^n$ to the stable eigenspace E^s as suggested above. On the Hilbert space $L^2(X,E^s)$ there are commuting operators $(\hat{U}f)(x) = f(Ux)$ and $(Sf)(x) = S(f(x))$. Then $\hat{U}$ is an isometry while $\|S\| < 1$. We are to solve

$$(\hat{U}-S)\beta_s = \alpha_s.$$

Thus

$$\begin{aligned}\beta_x &= (\hat{U}-S)^{-1}\alpha_s \\ &= \hat{U}^{-1}(I-\hat{U}^{-1}S)^{-1}\alpha_s \\ &= \hat{U}^{-1}\sum_{j=0}^{\infty}(\hat{U}^{-1}S)^j\alpha_s \\ &= \sum_{j=0}^{\infty}\hat{U}^{-j-1}S^j\alpha_s.\end{aligned}$$

Evaluating at x gives the definition of $\beta_s(x)$ above. A similar idea works for β_u by expanding

$$(\hat{U}-S)^{-1} = -S^{-1}(I-\hat{U}S^{-1})^{-1},$$

where $\|\hat{U}S^{-1}\|_{L^2(X,E^u)} < 1$.

References

1. M. Denker and M. Keane, Almost topological dynamical systems, Israel J. Math. 34 (1979), 139-160.

2. Harry Furstenberg, Rigidity and cocycles for ergodic actions of semi-simple Lie groups, Seminaire Bourbaki, No. 559, 1-20.

3. D. A. Lind, The structure of skew products with ergodic group automorphisms, Israel J. Math. 28 (1977), 205-248.

4. _____, Split skew products, a related functional equation, and specification, Israel J. Math. 30 (1978), 236-254.

5. _____, Ergodic group automorphisms and specification, Springer Lecture Notes in Math, Vol. 729.

6. Daniel J. Rudolph, A characterization of those processes finitarily isomorphic to a Bernoulli shift, these Proceedings, 1-64.

7. R. Zimmer, Strong rigidity for ergodic actions of semisimple Lie groups, Annals of Math., to appear.

D.S. Lind
Department of Mathematics
University of Washington
Seattle, Washington 98195

DISJOINTNESS OF MEASURE-PRESERVING TRANSFORMATIONS, MINIMAL SELF-JOININGS AND CATEGORY

Andres del Junco

Abstract. In the coarse topology on the group of measure-preserving transformations of a Lebesgue probability space, the class of transformations disjoint from a given ergodic transformation is a dense G_δ. The class of transformations T such that the family $\{T^i : i \in \mathbb{Z}\}$ is disjoint is also a dense G_δ. As a corollary there exists an uncountable family $\{T_\alpha : \alpha \in A\}$ of weakly-mixing transformations such that the family $\{T_\alpha^i : \alpha \in A, i \in \mathbb{Z} - \{0\}\}$ is disjoint.

§1. Definitions and statement of results:

We deal throughout with a Lebesgue probability space (Ω, F, μ). G denotes the group of invertible measure-preserving transformations (transformations, for short) of Ω. Following [3] we call the coarse topology (also known as the weak topology) on G that topology for which a net $\{T_\alpha\}$ converges to T if and only if $\mu(T_\alpha A \Delta TA) \to 0$ for all $A \in F$. This topology is induced by the metric

$$\rho(T,S) = \sum_{i=1}^{\infty} 2^{-i}(\mu(TA_i \Delta SA_i) + \mu(T^{-1}A_i \Delta S^{-1}A_i)),$$

where $\{A_i\}$ is a sequence of sets dense in F, and this metric is complete [6, p. 62 ff].

As defined in [5], transformations T and S are disjoint $(T \perp S)$ if whenever π is a measure on $(\Omega^2, F \times F)$ such that both of the marginals of π are μ and π is $T \otimes S$ invariant then $\pi = \mu \otimes \mu$. Note that the relation $S \perp T$ depends only on the isomorphism classes of S and T. The set of transformations disjoint from S is denoted $S^\perp$. A consequence of disjointness is that T and S have no common factors. More generally we say a countable family $\{T_i : i \in I\}$ is disjoint if any measure π on Ω^I with marginals μ which is invariant under $\bigotimes_{i \in I} T_i$ must be μ^I (product measure).

This is clearly equivalent to the disjointness of every finite subfamily. We do not know whether it is equivalent to pairwise disjointness. A positive answer to problem B of [5, p. 7] would certainly imply that this is the case. An arbitrary family is disjoint if each finite subfamily is.

Theorem 1: If S is ergodic $S^{\perp}$ is a dense G_{δ} in the coarse topology.

Corollary 2(a): If S is ergodic, $S^{\perp}$ contains weakly mixing transformations.

Corollary 2(b): There exists an uncountable family of weakly mixing transformations which is disjoint.

Obviously T can never be disjoint from itself - as well as product measure the off-diagonal measures μ_i^T defined by

$$\mu_i^T(E) \quad = \quad \mu\{x \in \Omega : (x, T^i x) \in E\}.$$

For $E \in F \times F$ are $T\times T$-invariant with marginals μ. However Rudolph [10] has shown that there exist transformations T such that these are the only $T \times T$ invariant and ergodic measures with marginals μ. This property implies, among other things, that the centralizer of T in G is reduced to the powers of T and that T has only the trivial invariant σ-algebras. In fact, Rudolph does much more - he constructs a T such that, for any countable I, any measure on Ω^I which has marginals μ and is $\underset{i\in I}{\otimes} T^{\ell(i)}$ invariant and ergodic for some $\ell : I \to \mathbb{Z} - \{0\}$ must be a product of off-diagonal measures, where an off diagonal measure of Ω^J, $j \subset I$, is a measure of the form $\underset{j\in J}{\otimes} T^{k(j)}\mu_d$, μ_d diagonal measure on Ω^J. Such a T is said to have minimal self-joinings. Another simpler example of minimal self-joinings is given in [7].

A consequence of minimal self-joinings is that the family $\{T^i : i \in \mathbb{Z}\}$ is disjoint. (One can include T^o here because the identity transformation is disjoint from anything ergodic, and minimal self-joinings implies weak-mixing.) This property is generic in the sense of category:

Theorem 3: The class of transformations T such that $\{T^i : i \in \mathbb{Z}\}$ is disjoint is a dense G_δ in the coarse topology.

As a consequence, for example, a transformation is in general not isomorphic to its inverse. We can apply Theorems 1 and 3 to obtain

Corollary 4: There exists an uncountable family $\{T_\alpha : \alpha \in A\}$ of weakly-mixing transformations such that the family $\{T_\alpha^i : \alpha \in A, i \in \mathbb{Z}\}$ is disjoint.

In particular the T_α are pairwise non-isomorphic and each is not isomorphic to its inverse. The K-automorphisms of Ornstein and Shields [8] are an explicit family with this property. Fieldsteel [4] has given a simpler construction of a family in entropy zero having this property. Our method gives such a family almost automatically once we have the existence of just one transformation which is disjoint from its inverse. It should be pointed out that the family which our approach yields is in principle also in entropy zero, because the entropy zero class is residual. Corollary 4 includes Corollary 2(b), but we have stated them separately because Corollary 4 depends on the existence of a T such that $\{T^i : i \in \mathbb{Z}\}$ is disjoint, which is not at all obvious, whereas Corollary 2 depends only on the fact that $S^\perp$ is non-empty for S ergodic - a relatively easy result.

We remark that the property of minimal self-joinings is not generic - in fact, even its weak consequence of trivial centralizer defines a meager set. This can be seen by looking at the class of transformations admitting "good approximation by partitions" as defined in [13, Chapter 5, 2.5.1] (see also [2]). As noted in [13, Chapter 5, 2.5.4] such a transformation has an uncountable centralizer. Moreover, it is shown in [12, Theorem 1.1] that the class of transformations admitting a "cyclic approximation by periodic transformations with speed $O(\frac{1}{n})$ contains a dense G_δ. Since this class is contained in the class admitting good approximation the latter class is residual. (In fact, it is not hard to see that it is exactly a dense G_δ.)

§2. Proofs

Proof of Theorem 1: If Q is a finite partition of Ω and T, T' are transformations we set

$$d(T,T';Q) = \sum_{q\in Q} \mu(Tq\Delta T'q).$$

If π and π' are measures on Ω^2 set

$$d(\pi,\pi';Q) = \sum_{q_1,q_2\in Q} |\pi(q_1\times q_2) - \pi'(q_1\times q_2)|.$$

Let $\mathcal{O}(P,\varepsilon,Q,\delta)$ consist of those transformations T such that if π is any measure on Ω^2 with marginals μ then

$$d(\pi,(S\otimes T)\pi;Q) < \delta \Longrightarrow d(\pi,\mu\otimes\mu;P) < \varepsilon.$$

We claim $\mathcal{O}(P,\varepsilon,Q,\delta)$ is contained in the interior of $\mathcal{O}(P,\varepsilon, Q,\delta/2)$. To see this suppose $T \in \mathcal{O}(P,\varepsilon,Q,\delta)$, T' is such that $d(T,T';Q) < \delta/2\#Q$ (this defines a coarse neighborhood of T) and π is a measure with marginals μ such that $d(\pi,(S\otimes T')\pi;Q) < \delta/2$. Now

$$\begin{aligned} & d((S\otimes T)\pi,(S\otimes T')\pi) \\ = & \sum_{q_1,q_2\in Q} |\pi(Sq_1\times Tq_2) - \pi(Sq_1\times T'q_2) \\ \le & \sum_{q_1,q_2\in Q} \pi(\Omega\times(Tq_2\Delta T'q_2)) \\ = & \#Q \sum_{q_2\in Q} \mu(Tq_2\Delta T'q_2) \le \#Q\ \delta/2\ \#Q. \end{aligned}$$

Thus $d(\pi(S\otimes T)\pi;Q) < \delta/2 + \delta/2 = \delta$ so we can conclude that $d(\pi,\mu\otimes\mu;P) < \varepsilon$. We have shown $T' \in \mathcal{O}(P,\varepsilon,Q,\delta/2)$ so the claim is established, It follows that

$$\mathcal{O}(P,\varepsilon,Q) = \bigcup_{\delta>0} \mathcal{O}(P,\varepsilon,Q,\delta)$$

is open.

Now choose an increasing sequence $\{P_n\}$ of finite partitions which generate F up to null sets. We further require of $\{P_n\}$ that, if $\mathfrak{A}$ is the algebra $\mathfrak{A} = \bigcup_n \sigma(P_n)$, then the countable algebra generated by rectangles $A \times B$, $A,B \in \mathfrak{A}$, which we denote $\mathfrak{A} \times \mathfrak{A}$, has the property that any decreasing sequence of sets in $\mathfrak{A} \times \mathfrak{A}$ which has empty intersection is eventually empty. This will ensure that any finitely additive measure on $\mathfrak{A} \times \mathfrak{A}$ is continuous at 0 and can be extended to $\sigma(\mathfrak{A}) \times \sigma(\mathfrak{A})$. This condition can be achieved, for example, by representing Ω as $\{0,1\}^{\mathbb{N}}$ and taking P_n to be the partition of Ω according to the first n co-ordinates. Set

$$\mathcal{O} = \bigcap_{m,n\in\mathbb{N}} \bigcup_{\ell\in\mathbb{N}} \mathcal{O}(P_n,1/n,P_\ell).$$

We claim $\mathcal{O} = S^\perp$. To see this first suppose $T \in \mathcal{O}$, so for each m,n there is an ℓ and $\delta > 0$ such that $T \in \mathcal{O}(P_m,1/n,P_\ell,\delta)$. If π is $S \otimes T$ invariant with marginals μ then certainly $d(\pi,(S\otimes T)\pi;P_\ell) = 0 < \delta$, so $d(\pi,\mu\otimes\mu;P_m) < 1/n$. Since this is true for all m,n, $\pi = \mu\otimes\mu$ and we've shown $T \in S^\perp$. Now suppose that $T \notin \mathcal{O}$. Thus there exist m,n such that for all ℓ, $T \notin \mathcal{O}(P_m,1/n,P_\ell,1/\ell)$. This means that for each ℓ there is a measure π_ℓ with marginals μ such that

(1) $$d(\pi_\ell,(S\otimes T)\pi_\ell;P_\ell) < 1/\ell$$

but

(2) $$d(\pi_\ell,\mu\otimes\mu;P_m) \geq 1/n.$$

Choose a subsequence ℓ_i such that $\pi_{\ell_i}(E)$ converges to $\pi(E)$, say, for each E in $\mathfrak{A} \times \mathfrak{A}$. π is certainly finitely additive on $\mathfrak{A} \times \mathfrak{A}$ so π can be extended to a measure on the σ-algebra $\sigma(\mathfrak{A}) \times \sigma(\mathfrak{A})$, which has marginals μ, since each π_ℓ does. Since all the π_ℓ and π have marginals μ it is easy to see that $\pi_{\ell_i}(A \times B) \to \pi(A \times B)$ for all $A, B \in \sigma(\mathfrak{U})$ (not just $A, B \in \mathfrak{U}$). Thus (1) implies that π is $S \otimes T$ invariant. Since π has marginals μ the completion of π is defined (at least) on $F \times F$ and continues to have marginals μ and be $S \otimes T$ invariant. But it is not product measure, because of (2), so $T \notin S^{\perp}$. Thus $S^{\perp} = \mathcal{O}$ and $S^{\perp}$ is a G_δ.

To complete the proof it suffices to show that $S^{\perp}$ contains at least one ergodic transformation, because $S^{\perp}$ is invariant under conjugation by elements of G and the conjugates of any ergodic (even aperiodic) transformation are dense in G [6, p. 77]. In [5, Theorem I.4] it is shown that if S is weakly mixing and T is ergodic with discrete spectrum then $S \perp T$. In fact, the same argument shows that one can drop the weak mixing assumption as long as $S \otimes T$ is ergodic (for $\mu \otimes \mu$). Now, if S is any ergodic transformation and we take T to be translation by α (mod 1) on $[0,1)$ where α is irrational and $e^{2\pi i n\alpha}$ is not an eigenvalue of S for any n, then it is easy to see that $S \otimes T$ is ergodic hence $S \perp T$ by the above remarks.

Proof of Corollary 2(a): This is immediate since the weakly-mixing transformations are a dense G_δ, [6, p. 77].

Proof of Corollary 2(b): Choose a maximal disjoint family $\{T_\alpha : \alpha \in A\}$ of weakly mixing transformations. If A is only countable form $S = \bigotimes_{\alpha \in A} T_\alpha$, considered as a transformation of (Ω^A, μ^A), which is again a Lebesgue space. Thus by Corollary 2(a) there is a weakly mixing $T \in S^{\perp}$, which contradicts the maximality of $\{T_\alpha : \alpha \in A\}$.

Proof of Theorem 3: It suffices to show that for each finite $I = [-n,n] \subset \mathbb{Z}$, the set $\mathcal{D}_I$ consisting of those T such that $\{T^i : i \in I\}$ is disjoint is a dense G_δ, for then we can intersect over n. Fixing n, let $\hat{T} = \bigotimes_{i=-n}^{n} T^i$. For a finite partition Q of Ω and measures π and π' on Ω^I we set

$$d(\pi,\pi';Q) = \sum_{p \in P^I} |\pi(p) - \pi'(p)|.$$

Let $\mathcal{O}(P,\varepsilon,Q,\delta)$ be the set of transformations T of Ω such that if π is any measure on Ω^I with marginals μ then

$$d(\hat{T}\pi,\pi;Q) < \delta \Longrightarrow d(\pi,\mu^I;P) < \varepsilon.$$

We claim

$$\mathcal{O}(P,\varepsilon,Q) = \bigcup_{\delta>0} \mathcal{O}(P,\varepsilon,Q,\delta)$$

is open. To see this suppose $T \in \mathcal{O}(P,\varepsilon,Q,\delta)$ and $d(T^i,S^i;Q) < \delta/2\ (2n+1)(\#Q)^{2n}$ for $-n \le i \le n$. These inequalities define a coarse neighborhood of T because G is a topological group. We want to show $S \in \mathcal{O}(P,\varepsilon,Q,\delta/2)$, so suppose π has marginals μ and

$$d(\hat{S}\pi,\pi;Q) < \delta/2.$$

Now

$$d(\hat{T}\pi,\hat{S}\pi;Q)$$

$$= \sum_{q_n \times \cdots \times q_n \in Q^I} \left| \pi\left(\bigotimes_{i=-n}^{n} T^i q_i\right) - \pi\left(\bigotimes_{i=-n}^{n} S^i q_i\right) \right|$$

$$\le \sum_{q_{-n} \times \cdots \times q_n \in Q^I} \sum_{h=-1}^{n} \left| \pi\left(\bigotimes_{i=-n}^{j-1} S^i q_i \otimes \bigotimes_{i=j}^{n} T^i q_i\right) - \pi\left(\bigotimes_{i=-n}^{j} S^i q_i \otimes \bigotimes_{i=j+1}^{n} T^i q_i\right) \right|$$

$$\le \sum_{q_{-n} \times \cdots \times q_n \in Q^I} \sum_{j=-n}^{n} \mu(T^j q_j \Delta S^j q_j)$$

$$\le \sum_{j=n}^{n} (\#Q)^{2n}\ \delta/2(2n+1)(\#Q)^{2n} = \delta/2.$$

Thus $d(\hat{T}\pi,\pi;Q) < \delta$ so $d(\pi,\mu^I;P) < \varepsilon$, and $S \in \mathcal{O}(P,\varepsilon,Q,\delta/2)$ which establishes the claim. We now choose a sequence of partitions P_n as in the proof of Theorem 1 and set $\mathcal{O} = \bigcap_{m,n} \bigcup_{\ell} \bigcup_{\delta>0} \varphi(P_m,1/n,P_\ell,\delta)$ and show as in Theorem 1 that $\mathcal{O} = \mathcal{D}_I$. Thus $\mathcal{D}_I$ is a G_δ and since $\mathcal{D}_I$ contains ergodic transformations (e.g. any T with minimal self-joinings) and is conjugation invariant it follows that $\mathcal{D}_I$ is dense.

Proof of Corollary 4: Choose a maximal family $\{T_\alpha : \alpha \in A\}$ such that $\{T_\alpha^i : i \in \mathbb{Z} - \{0\}, \alpha \in A\}$ is disjoint. If A is only countable form $S = \otimes \{T_\alpha^i : i \in \mathbb{Z} - \{0\}, \alpha \in A\}$. For each finite $I \subset \mathbb{Z} - \{0\}$ let $\mathcal{D}_I(S)$ consist of those T such that $\underset{i\in I}{\otimes} T^i$, as a transformation of (Ω^I,μ^I), is disjoint from S. Because of Theorem 1 and the fact that the map $T \to \underset{i\in I}{\otimes} T^i$ is continuous with respect to the coarse topologies $\mathcal{D}_I(S)$ is a G_δ. Moreover $\mathcal{D}_I(S)$ contains ergodic transformations - for example, if T is ergodic with discrete spectrum then each ergodic component of $\underset{i\in I}{\otimes} T^i$ has discrete spectrum, hence is disjoint from S, so $\underset{i\in I}{\otimes} T^i$ is itself disjoint from S. Thus $\mathcal{D}_I(S)$ is a dense G_δ. Now if we take the intersection of all the $\mathcal{D}(I)$, I finite and the class of T such that T is weakly mixing and $\{T^i : i \in \mathbb{Z} - \{0\}\}$ is disjoint, this intersection is a dense G_δ and any transformation in it allows us to enlarge $\{T_\alpha : \alpha \in A\}$, a contradiction.

References

1. K. Berg, Mixing, Cyclic Approximation and Homomorphisms.
2. R. V. Chacon and T. Schwartzbauer, Commuting point transformations. Z. Wahrscheinbichkeitstheorie verw. Geb. 11 (1969), 277-287.
3. J. R. Choksi and S. Kakutani, Residuality or ergodic measure-preserving transformations and of ergodic transformations which preserve an infinite measure. Preprint.
4. A. Fieldsteel, An uncountable family of prime transformations not isomorphic to their inverses. Unpublished.

5. H. Furstenberg: Disjointness in ergodic theory, minimal sets, and a problem in Diophantine approximation. Math. Systems Theory 1 (1967), 1-49

6. P. R. Halmos, Ergodic Theory. Chelsea, New York, 1956.

7. A. del Junco, M. Rahe and L. Swanson, Chacon's automorphism has minimal self-joinings, to appear J. d'Analyse Mathématique.

8. D. S. Ornstein and P. C. Shields, An uncountable family of k-automorphisms. Advances in Math. 10 (1973), 63-88.

9. V. A. Rohlin, In general a measure-preserving transformation is not mixing. Dok. Akad. Nauk 60 (1948), 349-351.

10. D. Rudolph, An example of a measure-preserving map with minimal self-joinings and applications, J. d'Analyse Mathématique 35 (1980), 97-122.

11. J. Neveu, Mathematical Foundations of the calculus of probability. Holden-Day, San Francisco, 1965.

12. A. B. Katok and A. M. Stepin, Approximations in Ergodic Theory. Russian Math Surveys 22 (1967), 77-102.

13. A. B. Katok, Ya. G. Sinai and A. M. Stepin, Theory of Dynamical Systems and General Transformation Groups with Invariant Measure. J. Soviet Math. 7 (1977), 974-1065. Translated from Itogi Nauki Teh. Ser. Mat. Anal. 13 (1975), 129-262.

The Ohio State University
Columbus, Ohio 43210

CONTINUOUS HOMOMORPHISMS OF BERNOULLI SCHEMES

A. del Junco   M. Keane*   B. Kitchens

B. Marcus†   L. Swanson

§1. Introduction

Let m and n be integers greater than one. We set

$$A = \{0,1,\ldots,m-1\} \qquad B = \{0,1,\ldots,n-1\}$$

$$X = A^{\mathbb{Z}} \qquad Y = B^{\mathbb{Z}}$$

$$S = \text{left shift on } X \qquad T = \text{left shift on } Y.$$

Then S and T are homeomorphisms of the compact spaces S and Y.

Let k be a positive integer. A *finite code of length* k is a map

$$\varphi : A^k \longrightarrow B.$$

It gives rise to a continuous equivariant map from (X,S) to (Y,T), which we shall also denote by φ, defined by putting

$$\varphi x = y$$

if

$$y_i = \varphi(x_i x_{i+1} \cdots x_{i+k-1}) \qquad (i \in \mathbb{Z}).$$

Moreover, any continuous equivariant map Φ from (X,S) to (Y,T) can be written as

$$\Phi = \varphi \circ S^t = T^t \circ \varphi \tag{1}$$

*Supported in part by NSF Grant MCS-78-01858.

†Supported in part by NSF Grants MCS-78-01244, MCS-80-01796.

for a suitable integer t and a suitable finite code φ [4].

Next let $p = (p_0,\ldots,p_{m-1})$ and $q = (q_0,\ldots,q_{n-1})$ be probability vectors with strictly positive entries over the respective alphabets A and B, and let μ and ν denote the probability measures on X and Y defined by

$$\mu = p^{\mathbb{Z}}, \qquad \nu = q^{\mathbb{Z}}.$$

Thus $B(p) = (X,\mu,S)$ and $B(q) = (Y,\nu,T)$ are the *Bernoulli schemes* defined by p and q.

A *continuous homomorphism*

$$\Phi: B(p) \longrightarrow B(q)$$

is a continuous equivariant map from (X,S) to (Y,T) which takes μ to ν. Up to a power of S or T (see (1)), any continuous homomorphism is given by a finite code. If Φ is invertible as a map from X to Y, then its inverse is also continuous and Φ is called a *continuous isomorphism*.

In this paper, we discuss the following problems, which arise naturally in the above context.

Problem #1. Given p and q, decide if there is a continuous homomorphism from $B(p)$ to $B(q)$.

Problem #2. Describe, in some sense, the continuous homomorphisms from $B(p)$ to $B(q)$.

By the remarks above, we may restrict our attention to those continuous homomorphisms arising from finite codes. The simplest such codes are codes of length one, i.e., maps

$$\psi: A \longrightarrow B.$$

A code ψ of this type yields a continuous homomorphism if and only if ψ sends p to q, that is, if and only if there is a partition of A into n disjoint subsets $A_0,\ldots,A_{n-1}$ such that for any $b \in B$,

$$q_b = \sum_{a \in A_b} p_a.$$

This motivates the following definition. The probability vector q is a *clustering* of the probability vector p if there is a finite code of length one which sends p to q.

Conjecture #1. There is a continuous homomorphism from $B(p)$ to $B(q)$ if and only if q is a clustering of p.

We have only been able to settle a special case of Conjecture #1, as follows. By entropy theory (see e.g. [1]), the existence of a continuous (or even measurable) homomorphism from $B(p)$ to $B(q)$ implies that

$$h(p) \geq h(q),$$

where

$$h(p) = -\sum_{a \in A} p_a \log p_a$$

and

$$h(q) = -\sum_{b \in B} q_b \log q_b.$$

(In this paper, log = logarithm to the base two.) Now if q is a clustering of p and if we suppose in addition that $h(p) = h(q)$, then it follows easily that this clustering must be trivial, i.e., $m = n$ and ψ is one-to-one. We then say that q is a *rearrangement* of p.

Theorem 1. If there is a continuous homomorphism from $B(p)$ to $B(q)$ and if $h(p) = h(q)$, then q is a rearrangement of p.

We give two different proofs of this result. A third proof, similar to our first one, has been discovered by S. Tuncel (oral communication). It contains a very nice use of the idea of pressure of a continuous function and can be applied to other classes of codes which will not be discussed here.

Two corollaries of Theorem 1 are worth mentioning. First, if $B(p)$ and $B(q)$ are continuously isomorphic, then q is a rearrangement of p. (This is an easier statement to prove; see §2 below.) Second, weak isomorphism implies isomorphism in the category of continuous homomorphisms of Bernoulli schemes.

In the general case of unequal entropy, we know only the following result:

<u>Theorem 2</u>. If $p = (\frac{1}{m},\frac{1}{m},\ldots,\frac{1}{m})$ and if there is a continuous homomorphism from $B(p)$ to $B(q)$, then q is a clustering of p.

Next we consider Problem #2, restricting our attention to the equal entropy case in light of the above. If

$$p = q = (\frac{1}{m},\frac{1}{m},\ldots,\frac{1}{m})$$

are both uniform distributions, then (see Hedlund [4]) a finite code φ is measure-preserving if and only if φ is onto. It is also shown in [4] that many such codes exist, both invertible and non-invertible, and that there is little hope for classifying them (e.g. [4], p. 336, Theorem 6.13).

On the other hand, if $p = q$ with

$$0 < p_0 < p_1 < \cdots < p_{m-1},$$

the situation can be different, as the next result indicates.

<u>Theorem 3</u>. If $p = q$ and $0 < p_0 < \cdots < p_{m-1}$, then any homomorphism from $B(p)$ to $B(q)$ arising from a finite code of length two is an isomorphism.

This leads us to:

<u>Conjecture #2</u>. If $p = q$ and $0 < p_0 < \cdots < p_{m-1}$, then any homomorphism from $B(p)$ to $B(q)$ is an isomorphis

In the special case $m = 2$ of this conjecture, the hypothesis implies that the homomorphism preserves all measures $(p_0,p_1)^{\mathbb{Z}}$, $0 \le p_0 \le 1$, and this is the case if and only if the image of each periodic point contains the same number of 0 (and 1) as the original point, counted over a period. We have not been able to verify the conjecture in this case.

The structure of our arguments is as follows. After proving a relation between periodic points and their images in §2 and using it in §3 to construct a coboundary function relating the information cocycles, we present our proofs of Theorem 1 in §4. Then §5 and §6 are devoted to the proofs of Theorem 2 and Theorem 3.

This work was begun while most of the authors were guests at the special year at the University of Maryland, and continued while all were guests for a week at the University of North Carolina, Chapel Hill. We thank both institutions for their hospitality.

§2. Periodic points and their images

In this section we assume that φ is a finite code of length k giving rise to a continuous homomorphism from $B(p)$ to $B(q)$. We sometimes write $p(a)$ for p_a and $q(b)$ for q_b. For $x \in X$, $y \in Y$ and $t \ge 1$ define

$$\pi_t(x) = p(x_1) \cdots p(x_t)$$

and

$$\pi_t(y) = q(y_1) \cdots q(y_t).$$

Note that by equivariance of φ, periodic points under S are mapped to periodic points under T, and the minimal period of the image point divides the minimal period of the point mapped.

Lemma 1. If $\bar{x} \in X$ with $S^u\bar{x} = \bar{x}$ and $\bar{y} = \varphi\bar{x}$, then $\pi_u(\bar{x}) \le \pi_u(\bar{y})$.

Proof: For any $t \geq 1$ set

$$[\bar{x}_1 \cdots \bar{x}_t] = \{x \in X: x_i = \bar{x}_i \text{ for } 1 \leq i \leq t\}$$

and

$$[\bar{y}_1 \cdots \bar{y}_t] = \{y \in Y: y_j = y_j \text{ for } 1 \leq j \leq t\}.$$

Then

$$[\bar{x}_1 \cdots \bar{x}_{t+k-1}] \subseteq \varphi^{-1}[\bar{y}_1 \cdots \bar{y}_t],$$

and since φ sends μ to ν, we obtain

$$\pi_{t+k-1}(\bar{x}) \leq \pi_t(\bar{y}) \qquad (t \geq 1).$$

Now set $t = t'u$ and note that since u is a period of $\bar{x}$ and $\bar{y}$,

$$\pi_{t+k-1}(\bar{x}) = \pi_{k-1}(\bar{x})(\pi_u(\bar{x}))^{t'}$$

and

$$\pi_t(\bar{y}) = (\pi_u(\bar{y}))^{t'}.$$

Therefore

$$\pi_u(\bar{x})(\pi_{k-1}(\bar{x}))^{1/t'} \leq \pi_u(\bar{y}),$$

and the lemma follows by letting t' tend to infinity. ∎

Corollary 1. If φ is a continuous isomorphism, then q is a rearrangement of p.

Proof: If φ is an isomorphism, then it establishes a one-to-one correspondence between the fixed points of S and the fixed points of T. Let $\bar{x} = (\ldots,a,a,a,\ldots)$ and $\bar{y} = \varphi\bar{x} = (\ldots,b,b,b,\ldots)$, and set $u = 1$ in Lemma 1 to obtain

$$p_a = \pi_1(\bar{x}) \leq \pi_1(\bar{y}) = q_b.$$

Since p and q are probability vectors, it follows that $p_a = q_b$ and the proof is finished. ■

The following corollary weakens the hypothesis of Lemma 1. It will be used to prove Lemma 2 below.

Corollary 2. Let $t < t'$ be integers with $u = t' - t$. If $\tilde{x} \in X$ with $\tilde{y} = \varphi\tilde{x}$ and

$$\tilde{x}_{t+1} \cdots \tilde{x}_{t+k-1} = \tilde{x}_{t'+1} \cdots \tilde{x}_{t'+j-1}$$

then

$$\pi_u(S^t\tilde{x}) \leq \pi_u(T^t\tilde{y}).$$

Proof: Define a point $\bar{x} \in X$ with $S^u\bar{x} = \bar{x}$ by setting

$$\bar{x}_1 \cdots \bar{x}_u = \tilde{x}_{t+1} \cdots \tilde{x}_{t'}$$

and continuing $\bar{x}$ periodically. The hypothesis, together with the assumption that the code length of φ is k, imply that if $\bar{y} = \varphi\bar{x}$, then

$$\bar{y}_1 \cdots \bar{y}_u = \tilde{y}_{t+1} \cdots \tilde{y}_{t'}.$$

Therefore

$$\pi_u(S^t\tilde{x}) = \pi_u(\bar{x}) \leq \pi_u(\bar{y}) = \pi_u(T^t\tilde{y})$$

by Lemma 1. ■

We can now state the main result of this section.

Lemma 2. If $h(p) = h(q)$, then under the hypothesis of Lemma 1 ($S^u\bar{x} = \bar{x}$ and $\bar{y} = \varphi\bar{x}$) we have

$$\pi_u(\bar{x}) = \pi_u(\bar{y}).$$

Proof: By the ergodic theorem ([1]), there is a subset $\tilde{X}$ of X with $\mu(\tilde{X}) = 1$ such that for any finite sequence $a_1 \ldots a_\ell$

of symbols of A and for any point $\tilde{x}$ of $\tilde{X}$, the frequency of occurrence of $a_1 \cdots a_\ell$ in $\tilde{x}_1\tilde{x}_2\tilde{x}_3\cdots$ exists and is equal to $p(a_1)\cdots p(a_\ell)$. In other words, almost all $\tilde{x} \in X$ are "normal" in $B(p)$. This statement is also valid for $B(q)$ with the appropriate change in wording. In particular, let $\tilde{Y}$ denote the set of sequences which are "normal" in $B(q)$. Then, since φ is measure preserving, we have

$$\mu(\varphi^{-1}(\tilde{Y}) \cap \tilde{X}) = 1,$$

which shows that this set is non-empty. We pick and fix a point

$$\tilde{x} \in \tilde{X} \cap \varphi^{-1}\tilde{Y}$$

and set

$$\tilde{y} = \varphi\tilde{x} \in \tilde{Y}.$$

Since any given symbol $a \in A$ occurs with frequency p_a in $\tilde{x}$, it follows that

$$\lim_{t\to\infty} \frac{1}{t} \log \pi_t(\tilde{x}) = -h(p);$$

similarly

$$\lim_{t\to\infty} \frac{1}{t} \log \pi_t(\tilde{y}) = -h(q).$$

Moreover, $h(p) = h(q)$ implies

(2) $$\lim_{t\to\infty} \frac{1}{t} (\log \pi_t(\tilde{y}) - \log \pi_t(\tilde{x})) = 0.$$

With these preparations, let $\bar{x} \in X$ with $S^u\bar{x} = \bar{x}$ and $\bar{y} = \varphi\bar{x}$. It is trivial to see that the conclusion of Lemma 2 holds for any period of $\bar{x}$ if and only if it holds for the minimal period of $\bar{x}$. Denote this minimal period by v, and let w be a multiple of v with $w \geq k-1$. Consider the set of positive integers t for which

$$\tilde{x}_{t+1} \cdots \tilde{x}_{t+w} = \bar{x}_1 \cdots \bar{x}_w,$$

and arrange it in increasing order:

$$0 < t_1 < t_2 < t_3 < \cdots .$$

Define $t_0 = 0$, and for any $\ell \geq 0$, set

$$P_\ell = \log[\pi_{t_{\ell+1}-t_\ell}(S^{t_\ell}\tilde{x})] = \sum_{i=t_\ell+1}^{t_{\ell+1}} \log p(\tilde{x}_i)$$

and

$$Q_\ell = \log[\pi_{t_{\ell+1}-t_\ell}(T^{t_\ell}\tilde{y})] = \sum_{j=t_\ell+1}^{t_{\ell+1}} \log q(\tilde{y}j).$$

By Corollary 2, $P_\ell \leq Q_\ell$ for all $\ell > 1$, and by (2),

$$\lim_{L\to\infty} \frac{1}{t_L} \sum_{\ell=0}^{L-1} (Q_\ell - P_\ell) = 0.$$

Further, consider the set of positive integers t for which

$$\tilde{x}_{t+1} \cdots \tilde{x}_{t+w+v} = \bar{x}_1 \cdots \bar{x}_w \bar{x}_1 \cdots \bar{x}_v.$$

If t is such an integer, then clearly there exists an index ℓ such that $t_\ell = t$ and $t_{\ell+1} = t + v$, hence

$$Q_\ell - P_\ell = \log \pi_v(\bar{y}) - \log \pi_v(\bar{x}).$$

Moreover, since $\tilde{x}$ is "normal," this set of integers has positive density δ $(=\pi_w(\bar{x})\pi_v(\bar{x}))$ in the non-negative integers. Therefore

$$0 = \lim_{L\to\infty} \frac{1}{t_L} \sum_{\ell=0}^{L-1} (Q_\ell - P_\ell) \geq \delta(\log \pi_v(\bar{y}) - \log \pi_v(\bar{x})) \geq 0$$

and $\pi_v(\bar{x}) = \pi_v(\bar{y})$, as desired. ∎

Corollary 3. If $h(p) = h(q)$, then under the hypotheses of Corollary 2 ($t < t'$, $u = t'-t$, $\tilde{y} = \varphi\tilde{x}$ and $\tilde{x}_{t+1}\cdots\tilde{x}_{t+k-1} = \tilde{x}_{t'+1}\cdots\tilde{x}_{t+k-1}$) $\pi_u(S^t\tilde{x}) = \pi_u(T^t\tilde{y})$.

Proof: Same as the proof of Corollary 2, using Lemma 2 instead of Lemma 1. ∎

§3. The information cocycle

In this paragraph we assume that φ is a finite code of length k giving rise to a continuous homomorphism from $B(p)$ to $B(q)$ and that $h(p) = h(q)$. The functions

$$I(x) = \log p(x_1) = \log \pi_1(x) \qquad (x \in X)$$

and

$$J(y) = \log q(y_1) = \log \pi_1(y) \qquad (y \in Y)$$

are called the information cocycles of $B(p)$ and $B(q)$. The articles Fellgett-Parry [3], Parry [6], [7], [8], study the relation between the functions I and $J \circ \Phi$ on the space X, under various conditions on the (not necessarily continuous) homomorphism Φ. We shall not make use of their results directly, but of course the ideas have guided us in an essential manner. In our setting, for example, [3] implies that there is a measurable function

$$C: X \longrightarrow \mathbb{R}$$

such that

$$I(x) = J(\varphi x) + C(Sx) - C(x)$$

μ-almost everywhere, and in §4 of [6] it is shown that $C \in \mathbb{L}^p(X,\mu)$ for all $1 \leq p < \infty$, under more general assumptions. In cohomology language, $C \circ S - C$ is called a coboundary and the functions I and $J \circ \varphi$ are said to be cohomologous. In the case we are considering, we shall need a stronger result, which follows easily from Corollary 3.

This lemma can be found elsewhere. The continuity of C is shown in [6]; the fact that Corollary 3 implies Lemma 3 is essentially contained in [2], Proposition 4.5.

Lemma 3. There exists a continuous real-valued function C on X such that for all $x \in X$,

$$I(x) = J(\varphi x) + C(Sx) - C(x). \tag{3}$$

Proof: Assume for a moment that (3) is valid for some function C, and replace x by $S^i x$ to obtain

$$C(S^{i+1}x) = C(S^i x) = I(S^i x) - J(T^i \varphi x).$$

Then let $t < t'$ and sum from t to $t' - 1$, recalling the definitions of I and J, to get

$$C(S^{t'}x) - C(S^t x) = \log \frac{p(x_{t+1}) \cdots p(x_{t'})}{q(y_{t+1}) \cdots q(y_{t'})} = \log \pi_u(S^t x) - \log \pi_u(T^t y),$$

where we have set $y = \varphi x$ and $u = t' - t$. We conclude that:

1) the value of C at a point determines its values on the orbit of the point, and

2) if $S^t x$ and $S^{t'} x$ agree in the first $k - 1$ coordinates, then $C(S^t x) = C(S^{t'} x)$ by Corollary 3.

Laying these remarks aside, we now proceed in the following manner. Fix a point $\tilde{x} \in X$ whose S-orbit is dense in X and define a function $\tilde{C}$ on the orbit of $\tilde{x}$ by setting $C(\tilde{x}) = 0$ and

$$\tilde{C}(S^t \tilde{x}) = \begin{cases} \log \dfrac{p(\tilde{x}_1) \cdots p(\tilde{x}_t)}{q(\tilde{y}_1) \cdots q(\tilde{y}_t)} & (t \geq 1) \\[2ex] -\log \dfrac{p(\tilde{x}_{t+1}) \cdots p(\tilde{x}_0)}{q(\tilde{y}_{t+1}) \cdots q(\tilde{y}_0)} & (t \leq -1), \end{cases}$$

where $\tilde{y} = \varphi \tilde{x}$. Then if $t < t'$ and $u = t' - t$ we have

$$\tilde{C}(S^{t'}\tilde{x})-\tilde{C}(S^{t}\tilde{x}) = \log \frac{p(\tilde{x}_{t+1})\cdots q(\tilde{x}_{t'})}{q(\tilde{y}_{t+1})\cdots q(\tilde{y}_{t'})} = \log \pi_u(S^t\tilde{x})-\log \pi_u(T^t\tilde{y}).$$

Corollary 3 now shows that the function $\tilde{C}$ is uniformly continuous on the orbit of $\tilde{x}$ and since this orbit is dense in X, we can extend $\tilde{C}$ to a continuous function C on all of X by continuity. Setting $t' = t + 1$ in the above equation shows that (3) holds for all x in the orbit of $\tilde{x}$, and hence by continuity of I and $J \circ \varphi$, for all $x \in X$. ∎

Corollary 4. The function $C(x)$ depends at most on the coordinates $x_1, x_2, \ldots, x_{k-1}$.

Proof: This is remark 2) above. ∎

Corollary 5. There exist constants $0 < d < D < \infty$ such that for any $x \in X$ and $t \geq 1$, $y = \varphi x$,
$dp(x_1)\cdots p(x_t) \leq q(y_1)\cdots q(y_t) \leq Dp(x_1)\cdots p(x_t)$.

Proof: As in the proof of Lemma 3, iteration of the equation (3) yields

$$\log \frac{p(x_1)\cdots p(x_t)}{q(y_1)\cdots q(y_t)} = C(S^t x) - C(x),$$

so that the corollary follows immediately from the continuity of C. ∎

§4. Two proofs of Theorem 1

If $r = (r_0, \ldots, r_{\ell-1})$ is any vector with positive real entries, we call the function $F_r(s)$, defined for $s > 0$ by

$$F_r(s) = \sum_{a=0}^{\ell-1} r_a^s,$$

the transform of r. It is easy to verify that

(i) the transforms of two vectors are equal if and only if one vector is a rearrangement of the other, and

(ii) $F_{r\otimes r'} = F_r \cdot F_{r'}$, where $r \otimes r' = (r_a r'_b)$.

Note that F_r is just the Laplace transform of the measure obtained from r by assigning mass one to each point $-\log_e r_a$ (with multiplicity).

We repeat the statement of the theorem.

Theorem 1. If there is a continuous homomorphism form $B(p)$ to $B(q)$ and if $h(p) = h(q)$, then q is a rearrangement of p.

Proof #1: We may assume that the homomorphism is given by a finite code φ. Let $y = \varphi x$. From Corollary 5 we have

$$dp(x_1) \cdots p(x_t) \leq q(y_1) \cdots q(y_t) \leq Dp(x_1) \cdots p(x_t)$$

for all $t \geq 1$. Since these expressions are products of independent random variables, we can raise to the power $s > 0$ and integrate to obtain

$$d^s (F_p(s+1))^t \leq (F_q(s+1))^t \leq D^s (F_p(s+1))^t.$$

Dividing by $(F_p(s+1))^t$ and letting t tend to infinity yields

$$F_p(s) = F_q(s) \qquad (s > 1),$$

so that by (i), q is a rearrangement of p. ∎

Our second proof is more combinatorial in nature and, although a bit longer, seems to yield an interesting method (see §6). We preface the proof with two remarks:

Remark 1. It is sufficient to show that Theorem 1 holds for codes of length two.

To see this, suppose φ is a code of length k. By grouping symbols in $B(p)$ and $B(q)$ into groups of length ℓ, where $\ell \geq 1$ is arbitrary, φ induces a homomorphism $\tilde{\varphi}$ from $B(p^\ell)$ to $B(q^\ell)$, where

$$p^\ell = p \otimes \cdots \otimes p \quad (\ell \text{ times})$$

and q^ℓ is defined similarly. Now if $\ell \geq k - 1$, then $\tilde{\varphi}$ is a finite code of length two, and if

$$F_{p^\ell} = F_{q^\ell}$$

for some ℓ, then (ii) shows that $F_p = F_q$.

Remark 2. The hypotheses of Theorem 1 imply that $m = n$ and that Φ preserves the probability measure λ arising from equidistribution $(\frac{1}{m},\ldots,\frac{1}{m})$ on A and B.

This statement follows easily from entropy theory and the results of [4], [9], since $h(p) = h(q)$ implies that the topological entropy cannot be decreased by φ. Thus $m = n$, and the rest follows from [4]. Nevertheless, we shall outline a simple counting argument giving $m = n$, and an argument different from that of [4] which yields $\varphi\lambda = \lambda$.

Obviously φ maps X onto Y, since φX is compact and support (ν) = support $(\varphi\mu)$ = Y. For each $t \geq 1$, consider the map, which we shall also denote by φ, induced by φ from A^{t+k-1} to B^t. These maps are also all surjective, and since

$$\text{card } A^{t+k-1} = m^{t+k-1}$$

$$\text{card } B^t = n^t,$$

we conclude that

$$n^t \leq m^{t+k-1} \qquad (t \geq 1).$$

Letting t go to infinity, we obtain $n \leq m$. If $n < m$, then there exists $b' \in \bigcup B^t$ such that

$$\text{card } \varphi^{-1}b' \geq \frac{m^{t+k-1}}{n^t} \longrightarrow \infty$$

as $t \to \infty$. Therefore we can find $t' \geq 1$, $b' \in B^{t'}$, and $a', a'' \in \varphi^{-1}b'$ such that a' and a'' are not equal but do agree in the first and last $k-1$ places. Now, if $t \geq 1$ is arbitrary and if $b \in B^t$ such that a' occurs disjointly in one element of $\varphi^{-1}(b)$, say ℓ times, then $\varphi^{-1}b$ contains at least 2^ℓ different elements of the same measure (by Corollary 3), obtained by arbitrarily replacing occurrences of a' with a" and vice-versa. Since for large t and most $b \in B^t$, $\frac{\ell}{t}$ is at least equal to $\delta = \frac{1}{2}\frac{\mu[a']}{t'+k-1} > 0$, it follows that in the limit,

$$h(q) \leq h(p) - \delta < h(p).$$

This contradiction implies $m = n$.

To show that φ maps λ to λ, note that the surjectivity of φ implies that (first for any cylinder set and then) for any measurable set $U \subseteq Y$,

$$\lambda(U) \leq n^{k-1}\lambda(\varphi^{-1}U).$$

Therefore λ is absolutely continuous with respect to $\varphi\lambda$, and ergodicity of $\varphi\lambda$ under T and invariance of λ then imply that $\lambda = \varphi\lambda$.

Finally we remark that $\lambda = \varphi\lambda$ means simply that

$$\text{card } \lambda^{-1}b = m^{k-1}$$

for all $b \in B^t$ and $t \geq 1$. Now we can present our second proof of Theorem 1.

Proof #2: Assume by Remark 1 that

$$\varphi: A \times A \longrightarrow B$$

is a finite code of length two. Let $C(x)$ be the function constructed in Lemma 3 which, by Corollary 4, depends only on x_1, and define $c = (c_0, \ldots, c_{m-1})$ by setting

$$c_a = 2^{-C(x)} \quad \text{iff} \quad x_1 = a.$$

Then the cohomology equation (3) exponentiates to

$$q_\varphi(a,a') = \frac{p_a}{c_a} \cdot c_{a'} \qquad (a,a' \in A)$$

(obtained by selecting $x \in X$ with $x_1 = a$ and $x_2 = a'$). We get

$$q_\varphi(a',a") = \frac{p_{a'}}{q_\varphi(a,a')} \cdot q_\varphi(a,a")$$

and

$$q_\varphi(a,a')q_\varphi(a',a") = p_{a'}q_\varphi(a,a"),$$

valid for all $a,a',a" \in A$. If $s > 0$, then

$$\sum_{a,a',a"} q^s_\varphi(a,a')q^s_\varphi(a',a") = \sum_{a,a',a"} p^s_{a'}q^s_\varphi(a,a"),$$

and recalling that Remark 2 $(\varphi\lambda=\lambda)$ yielded

$$\text{card}\{(a,a"): \varphi(a,a") = b\} = m,$$

$$\text{card}\{(a,a',a"): \varphi(a,a') = b,\ \varphi(a',a") = b'\} = m$$

for all $b,b' \in B$, the above equation becomes

$$m(F_q(s))^2 = mF_p(s)F_q(s) \qquad (s > 0).$$

Therefore $F_p(s) = F_q(s)$ and by (i), q is a rearrangement of p. ■

§5. The case of equal probabilities

Theorem 2. If $p = (\frac{1}{m},\dots,\frac{1}{m})$ and if there is a continuous homomorphism from $B(p)$ to $B(q)$, then q is a clustering of p.

Proof: For $b \in B$ and $t \geq 1$ set

$$[b^t] = \{y \in Y: y_j = b \quad \text{for} \quad 1 \le j \le t\}.$$

Suppose that the code length of φ is k. Then

$$\varphi^{-1}[b^t]$$

is a finite union of cylinder sets in X depending on the coordinates 1 to $t + k - 1$, and hence this set has measure of the form

$$\frac{e}{m^{t+k-1}},$$

where e is an integer. In particular, for $t = 1$, since

$$\nu([b]) = \mu(\varphi^{-1}[b]),$$

we see that q must have the form

$$q = (q_0,\ldots,q_{n-1}) = \left(\frac{e_0}{m^k},\ldots,\frac{e_{n-1}}{m^k}\right).$$

Then

$$\nu([b^t]) = \mu(\varphi^{-1}[b^t])$$

implies that

$$\left(\frac{e_b}{m^k}\right)^t = \frac{e}{m^{t+k-1}},$$

so that

$$m^{t+k-1}\left(\frac{e_b}{m^k}\right)^t = m^{k-1}\left(\frac{e_b}{m^{k-1}}\right)^t$$

must be an integer for all $t \ge 1$. This is only possible if m^{k-1} divides e_b, and hence

$$q_b = \frac{e'_b}{m} .$$

Therefore q is a clustering of p. ∎

§6. Proof of Theorem 3

Theorem 3. If $p = q$ and $0 < p_0 < \cdots < p_{m-1}$, then any homomorphism from $B(p)$ to $B(q)$ arising from a finite code of length two is an isomorphism.

Proof: Let $a_i, a'_i \in A$ for $1 \leq i \leq 3$. If we can show that

$$\varphi(a_1, a_2) = \varphi(a'_1, a'_2)$$

and

$$\varphi(a_2, a_3) = \varphi(a'_2, a'_3)$$

together imply

$$a_2 = a'_2 ,$$

then the theorem will be proved, since this allows us to construct an "inverse" ψ by setting

$$\psi(b_1, b_2) = a_2$$

iff

$$\varphi(a_1, a_2) = b_1$$

and

$$\varphi(a_2, a_3) = b_2$$

for some choice of $a_1, a_3 \in A$. (It is trivial to check that $S^{-1} \circ \psi \circ \varphi$ is the identity map on X.)

Consider the cohomology equation (see Proof #2 of Theorem 1 in §4)

$$q_\varphi(a,a') = \frac{p_a}{c_a} c_{a'} \qquad (a,a' \in A).$$

Set $r = (r_0,\ldots,r_{m-1}) = (\frac{p_0}{c_0},\ldots,\frac{p_{m-1}}{c_{m-1}})$. By raising to the sth power and summing we get (as in §4)

$$(4) \qquad mF_p(s) = F_r(s)F_c(s) \qquad (s > 0).$$

Setting $z = e^s$ in (4), we apply the Gauss lemma ([5]) Chapter IV; the proof is the same for nonintegral exponents) to the "polynomials" in z corresponding to F_p, F_r and F_e, and obtain integers u and v with $uv = m$ such that each value which is assumed by a component of r is assumed with a multiplicity which is a multiple of u, and similarly for c and v. In other words, if we define two equivalence relations on A by

$$a \sim a' \quad \text{iff} \quad \frac{p_a}{c_a} = \frac{p_{a'}}{c_{a'}}$$

and

$$a \sim a' \quad \text{iff} \quad c_a = c_{a'},$$

then $\sim$-equivalence classes have cardinalities which are multiples of u and $\sim$-equivalence classes have cardinalities which are multiples of v. Recalling now that for any $b \in B$,

$$\text{card}\{(a,a'): \varphi(a,a') = b\} = m,$$

it follows from the cohomology equation and the fact that the p_a are distinct that the cardinalities of $\sim$-equivalence classes and $\sim$-equivalence classes must all be equal to u and v respectively, and that each pair of equivalence classes, say with representatives a and a', produces a value

$$q_\varphi(a,a') = \frac{p_a}{c_a} c_{a'}$$

which is distinct from the value produced by a different pair of classes.

Now suppose that

$$\varphi(a_1,a_2) = \varphi(a_1',a_2')$$

and

$$\varphi(a_2,a_3) = \varphi(a_2',a_3').$$

The first equation implies that $a_2 \sim a_2'$ and the second that $a_2 \sim a_2'$. Therefore

$$\varphi(a_2,a_2) = \varphi(a_2',a_2'),$$

and the fixed points $(\ldots,a_2.a_2,a_2,\ldots)$ and $(\ldots,a_2',a_2',a_2',\ldots)$ have the same image under φ. By Lemma 2, we then have

$$p_{a_2} = p_{a_2'},$$

and since the p_a are all distinct, it follows that $a_2 = a_2'$. ∎

References

[1] Billingsley, P., Ergodic Theory and Information, John Wiley & Sons, New York (1965).

[2] Bowen, R., Equilibrium States in the Ergodic Theroy of Anosov Diffeomorphisms, Springer Verlag Lecture Notes in Mathematics #470.

[3] Fellgett, R. and W. Parry, Endomorphisms of a Lebesgue Space II, Bull. London Math. Soc. 7 (1975), 151-158.

[4] Hedlund, G. A., Endomorphisms and Automorphisms of the Shift Dynamical System, Math. Systems Theory 3 (1969), 320-375.

[5] Jacobson, N., Lectures in Abstract Algebra, van Nostrand, New York (1966).

[6] Parry, W., Endomorphisms of a Lebesgue Space III, Israel J. Math. 21 (1975), 167-172.

[7] _____, The Information cocycle and ε-bounded codes, Israel J. Math. 29 (1978), 205-220.

[8] _____, Finitary Isomorphisms with Finite Expected code Lengths, Bull. London Math. Soc. 11 (1979), 170-176.

[9] Parry, W., Finitary Classification of Topological Markow Chains and Sofic Systems, Bull. of London Math. Soc. 9 (1977), 86-92.

A. del Junco
Mathematics Department
The Ohio State University
Columbus, Ohio 43210

M. Keane
Mathematics Department
Julianalaan 134
2600 AA Delft
The Netherlands

B. Kitchens
Mathematics Department
University of North Carolina
Chapel Hill, North Carolina
27514

B. Marcus
Mathematics Department
University of North Carolina
Chapel Hill, North Carolina
27514

L. Swanson
Mathematics Department
Texas A & M University
College Station, Texas 77843

PROJECTIVE SWISS CHEESES AND UNIQUELY ERGODIC INTERVAL EXCHANGE TRANSFORMATIONS

William A. Veech*

1. Introduction. Recall that an *interval exchange* on a finite (left closed-right open) interval, $J \subseteq \mathbb{R}$ is a transformation, T, of J which results from decomposing J into a finite number of (left closed-right open) subintervals and translating these subintervals in such a way that their union is again J. T is determined by J and three additional entities:

1) $m > 0$, the number of subintervals
2) $\lambda \in \Lambda_m$ (the cone of positive vectors in $\mathbb{R}^m$), giving the lengths of the subintervals, reading from left to right
3) $(i \to \pi i) \in \mathfrak{S}_m$ (the permutation group on $\{1,2,\ldots,m\}$), giving the order of the subintervals, reading from left to right, after T has been applied.

Given interval exchanges, (T,J) and (T',J'), which have associated the same m and (λ,π), J has the same length as J', and the unique translation $L: J \to J'$ conjugates T and T'. For this reason it is customary to replace J by $[0,|\lambda|) = I^\lambda$, $|\lambda| = \sum_{j=1}^{m} \lambda_j$, and speak of "the" (λ,π) interval exchange. Indeed, we shall often think of (λ,π) itself as the transformation and write (λ,π) in place of T or $T_{(\lambda,\pi)}$.

Remark. An argument similar to the above shows (λ,π) and $(t\lambda,\pi)$ are (continuously) isomorphic for any $t > 0$. Indeed, the map $x \to tx$ from I^λ to $I^{t\lambda}$ conjugates the two.

---

*Research supported by NSF-MCS-7801858

It is obvious from the definition that (λ,π) preserves Lebesgue measure on I^λ. One says (λ,π) is *uniquely ergodic* if every finite (λ,π)-invariant Borel measure is a multiple of Lebesgue measure.

Recall that $\pi \in \mathfrak{S}_m$ is *reducible* if there exists $i < m$ such that $\pi\{1,2,\ldots,i\} = \{1,2,\ldots,i\}$. Otherwise π is *irreducible*. If π is reducible, then I^λ, $\lambda \in \Lambda_m$, decomposes naturally into two nonempty (λ,π)-invariant subintervals, and (λ,π) cannot be uniquely ergodic. It is therefore natural, for purposes of studying unique ergodicity, to restrict attention to $\mathfrak{S}_m^0 \subseteq \mathfrak{S}_m$ the set of irreducible permutations. We shall prove

1.1 Theorem. $(m = 4)$ *Let* $\pi \in \mathfrak{S}_4$ *be an irreducible permutation. For Lebesgue almost all* $\lambda \in \Lambda_4$ *the* (λ,π) *interval exchange is uniquely ergodic.*

Recall the result of Keane [6]. If $\pi \in \mathfrak{S}_m^0$, and if the orbits of the discontinuities of (λ,π) are infinite (i.e., each orbit is infinite) and pairwise distinct, then (λ,π) is *minimal*; that is, *every* (λ,π) orbit is dense in I^λ. In particular, when the components of λ are rationally independent and $\pi \in \mathfrak{S}_m^0$), (λ,π) is minimal.

When (λ,π) is minimal, the only obvious finite invariant measures are the multiples of Lebesgue measure. In view of this as well as the fact that minimality implies unique ergodicity when $m = 2,3$ ([5], [6]), Keane was led to conjecture that, in general, (λ,π) is uniquely ergodic once it is minimal. However Keynes and Newton [9] produced a counterexample with $m = 5$, by noting (effectively) that the minimal nonuniquely ergodic skew products constructed in [17] are isomorphic to interval exchang

Keynes and Newton in turn suggested that if $\pi \in \mathfrak{S}_m^0$, and if $\lambda \in \Lambda_m$ has rationally independent components, then (λ,π) is uniquely ergodic. But this time Keane produced the counter-example, with $m = 4$, the smallest possible.

In the wake of these counterexamples Keane set forth a final (one hopes) conjecture:

1.2. Conjecture. (Keane [7]). If $\pi \in \mathfrak{S}_m^0$, then for Lebesgue almost all $\lambda \in \Lambda_m$ the (λ,π) interval exchange is uniquely ergodic.

A topological analogue of Conjecture 1.2 (replace "Lebesgue almost all" by "residual set of") has been confirmed by Keane and Rauzy [8], but the first measure theoretic evidence for Keane's conjecture is contained in Theorem 1.1. (One suspects the intricacy of the proof of Theorem 1.1 is as much a reflection of its not being the "right" proof as it is of the existence of the rather delicate examples by Keane [7] for $m = 4$.)

We adopt the point of view of [18], [15] wherein it was shown that results such as Theorem 1.1 might arise as corollaries of theorems concerning the ergodic behavior of certain related transformations on the body of (irreducible) interval exchanges. This has proved to be so, and here we describe one of these theorems.

Define $\theta_m \in \mathfrak{S}_m$ by $\theta_m i = m - i + 1$, $1 \leq i \leq m$. Given $\lambda \in \Lambda_m$, denote $I_1^\lambda = [0,\lambda_1)$. Let $U\colon I_1^\lambda \to I_1^\lambda$ be the induced transformation (of (λ,θ_m)), $Ux = T^n x$, where $n = n(x) > 0$ is the least $n > 0$ such that $T^n x \in I_1^\lambda$. According to Rauzy [10], it is true for Lebesgue almost all $\lambda \in \Lambda_m$ that $U = (\alpha,\theta_m)$ for some $\alpha \in \Lambda_m$ such that $|\alpha| = \lambda_1$. If isomorphism properties are the only concern, one may restrict attention to $\lambda \in \Delta_{m-1}$, the standard (m-1)-simplex, $\Delta_{m-1} = \{\lambda \in \Lambda_m \mid |\lambda| = 1\}$, and normalize α above to obtain an a.e. defined map $u_1\colon \Delta_{m-1} \to \Delta_{m-1}$,

$$u_1\lambda = \frac{\alpha}{|\alpha|} .$$

u_1 is nonsingular relative to Lebesgue (m-1) measure and many-to-one. One of the principal results of this study is

1.3. Theorem. *Suppose* $m = 4$, *and let* $u_1 : \Delta_3 \to \Delta_3$ *be as defined above. There exists a α-finite Borel measure,* u, *on* Δ_3 *such that* a) u *is equivalent to Lebesgue* (m-1) *measure, and* b) u *is invariant and ergodic relative to* u_1.

We shall understand by "ergodic" what is usually called "ergodic and incompressible." T is incompressible relative to a measure, ν, if whenever E is measurable and $T^{-1}E \subseteq E$, then $\nu(E \Delta T^{-1}E) = 0$. The latter conditon holds, for example, when ν is a finite invariant measure (Poincaré Recurrence).

It is natural to ask whether the assumption that Lebesgue measure on I^λ be ergodic for (λ,π) has any bearing on the question of unique ergodicity. Strictly speaking it does not ([6]; see also [18]). However, in the hopes that one may be led somehow to an "easy" proof of Conjecture 1.2, we shall prove

1.4. Theorem. *Let* $\pi \in \mathfrak{S}_m^0$, *and suppose that for Lebesgue almost all* $\lambda \in \Lambda_m$ *Lebesgue measure on* I^λ *is ergodic for* (λ,π). *Then for Lebesgue almost all* $\lambda \in \Lambda_m$ (λ,π) *is uniquely ergodic.*

Remark. As is indicated in the remark preceding the statement of Theorem 1.4, the set of $\lambda \in \Lambda_m$ such that (λ,π) is uniquely ergodic is in general a *proper* subset of the set of λ such that (λ,π) is ergodic relative to Lebesgue measure on I^λ.

2. Interval exchanges. Let $m > 1$, and suppose $(i \to \pi i) \in \mathfrak{S}_m$. Following [15], [11] we associate to π an *alternating* $m \times m$ matrix, L^π, defined above the main diagonal by

$$L^\pi_{ij} = \begin{cases} 0 & \pi i < \pi j, \quad i < j \\ 1 & \pi i > \pi j, \quad i < j \end{cases} \quad . \tag{2.1}$$

As in the introduction Λ_m denotes the cone of positive vectors in $\mathbb{R}^m$, and we write $|\lambda| = \sum_{j=1}^{m} \lambda_j$, $\lambda \in \Lambda_m$. If $\lambda \in \Lambda_m$, the interval $I^\lambda = [0, |\lambda|)$ is partitioned into m left closed-right open subintervals I^λ_j, $1 \le j \le m$, whose lengths are successively $\lambda_1, \lambda_2, \ldots, \lambda_m$.

The (λ, π) interval exchange is conveniently expressed in terms of L^π as

$$Tx = x + (L^\pi \lambda)_j \quad (x \in I^\lambda_j,\ 1 \le j \le m) \tag{2.2}$$

(see [15], [11]).

Remark. Let $\vartheta(\lambda,\pi)$ be the cone of (λ,π)-invariant finite Borel measures; it is proved in [15] that if (λ,π) is *minimal* then

$$\dim \vartheta(\lambda,\pi) \le \tfrac{1}{2} \operatorname{Rank} L^\pi \quad . \tag{2.3}$$

For the equivalence between this result and a bound of Katok's [4] for the dimension of the cone of nonatomic foliation invariant measures for certain foliations of M^2, see [16].

We adopt below the convention that "interval" shall mean left closed-right open interval. $|J|$ denotes the length of an interval J.

Given an interval exchange (λ,π) and an interval $J \subseteq I^\lambda$, the induced transformation $U: J \to J$ is defined by $Ux = T^n x$, $x \in J$, where $n = n(x)$ is the least positive integer such that

$T^n x \in J$. As noted in [15], U is again an interval exchange, although with perhaps a different number of discontinuities. Taking into account the remarks in the Introduction, we write $U = (\alpha, \pi')$, $\alpha \in \Lambda_{m'}$, $\pi' \in \mathfrak{S}_{m'}$, $m' > 0$, with $|\alpha| = |J|$ We recall that if J is an "admissible" interval, in the sense of [15], then $m' \leq m$. With certain additional hypotheses it is guaranteed that $m' = m$, the only case of concern to us is what follows. As $m' = m$ will be clear in the context of the present study, we do not here elaborate on the general situation, but instead refer the reader to [15] for a complete discussion.

In what follows $U = (\alpha, \pi')$ is as above with the assumption that $m' = m$. As in [7], [15], [11] we define an $m \times m$ non-negative integer matrix, A, by

$$A_{ij} = \mathrm{Card}\{k \mid 0 \leq k < N_j,\ T^k I_j^\alpha \subseteq I_i^\lambda\} \tag{2.4}$$

where, for each j, N_j is the common value of $n(x)$ on I_j^α.

(Note the abuse of notation in (2.4), where I_j^α is regarded as a subset of J and not $[0, |\alpha|)$.) If (λ, π) is minimal, then A has the following properties, of which at least the first is obvious.

$$\lambda = A\alpha \tag{2.5}$$

$$A^t L^\pi A = L^{\pi'} \tag{2.6}$$

$$\det A = \pm 1 \tag{2.7}$$

See [15], [11] for the proofs.

We recall now from [15], [11] the important general principle, implicit in [7]:

2.8. Proposition. Let (λ, π) be a minimal interval exchange and let $\{J_n\}$ be a sequence of admissible intervals (nested or not) such that $\lim_{n\to\infty} |J_n| = 0$. Let $A^{(n)}$ be the corresponding sequence of matrices (2.4), and define

$$(2.9) \qquad \Sigma\,(\lambda,\pi) = \bigcap_{n=1}^{\infty} A^{(n)}\,\Lambda_m \quad .$$

Then $\Sigma\,(\lambda,\pi)$ is a cone, independent of the sequence $\{J_n\}$, which is naturally affinely isomorphić to the cone, $\vartheta(\lambda,\pi)$, of finite (λ,π) invariant Borel measures on $[0,|\lambda|)$.

Remark. Suppose the sequence $\{J_n\}$ above is nested, $J_n \subseteq J_{n+1}$. If $(\alpha^{(n)},\pi_n)$, $|\alpha^{(n)}| = |J_n|$, is the interval exchange corresponding to inducing (λ,π) on J_n, we may view $(\alpha^{(n+1)},\pi_{n+1})$ also as $(\alpha^{(n)},\pi_n)$ induced on (an interval corresponding to) J_{n+1}. In this interpretation there is a matrix A_{n+1} as in (2.5)-(2.7) such that $\alpha^{(n)} = A_{n+1}\alpha^{(n+1)}$ $(\lambda = A_1\alpha^{(1)})$. One finds then that $A^{(n)} = A_1A_2\ldots A_n$ for all $n \geq 1$. We have thus defined an "expansion" $(\lambda,\pi,\{J_n\}) \sim [A_1,A_2,\ldots]$. Later we shall fix π irreducible and choose a canonical sequence $\{J_n\}$ of nested intervals whose lengths decrease to 0 whenever (λ,π) is minimal. It follows that to (Lebesgue) almost all $\lambda \in \Lambda_m$ is associated an expansion

$$(2.10) \qquad \lambda \sim [A_1,A_2,\ldots] \qquad (A_n = A_n(\lambda),\ n \geq 1)$$

with $[\lambda] \approx \vartheta(\lambda,\pi)$, where

$$(2.11) \qquad [\lambda] = \bigcap_{n=1}^{\infty} (A_1A_2\ldots A_n)\Lambda_m \quad .$$

We recall from [15] two additional principles:

2.12. Proposition. With notations as above, if (λ,π) is minimal, there exists in (2.11) an n such that $B = A_1A_2\ldots A_n$ has all entries positive.

2.13. Proposition. If in (2.11) there exists a matrix with all entries positive such that for some n and infinitel man $k \geq 0$ $B = A_{k+1}A_{k+2}\cdots A_{k+n}$, then $\dim[\lambda] = 1$.

It is clear now that the unique ergodicity problem of the Introduction can be reduced to a "normality" question for expa sions such as (2.11). However, the present state of our knowledge of these expansions forces us to work with a reduction less direct (and more complicated) than the "simple" question of normality. This latter reduction is in the same spirit as the normality reduction, however. (And, indeed, normality wil be a consequence of the analysis of the specific expansions below.)

3. The (λ,θ_m) interval exchange, $m > 1$. Recall the definition of $\theta_m \in \mathfrak{S}_m$, $m > 1$, by $\theta_m i = m - i + 1$. The associated matrix L^{θ_m} (Section 2) is

$$L^{\theta_m}_{ij} = \begin{cases} 1 & i < j \\ 0 & i = j \\ -1 & i > j \end{cases} .$$

As the approach of this section is inductive, it will be useful to adopt conventions for dealing with indices m and m-1 in the same discussion. Accordingly, we define

$$\begin{aligned} &\theta = \theta_m \\ (3.1) \qquad &\hat{\theta} = \theta_{m-1} \\ &\hat{\lambda} = (\lambda_1-\lambda_m,\lambda_2,\ldots,\lambda_{m-1}) \qquad (\lambda \in \Lambda_m,\ \lambda_1 \geq \lambda_m) \end{aligned}$$

3.2. Lemma. Let $\lambda \in \Lambda_m$ be such that $\lambda_1 > \lambda_m$. With notations given by (3.1), one has

$$(3.3) \qquad (L^{\hat{\theta}}\hat{\lambda})_j = \begin{cases} (L^{\theta}\lambda)_1-\lambda_m & (j = 1) \\ (L^{\theta}\lambda)_j & (1 < j < m) \end{cases}$$

Proof. Computation.

When $\lambda \in \Lambda_m$ is such that $\lambda_1 > \lambda_m$, we define a subset, X, of I^{λ} by

$$X = I^{\lambda} \cap ([\lambda_1-\lambda_m,\lambda_1) \cup I^{\lambda}_m)^c$$

X has length $|\lambda| -2\lambda_m = |\hat{\lambda}|$, and we identify X with $I^{\hat{\lambda}}$ by means of $\varphi: X \to I^{\hat{\lambda}}$, where

$$(3.4)\qquad \varphi(x) = \begin{cases} x & (x \in [0,\lambda_1-\lambda_m)) \\ x-\lambda_m & (x \in [\lambda_1,|\lambda|-\lambda_m)) \end{cases} .$$

It is immediate from the definitions that

$$(3.5)\qquad \varphi^{-1}I_j^{\hat{\lambda}} = I_j^{\lambda} \cap X \quad (1 \le j \le m-1)$$

3.6. Lemma. Let $\lambda \in \Lambda_m$ be such that $\lambda_1 > \lambda_m$. With notations given by (3.1) and (3.4), one has

$$(3.7)\qquad \hat{T}\varphi(x) = Tx-\lambda_m \quad (x \in X)$$

where T and $\hat{T}$ refer to (λ,θ) and $(\hat{\lambda},\hat{\theta})$, respectively.

Proof. Immediate from (3.5), (2.2), (3.3) and (3.4).

Using (3.4), one obtains as corollary to (3.7)

3.8. Lemma. Let $\lambda \in \Lambda_m$ be such that $\lambda_1 > \lambda_m$. With all notations as above, one has

$$(3.9)\qquad \hat{T}^k\varphi(x) = \varphi(T^kx) \quad (x \in X,\ T^jx \notin I_1^{\lambda},\ 1 \le j \le k)$$

Proof: When $k = 1$ this follows from (3.7) and the definition (3.4). Argue then by induction.

In what follows U and $\hat{U}$ denote the transformations induced by T and $\hat{T}$ on I_1^{λ} and $I_1^{\hat{\lambda}}$, respectively. Recall $Ux = T^{n(x)}x$, $x \in I_1^{\lambda}$, $\hat{U}x = \hat{T}^{\hat{n}(x)}x$, $x \in I_1^{\hat{\lambda}}$, where $n(x)$ and $\hat{n}(x)$ are the "first return times" to I_1^{λ} and $I_1^{\hat{\lambda}}$, respectively.

If $x \in I_1^{\hat{\lambda}} \subseteq I_1^{\lambda}$, and if $1 \le k < n(x)$, then by (3.9) and

(3.5) $\hat{T}^k\varphi(x) = \hat{T}^k x \notin I_1^{\hat{\lambda}}$. Therefore, $\hat{n}(x) \geq n(x)$. On the other hand, if we set $k = n(x)-1$ in (3.9) and then apply (3.7) to $\hat{T}^k\varphi(x) = \hat{T}^k x$, we find

$$\hat{T}^{n(x)}x = Ux - \lambda_m \in I_1^{\hat{\lambda}}$$

and therefore $n(x) = \hat{n}(x)$. Noting that if $\lambda_1-\lambda_m \leq x < \lambda_1$, then $Ux = T^2x = x - (\lambda_1-\lambda_m)$, we have proved

$$(3.10) \qquad Ux = \begin{cases} x-(\lambda_1-\lambda_m) & (x \in [\lambda_1-\lambda_m,\lambda_1)) \\ \hat{U}x+\lambda_m & (x \in [0,\lambda_1-\lambda_m)) \end{cases}$$

In addition, if $x \in [0,\lambda_1-\lambda_m) = I_1^{\hat{\lambda}}$ the sequences $T^k x$ and $\hat{T}^k x$, $0 \leq k < n(x) = \hat{n}(x)$, wlll, by (3.5) and (3.9), visit corresponding sequences of subintervals, I_j^{λ} and $I_j^{\hat{\lambda}}$, $2 \leq j \leq m-1$. It follows that if A and $\hat{A}$ are the matrices associated to U and $\hat{U}$, respectively, by (2.4), then A and $\hat{A}$ are related by

$$(3.11) \qquad A = \left(\begin{array}{c|c} \hat{A} & \begin{matrix} 1 \\ 0 \\ \vdots \\ 0 \end{matrix} \\ \hline 0\ldots0 & 1 \end{array}\right).$$

Suppose now $\lambda \in \Lambda_m$, and $\lambda_m > \lambda_1$. We shall study the induced transformation, W, of T on the interval $[0,\lambda_m)$. To this end, define V to be the induced transformation of T on $I_m^{\lambda} = [|\lambda|-\lambda_m,|\lambda|)$, and note that V and W are conjugated by

$$(3.12) \qquad W = TVT^{-1}$$

We shall obtain another representation for V, based in part on the obvious symmetry of $\theta = \theta_m$. Given λ as above, i.e., $\lambda_1 < \lambda_m$, define a new vector $\eta \in \Lambda_m$ by $\eta = (\lambda_m, \lambda_{m-1}, \ldots$ Of course, $\eta_1 > \eta_m$, and so if U_η is the induced transformation of (η, θ) on I_1^η, (3.10) and (3.11) apply to U_η, with $\hat{\eta}$ to η what $\hat{\lambda}$ was to λ.

Let S be the orientation reversing isometry of $I^\lambda = I^\eta$ onto itself. We have

$$T_{(\lambda,\theta)} = ST_{(\eta,\theta)}S^{-1} \tag{3.13}$$

and therefore with notations as above

$$V = SU_\eta S^{-1} \quad . \tag{3.14}$$

Combining this with (3.12), one has (for $T = T_{(\lambda,\pi)}$)

$$W = TSU_\eta S^{-1}T^{-1} \quad .$$

$TS: [0, \lambda_m) \to [0, \lambda_m)$ is the orientation *reversing* isometry. But as U_η, it will develop, is an (α, θ) interval exchange for a.e. $\lambda, \eta = \eta(\lambda)$, it will follow that W is the (α', θ) interval exchange, $\alpha'_j = \alpha_{\theta j}$, $1 \leq j \leq m$.

If C is an $m \times m$ matrix, define a new $m \times m$ matrix, C^θ, $\theta = \theta_m$. by

$$C^\theta_{ij} = C_{\theta i, \theta j} \qquad (1 \leq i, j \leq m)$$

(While this notation does conflict with the notation $L^\theta = L^{\theta_m}$, the conflict will cause no confusion in what follows.) In terms of a corresponding notation for vectors, one has $\eta = \lambda^\theta$, $\alpha' = \alpha^\theta$ above.

3.15. Lemma. Let $\lambda \in \Lambda_m$ be such that $\lambda_1 < \lambda_m$, and let W and $U_\eta = U_{\lambda^\theta}$ be defined above. If A and B are the matrices (2.4) associated to W and U_{λ^θ} respectively, then

$$A = B^\theta \quad .$$

Proof. It follows from (3.12) and the definition (2.4) that V and W have the same associated matrix A. Then (3.13) - (3.14) combine to imply $A = B^\theta$. The lemma is proved.

If $\lambda \in \Lambda_m$, define intervals $J_1(\lambda)$, $J_2(\lambda) \subseteq I^\lambda$ by

$$J_1(\lambda) = I_1^\lambda$$

$$J_2(\lambda) = [0, \max(\lambda_1, \lambda_m)) \quad .$$

Also, let U_1 and U_2 be the induced transformations of (λ, π) on $J_1(\lambda)$ and $J_2(\lambda)$, respectively. We shall write $U_j = U_j(\lambda)$ (since θ is understood), $j = 1,2$. Of course, $U_1 = U_2$ when $J_1 = J_2$, i.e., when $\lambda_1 \geq \lambda_m$. In this situation (more precisely, when $\lambda_1 > \lambda_m$) (3.10) and (3.11) provide a connection between $U_1(\lambda) = U_2(\lambda)$ and $U_1(\hat{\lambda})$, $\hat{\lambda} = (\lambda_1 - \lambda_m, \lambda_2, \ldots, \lambda_{m-1})$. Note the mapping $\lambda \to \hat{\lambda}$ is nonsingular.

The following proposition extends a similar result by Rauzy [10], and could in fact also be proved by his techniques.

3.16. Lemma. Let $\lambda \in \Lambda_m$, and suppose the components of λ are rationally independent (or, more generally, suppose (λ, θ_m) satisfies the "infinite distinct orbit condition" [6]). With notations as above there exist $\alpha^{(j)} \in \Lambda_m$, $|\alpha^{(j)}| = |J_j(\lambda)|$, $j = 1,2$, such that $U_j(\lambda) = (\alpha^{(j)}, \theta)$, $j = 1,2$.

Proof. The lemma is straightforward for $m = 2$, and so we suppose it is known for all integers less than a given integer $m > 2$. Suppose also $\lambda \in \Lambda_m$ satisfies the hypothesis of the lemma. There are two cases to consider, $\lambda_1 > \lambda_m$ and $\lambda_1 < \lambda_m$ ($\lambda_1 = \lambda_m$ is not allowed).

Case 1. ($\lambda_1 > \lambda_m$). In this case $U_1(\lambda) = U_2(\lambda)$ are related to $U_1(\hat{\lambda})$, $\hat{\lambda} \in \Lambda_{m-1}$, by (3.10). By the induction hypothesis $U_1(\hat{\lambda}) = (\beta, \hat{\theta})$ for some $\beta \in \Lambda_{m-1}$. But then by (3.10), $U_1(\lambda) = U_2(\lambda) = (\alpha, \theta)$, where $\alpha = (\beta_1, \ldots, \beta_{m-1}, \lambda_m)$.

Case 2. ($\lambda_1 < \lambda_m$). In this case the analysis above of W, V, U_{λ^θ} combines with Case 1 to imply that $U_2(\lambda) = (\alpha^{(2)}, \theta)$, as claimed. Now $\alpha_1^{(2)} = \lambda_1$, and therefore $U_1(\lambda)$ and $U_1(\alpha^{(2)})$ are the same. Write $\alpha^{(2)} = \alpha^{(2)}(1)$, and suppose $\alpha^{(2)}(k)$, $1 \leq k \leq \ell$, have been constructed so that $U_2(\alpha^{(2)}(k)) = (\alpha^{(2)}(k+1), \theta)$, $1 \leq k < \ell$, and $J_1(\lambda) = J_1(\alpha^{(2)}(k))$, $1 \leq k \leq \ell$. If $J_1(\alpha^{(2)}(\ell)) \neq J_2(\alpha^{(2)}(\ell))$, the construction can be repeated, yielding $\alpha^{(2)}(\ell+1)$, and so on. The fact $U_1(\lambda)$ is *defined* implies $J_1(\lambda) \not\subseteq J_2(\alpha^{(2)}(\ell))$ cannot be true for all ℓ, and therefore there exists an ℓ such that $J_1(\alpha^{(2)}(\ell)) = J_2(\alpha^{(2)}(\ell))$ and $U_1(\lambda) = U_1(\alpha^{(2)}(\ell)) = U_2(\alpha^{(2)}(\ell)) = (\alpha^{(2)}(\ell+1), \theta)$. The lemma is proved.

With notations as in the lemma, let $\mathcal{F}_j(m)$, $m > 1$, be the set of matrices (2.4) which arise from $U_j(\lambda, \theta)$ as λ varies, $j = 1,2$. According to [15] if $A \in \mathcal{F}_j(m)$, then for every $\lambda \in A\Lambda_m$, $U_j(\lambda, \theta) = (A^{-1}\lambda, \theta)$. Working backwards from (3.11) one finds that for any $\hat{A} \in F_1(m-1)$, the matrix A *defined* by (3.11) belongs to $F_2(m)$. Combining this observation with Lemmas 3.15 and 3.16, we shall prove

3.17. Lemma. If $m > 2$, then $F_2(m)$ is the set of matrices, B, of the form $B = A$ or A^θ, where A is given by (3.11) for some $\hat{A} \in F_1(m-1)$. Also, $F_1(m)$ is the set of all matrices, B, of the form $B = A$ or $B = A_1^\theta \ldots A_k^\theta A$, where $A_1, A_2, \ldots, A_k$, and A are given by (3.11) for some $\hat{A}_1, \ldots, \hat{A}_k$, $\hat{A} \in F_1(m-1)$.

Proof. The first statement follows from the discussion preceding the lemma. As for the second, suppose $B \in F_1(m)$ is associated to (λ, θ) by (2.4) (for $U_1(\lambda,\theta)$). If $\lambda_1 > \lambda_m$, then $B = A \in F_2(m)$, where A is given by (3.11). If $\lambda_1 > \lambda_m$, let $\alpha^{(2)}(1), \alpha^{(2)}(2), \ldots, \alpha^{(2)}(\ell)$, $\alpha^{(2)}(\ell+1)$ be the sequence constructed in the proof of Lemma 3.16. By Lemma 3.15 the matrices (2.4) associated to $U_2(\lambda)$, $U_2(\alpha^{(2)}(1)), \ldots, U^{(2)}(\alpha^{(2)}(\ell-1))$ each have the form $B = A^\theta$ where A is as in (3.11) for some $\hat{A} \in F_1(m-1)$. Writing these matrices out in order as $A_1^\theta, A_2^\theta, \ldots, A_\ell^\theta$, the matrix associated by (2.4) to the induced transformation of (λ,θ) on $J_2(\alpha^{(2)}(\ell-1) = J_1(\alpha^{(2)}(\ell)) = J_1(\lambda)$, $U_1(\alpha^{(2)}(\ell))$ have to it by (2.4) an A of the form (3.11). Therefore $U_1(\lambda)$ has associated matrix $B = A_1^\theta A_2^\theta \ldots A_\ell^\theta A$, as claimed. On the other hand, given any matrix, B, of this form it is true, as above or in [15], that $\lambda \in B\Lambda_m$ has B associated to $U_1(\lambda)$. That is, $B \in F_1(m)$. The lemma is proved.

Suppose $m > 1$, and let (λ,θ) be such that $U_j(\lambda,\theta) = (\alpha^{(j)},\theta)$, $j = 1,2$ hold. Assume also that $\lambda \in \Delta_{m-1}$, and then define $u_j\lambda \in \Delta_{m-1}$ by

(3.18) $$u_j\lambda = \frac{\alpha^{(j)}}{|\alpha^{(j)}|} \qquad (j = 1,2) \quad .$$

If $A \in F_j(m)$, $j = 1,2$ then on $A\Lambda_m \cap \Delta_{m-1}$ u_j is given by the formula

(3.19) $$u_j\lambda = \frac{A^{-1}\lambda}{|A^{-1}\lambda|} \quad .$$

Much of the rest of this paper is devoted to proving the following theorem (which contains Theorem 1.3).

3.20. Theorem. *Suppose* $m = 4$ *above, and let* $u_j\colon \Delta_3 \to \Delta_3$ *be defined by* (3.18)-(3.19) *and the discussion which precedes. There exist* σ*-finite Borel measures* μ_j, $j = 1,2$, *each of which is equivalent to Lebesgue measure on* Δ_3, *such that* μ_j *is invariant and ergodic for* u_j, $j = 1,2$.

Remark. It is very probable that $\mu_j \times \mu_j$ is ergodic on $\Delta_3 \times \Delta_3$. In that event one has also that if $B(m)$ is the Borel σ-algebra of Δ_3, then

(3.21) $$B_\infty^{(j)}(m) = \bigcap_{n=1}^{\infty} u_j^{-n} B(m) \qquad (j = 1,2)$$

is trivial (Proposition 6.32).

We shall conclude this section with an explicit calculation of the set $F_2(4)$. The computation is aided by Lemma 3.17 which implies a corresponding description of $F_1(3)$ is sufficient. The latter, which was given in [15], will be recalled here.

Let $G \subseteq SL(2,\mathbb{Z})$ be the set of matrices, $g = \begin{pmatrix} \ell & 1 \\ -1 & 0 \end{pmatrix}$, $\ell \geq 1$, or $g = \begin{pmatrix} a & b \\ c & d \end{pmatrix}$, $a,b,d \geq 1$, $c \geq 0$. To each $g = \begin{pmatrix} a & b \\ c & d \end{pmatrix} \in G$ we associate a 3×3 matrix, $M(g)$, where

$$(3.22) \qquad M(g) = \begin{pmatrix} 1 & 1 & 1 \\ a-1 & a+b-1 & b-1 \\ c+1 & c+d+1 & d+1 \end{pmatrix} .$$

Also, define a 4×4 matrix, $N(g)$, by

$$(3.23) \qquad N(g) = \left(\begin{array}{c|c} M(g) & \begin{matrix} 1 \\ 0 \\ 0 \end{matrix} \\ \hline 0\ 0\ 0 & 1 \end{array}\right) .$$

3.24. Lemma. With notations as above, one has

$$\mathcal{F}_1(3) = \{M(g) \mid g \in G\}$$

(3.25)

$$\mathcal{F}_2(4) = \{A \mid A = N(g) \quad \text{or} \quad N(g)^{\theta},\ g \in G\} .$$

Proof. The first line of (3.25) is proved in [15]. The second line follows from the first and Lemma 3.17.

A basic fact used in calculating $\mathcal{F}_1(3)$ in [15] was that if $u = (1,-1,1)$, then $Au = u$, $A \in \mathcal{F}_1(3)$. If we let v be the vector $v = (1,-1,1,0)$, it follows from this (and is in any case obvious from (3.22)-(3.23)) that $N(g)v = v$, $g \in G$. Now let $\theta = \theta_4$, $w = v^{\theta} = (0,1,-1,1)$. If $N(g)$ is given by (3.22)-(3.23), then

$$N(g)w = \begin{pmatrix} 1 \\ a \\ c \\ 1 \end{pmatrix} \geq \begin{pmatrix} 0 \\ 1 \\ -1 \\ 1 \end{pmatrix} = w .$$

As

$$N(g)w = N(g)v^{\theta} = (N(g)^{\theta} v)^{\theta}$$

it follows that $N(g)^{\theta}v \geq v$. Collecting results (and arguing by symmetry) we have

$$(3.26)\qquad \begin{matrix} Av \geq v \\ Aw \geq w \end{matrix} \qquad (A \in \mathcal{F}_2(4)) \quad .$$

Of course, (3.26) must also hold for any matrix A which is a product of elements of $\mathcal{F}_2(4)$.

In what follows, if A is a matrix, a_j will denote the sum of the entries of the jth column of A, $1 \leq j \leq 4$. Inequalities (3.26) imply the following inequalities between the various entries and various column sums of a matrix which is a product of elements of $\mathcal{F}_2(4)$:

3.27. Lemma. Suppose $A = \{A_{ij}\}$ is a matrix which is a product of elements of $\mathcal{F}_2(4)$. We have

$$(3.28)\qquad \begin{matrix} A_{i1} + A_{i3} \geq A_{i2} \quad (i \neq 2) \\ \\ A_{21} + A_{23} \geq A_{22} - 1 \end{matrix}$$

$$(3.29)\qquad \begin{matrix} A_{i2} + A_{i4} \geq A_{i3} \quad (i \neq 3) \\ \\ A_{32} + A_{34} \geq A_{33} - 1 \end{matrix}$$

$$(3.30)\qquad \begin{matrix} a_1 + a_3 \geq a_2 \\ \\ a_2 + a_4 \geq a_3 \end{matrix}$$

Analogous arguments using the vectors $z = (-1,1,0,0)$ and $z^\theta = (0,0,1,-1)$ show $Az \geq z$, $Az^\theta \geq z^\theta$ for any matrix as in the lemma. Moreover, if A has all entries positive, meaning in particular that the representation of A as a product of elements of $F_2(4)$ contains elements of both forms A, A^θ from (3.11), then $Az \geq z + (1,0,0,0)$, $Az^\theta \geq z^\theta + ((0,0,0,1)$. Thus, we have

3.31. Lemma. Let A be a matrix satisfying the hypotheses of Lemma 3.27. Then

$$a_2 \geq a_1$$

(3.32)

$$a_3 \geq a_4 \quad .$$

If in addition all entries of A are positive, then

$$A_{i2} \geq A_{i1}$$

(3.33) $(1 \leq i \leq 4)$

$$A_{i3} \geq A_{i4} \quad .$$

4. The "Rauzy Class" of an irreducible permutation. If $m > 1$, we define $\mathfrak{S}_m^0$ to be the set of irreducible permutations $\pi \in \mathfrak{S}_m$. Following Rauzy [10] we define for $\lambda \in \Lambda_m$ the interval $J_3(\lambda) = [0, |\lambda|-b)$, where $b = b(\lambda) = \min(\lambda_m, \lambda_{\pi^{-1}m})$. Rauzy observes $J_3(\lambda)$ is the longest interval with left end-point 0 which is admissible in the sense of [15].

Rauzy defines one-one maps a and b from $\mathfrak{S}_m^0$ to itself by

$$(4.1) \qquad (a\pi)(i) = \begin{cases} \pi i & i \le \pi^{-1}m \\ \pi m & i = \pi^{-1}m + 1 \\ \pi(i-1) & i > \pi^{-1}m + 1 \end{cases}$$

and

$$(4.2) \qquad (b\pi)(i) = \begin{cases} \pi i & \pi i < m \\ \pi i+1 & \pi m < \pi i < m \\ \pi m+1 & \pi i = m \end{cases}$$

In terms of $(\pi 1, \pi 2, \ldots, \pi m)$, $a\pi$ is constructed as follows. Reading from left to right all terms up to and including $m = \pi\pi$ are unchanged. The remaining terms are then permuted cyclicall with πm moving to the $(\pi^{-1}m+1)^{st}$ place, etc. As for $b\pi$, all terms in $(\pi 1, \pi 2, \ldots, \pi m)$ which are less than the last term (i.e., πm) are unchanged. Then add one to each remaining term save th $(\pi^{-1}m)^{th}$ which is replaced by $(\pi m) + 1$. Now let I_k be the $k \times k$ identity matrix, and define matrices $A_{\pi,a}$, $A_{\pi,b}$, $\pi \in \mathfrak{S}$

by

$$A_{\pi,a} = \left(\begin{array}{c|c} I_{\pi^{-1}m} & \\ & 1\ 0 \ldots 0 \\ \hline & \begin{matrix} 0 & 1 & 0 & \ldots & 0 \\ . & 0 & 1 & \ldots & 0 \\ \cdot & \cdot & & & \vdots \\ \cdot & \cdot & & & 0 \\ 0 & \cdot & & & 1 \\ 1 & 0 & \ldots & & 0 \end{matrix} \end{array}\right)$$

$$A_{\pi,b} = \left(\begin{array}{c|c} I_{m-1} & \begin{matrix} 0 \\ \cdot \\ \cdot \\ \cdot \\ 0 \end{matrix} \\ \hline 0 \ldots 010 \ldots 0 & 1 \end{array}\right)$$

The first 1 in the last row of $A_{\pi,b}$ occurs at the $(\pi^{-1}m)^{st}$ position.

4.3. Proposition. (Rauzy [10]) With notations as above, assume the orbits of the left endpoints of I_j^λ, $2 \leq j \leq m$, are pairwise infinite and distinct (of course $|\lambda| \notin I^\lambda$). If $U_3 = U_3(\lambda,\pi)$ is the induced transformation of (λ,π) on $J_3(\lambda)$, then

$$U_3 = (A^{-1}\lambda, \pi^*) \tag{4.4}$$

where $A = A_{\pi,a}$ or $A_{\pi,b}$ and $\pi^* = a\pi$ or $b\pi$ as $\lambda_m < \lambda_{\pi^{-1}m}$ or $\lambda_m > \lambda_{\pi^{-1}m}$.

Notation. If $\pi \in \mathfrak{S}_m^0$, we shall denote by $\mathcal{R}[\pi]$ the orbit of π under the group of transformations of $\mathfrak{S}_m^0$ generated by the maps a and b. We call $\mathcal{R}[\pi]$ the *Rauzy class* of π. If we normalize in (4.4), requiring $\lambda \in \Delta_{m-1}$ and replacing $A^{-1}\lambda$ by $\frac{A^{-1}\lambda}{|A^{-1}\lambda|}$, we see that there is defined a.e. on $\Delta_{m-1} \times \mathcal{R}[\pi]$ a map

$$u_3(\lambda,\pi) = \left(\frac{A^{-1}\lambda}{|A^{-1}\lambda|}, \pi^* \right) \quad . \tag{4.5}$$

In case $m = 4$, there is only one "interesting" Rauzy clas in $\mathfrak{S}_4^0$, the class of θ_4. This is because all other permutatio are either reducible or else give rise to interval exchanges with less than 3 discontinuities (where minimality already implies unique ergodicity). (Already for $m = 5$ there are tw "interesting" Rauzy classes.) Continuing to write $\pi \in \mathfrak{S}_m^0$ a $(\pi_1,\pi_2,\ldots,\pi_m)$, the Rauzy class, $\mathcal{R}[\theta_4]$, is given by the list [10]

$$\mathcal{R}[\theta_4] = \begin{matrix} (4,3,2,1) & \\ & (2,4,3,1) \\ (4,1,3,2) & \\ & (3,2,4,1) \\ (4,2,1,3) & \\ & (2,4,1,3) \\ (3,1,4,2) & \end{matrix} \quad .$$

The corresponding "Rauzy diagram" of u_3 on $\Delta_3 \times \mathcal{R}[\theta_4]$ is (interpret each permutation as a copy of Δ_3):

$$\text{(4.10)} \qquad \pi = (2,4,3,1) \quad ba^k b\pi = (4,3,2,1)$$

$$ba^k b\lambda = (\lambda_1,\lambda_2,\lambda_3^*,\lambda_4^*)$$

$$\lambda_3^* = \lambda_3 - k(\lambda_4-\lambda_2)$$

$$\lambda_4^* = (k+1)(\lambda_4-\lambda_2)-\lambda_3$$

Consider now an element $\lambda \in \Lambda_3$ such that $\lambda_1 > \lambda_4$. If the components of λ are rationally independent (say), the first return map to $\Lambda_4 \times \{\theta_4\}$ is defined, and we write

$$U_3^N(\lambda,\theta_4) = (\mu,\theta_4) \qquad .$$

By (4.6) there exist integers $n \geq 0$, $\ell_1,\ldots,\ell_n \geq 0$, and $k \geq 0$ such that $(N = 3+k+2n+\ell_1+\ldots+\ell_n)$

$$\mu = (ba^k b)\left(\prod_{j=1}^{n} ab^{\ell_j}a\right)b\lambda \qquad .$$

Also, define

$$\mu^* = \left(\prod_{j=1}^{n} ab^{\ell_j}a\right)b\lambda \qquad .$$

We have from (4.10) that

$$|\mu| = \mu_1^* + \mu_4^* \ ,$$

and then from (4.9),

$$\mu_1^* + \mu_4^* = (b\lambda)_1 + (b\lambda)_4 \qquad .$$

Combining this with (4.8), $|\mu| = \lambda_4$. In other words,

$U_3^N(\lambda,\theta_4) = U_2(\lambda,\theta_4)$.

4.11. Proposition. u_2 is the induced transformation of u_3 on $\Delta_3 \times \{\theta_4\}$.

Proof. If $\lambda \in \Delta_3$ is as above, we have that $u_3^N(\lambda,\theta_4) = u_2(\lambda,\theta_4)$, where N is the first return time to $\Delta_3 \times \{\theta_4\}$. If now $\lambda \in \Delta_3$ has rationally independent components but $\lambda_1 > \lambda_4$, then by Rauzy [11] $u_3^N(\lambda,\theta_4) = u_1(\lambda,\theta_4)$, where N is the first return time. In this case we also have $u_1(\lambda,\theta_4) = u_2(\lambda,\theta_4)$, and the proposition obtained.

5. Regular partitions. Let F be a set of $m \times m$ matrices with nonnegative integer entries and determinants ± 1. F shall be called a *partition* (of Λ_m) if

(5.1) $$A\Lambda_m \cap A'\Lambda_m = \emptyset \qquad (A, A' \in F,\ A \neq A')$$

and if, modulo a null set,

(5.2) $$\Lambda_m = \bigcup_{A \in F} A\Lambda_m \quad .$$

The discussion in Section 3 implies the sets $F_j(m)$, $j = 1,2$, are partitions, for example.

If F is a partition, define $F^{(n)}$, $n \geq 1$, to be the set of n-fold products of elements of F; $A \in F^{(n)}$ if and only if $A = A_1 A_2 \ldots A_n$, where $A_1, A_2, \ldots, A_n \in F$. It is evident from (5.1) and (5.2) that $F^{(n)}$ is a partition for each n. Applying (5.1) and (5.2) to $F = F^{(n)}$ for all n it follows there exists for a.e. $\lambda \in \Lambda_m$ a sequence $A_j = A_j(\lambda) \in F$, $j \geq 1$, such that

$$\lambda \in A_1(\lambda)A_2(\lambda)\ldots A_n(\lambda)\Lambda_m \qquad (n \geq 1) \quad .$$

We set $A^{(n)}(\lambda) = A_1(\lambda)\ldots A_n(\lambda)$ and define

(5.3) $$[\lambda]_F = \bigcap_{n=1}^{\infty} A^{(n)}(\lambda)\Lambda_m \quad .$$

By (5.1) we have $[\lambda]_F = [\lambda']_F$ if and only if $[\lambda]_F \cap [\lambda']_F \neq \emptyset$ if and only if $A_j(\lambda) = A_j(\lambda')$, $j \geq 1$. Of course, $[\lambda]_F = [\tau\lambda]_F$, $\tau > 0$. Corresponding to (5.3) is a subset of Δ_{m-1},

(5.4) $$\Delta_F(\lambda) = [\lambda]_F \cap \Delta_{m-1} \quad .$$

5.5. Definition. A partition, F, is said to be *regular* if $[\lambda]_F = \mathbb{R}^+\lambda$ for Lebesgue almost all $\lambda \in \Lambda_m$, or, equivalently, if $\Delta_F(\lambda) = \{\lambda\}$ for Lebesgue almost all $\lambda \in \Delta_{m-1}$.

The result which follows is not used in the proof of Theorems 1.1 (and 1.3) or the related theorems, Theorem 3.20 and Theorem 4.7.

5.6. Proposition. If for Lebesgue almost all $\lambda \in \Delta_{m-1}$ λ is an extreme point of $\Delta_F(\lambda)$, then F is regular.

Later on we shall find Proposition 5.6 has a corollary for interval exchanges.

5.7. Corollary. *Let* $\pi \in \mathfrak{S}_m^0$ *(the set of irreducible permutations. If for Lebesgue almost all* $\lambda \in \Delta_{m-1}$ *Lebesgue measure on* I^λ *is ergodic for* (λ,π)*, then for Lebesgue almost all* $\lambda \in \Lambda_m$ *the* (λ,π) *interval exchange is uniquely ergodic.*

Of course, the corollary is Theorem 1.4 of the Introduction.

The proof of Proposition 5.6 will require some preliminary notations and discussion. As usual Δ_k denotes the standard k simplex, $\Delta_k = \{\lambda = (\lambda_1,\dots,\lambda_{k+1}) | \lambda_j > 0,\ |\lambda| = 1\}$, and we use $\omega = \omega_k$ to denote normalized (k-dimensional) Lebesgue measure on Δ_k. If Δ is an arbitrary k-dimensional simplex in $\mathbb{R}^m$ for some $m \geq k$, let $E(\Delta) = \{e_1,\dots,e_{k+1}\}$ be the set of extreme points of Δ in some order, and define an affine map $\rho: \Delta_k \to \Delta$ by $\rho(\lambda) = \sum_{j=1}^{k+1} \lambda_j e_j$. Then $\omega_\Delta = \rho\omega_k$ is normalized Lebesgue measure on Δ (and is independent of the ordering chosen for $E(\Delta)$).

It is useful to generalize the construction just made. Thus, we suppose $\dim \Delta \leq k$, and we suppose $E^* \subseteq \Delta$ is any set with k+1 elements which contains $E(\Delta)$. Define $\rho^*: \Delta_k \to$ in terms of some ordering chosen for E^*, and let $\omega^*_\Delta = \rho^*\omega_k$. Again ω^*_Δ does not depend upon the ordering chosen for E^*, and because ρ^* is onto and nonsingular, ω^*_Δ is *equivalent* to ω

Suppose now F is a partition, and let P_n, $n \geq 1$, be the partition of Δ_{m-1} whose elements are the simplices $\Delta_{m-1} \cap A\Lambda_m$, $A \in F^{(n)}$. Let $B^{(n)}$ be the σ-algebra generated by the elements of P_n, so that $B^{(n)} \subseteq B^{(n+1)}$ (mod 0) for each n. Also, define $B(F) = \bigvee_{n=1}^{\infty} B^{(n)}$.

The elements of P_n are (m-1) dimensional simplices, Δ; if $f \in L^1(\Delta_{m-1})$, then $\mathcal{E}(f|B^{(n)})(\cdot)$ is constant on $\Delta \in P_n$, defined there by

(5.8) $$\mathcal{E}(f|B^{(n)})(\lambda) = \int_{\Delta} f(\zeta)\omega_{\Delta}(d\zeta) \quad (\lambda \in \Delta \in P_n) \quad .$$

By the Martingale theorem

(5.9) $$\mathcal{E}(f|B(F))(\lambda) = \lim_{n\to\infty} \mathcal{E}(f|B^{(n)})(\lambda)$$

holds ω_{m-1}.- a.e. λ.

Fix f continuous with compact support in Δ_{m-1}, and suppose $\lambda \in \Delta_{m-1}$ is such that (5.9) holds for f, λ. Set $\Delta^{(n)}(\lambda) = A^{(n)}(\lambda)\Lambda_m \cap \Delta_{m-1}$, and for each n let $\rho_n: \Delta_{m-1} \to \Delta^{(n)}(\lambda)$ be affine and onto. If E^* is a cluster point of the sequence of (ordered) sets $E(\Delta^{(n)}(\lambda))$, then the convex hull of E^* is $\Delta_F(\lambda)$. With notations as above, $\rho^*: \Delta_{m-1} \to \Delta_F(\lambda)$ sends ω_{m-1} to some measure ω^* (we suppress the subscript $\Delta_F(\lambda)$). Now because f is continuous, the sequence of integrals $\int_{\Delta_{m-1}} f\circ\rho^*\omega_{m-1}$ has for a cluster point the integral $\int_{\Delta_{m-1}} f\circ\rho^*\omega_{m-1}$. That is, by (5.9)

$$(5.10) \qquad \mathcal{E}(f|\mathcal{B}(F))(\lambda) = \int_{\Delta_F(\lambda)} f(\zeta)\omega^*(d\zeta) \quad .$$

Moreover, if f is allowed to run through any countable set of continuous compactly supported functions on Δ_{m-1}, the set of $\lambda \in \Delta_{m-1}$ for which (5.10) will hold, with ω^* depending only upon λ (and F), has full Lebesgue measure. By the theory of Lebesgue spaces there exists for ω_{m-1} - a.e. λ a measure, ν_λ, on $\Delta_F(\lambda)$ such that for any $f \in L^1(\omega_{m-1})$

$$(5.11) \qquad \mathcal{E}(f|\mathcal{B}(F))(\lambda) = \int f(\zeta)\nu_\lambda(d\zeta)$$

holds for ω_{m-1} - almost all $\lambda \in \Delta_{m-1}$. For any λ ν_λ is determined if one knows the right hand side of (5.11) for a countable dense set of compactly supported continuous functions Combining (5.10) and (5.11), we have $\nu_\lambda = \omega^*_{\Delta_F(\lambda)}$ for a.e. $\lambda \in \Delta_{m-1}$, and in particular,

5.12. Lemma. With notations as above, we have for Lebesg almost all $\lambda \in \Delta_{m-1}$ that the conditional measure ν_λ is equi valent to $\omega_{\Delta_F(\lambda)}$.

We shall now prove Proposition 5.6: Consider in $\Delta_{m-1} \times \Delta_{m-1}$ the set

$$\Omega = \{(\lambda,\lambda')|\Delta_F(\lambda) = \Delta_F(\lambda')\}$$

$$= \bigcap_{n=1}^{\infty} \{(\lambda,\lambda')\}|A_n(\lambda) = A_n(\lambda')\}$$

where of course in the intersection it is understood we conside only those λ,λ' such that $A_n(\lambda)$ $A_n(\lambda')$ are defined for all n. Ω is a Borel set. Define a map $\Psi\colon \Omega \to \Delta_{m-1}$ by $\Psi(\lambda+\lambda') = \frac{\lambda + \lambda'}{2}$. If $\Omega' \subseteq \Omega$ is any Borel set, then $\Psi(\Omega')$

is an analytic set, hence Lebesgue measurable (see [13]). Let Ω^{ε}, $\varepsilon > 0$, be the complement of Ω of the ε-neighborhood of the diagonal in Ω, relative (say) to the Euclidean metric. We define

$$Z = \bigcup_{k=1}^{\infty} \Psi(\Omega^{\frac{1}{k}})$$

which is also Lebesgue measurable. Now we use the standard trick of Choquet theory, to wit that for any $\Delta_{F}(\lambda)$, $Z \cap \Delta_{F}(\lambda)$ contains precisely the complement of the set of extreme points of $\Delta_{F}(\lambda)$. If for ω_{m-1} - almost all $\lambda \in \Delta_{m-1}$ λ is an extreme point of $\Delta_{F}(\lambda)$, then $\omega_{m-1}(Z) = 0$. But then for almost all λ $\omega_{\Delta_{F}(\lambda)}(Z \cap \Delta_{F}(\lambda)) = 0$, because the conditional probabilities are absolutely continuous. In turn, this says $\Delta_{F}(\lambda) = E(\Delta_{F}(\lambda))$ for almost all λ, or $\Delta_{F}(\lambda) = \{\lambda\}$, a.e. λ. That is, F is regular.

Proof of Corollary 5.7: Let $\pi \in \mathfrak{S}_m^0$ be irreducible, and suppose for Lebesgue almost all $\lambda \in \Delta_{m-1}$ (λ,π) is ergodic relative to Lebesgue measure on $I^{\lambda} = [0,1)$. Rauzy's theory and the nonsingularity of u_3 implies the same statement is true for all $\pi' \in \mathcal{R}[\pi]$. (For any $\pi' \in \mathcal{R}[\pi]$ there exists an $n > 0$ and a simplex $\Delta \subseteq \Delta_{m-1}$ such that $u_3^n \Delta x\{\pi\} = \Delta_{m-1} x\{\pi'\}$. (λ,π) is ergodic if and only if $u_3^n(\lambda,\pi) = (\lambda',\pi')$ is.) Let F be the partition of $\Delta_{m-1} x \mathcal{R}[\pi]$ arising from the matrices $A_{\pi,a}$, $A_{\pi,b}$ (we here enlarge the definition of "partition" in an obvious way). For Lebesgue almost all $(\lambda',\pi') \in \Delta_{m-1} x \mathcal{R}[\pi]$ we have that (λ',π') is an extreme point of $\Delta_{F}(\lambda',\pi')$ because a) (λ',π') is ergodic relative to Lebesgue measure on $[0,1)$, b) $\Delta_{F}(\lambda',\pi')$ is affinely isomorphic to the set of (λ',π') invariant Borel probability measures on $[0,1)$, with λ itself paired with Lebesgue measure under this correspondence (see [15]), and c) an invariant probability measure is ergodic if and only if it is an extreme point of the set of all invariant probability measures ([1]). The corollary now follows because by Proposition 5.6 $\Delta_F(\lambda',\pi') = \{(\lambda',\pi')\}$, a.e. (λ',π') .

6. Existence of invariant measures. The discussion which follows is more general than is necessary for the purposes of the present paper. It is possible it will be useful in later work, however.

$\chi = (X, \mathcal{B}, \omega)$ represents a probability triple in which X is a locally compact, σ-compact metric space and $\mathcal{B}$ is the Borel σ-algebra of X.

In addition there will be given objects $\mathcal{P}$ and $\mathcal{T}(\mathcal{P}) = \mathcal{T} = \{T_\Delta, D_\Delta) \mid \Delta \in \mathcal{P}\}$ of the following description. $\mathcal{P}$ is a partition of X into measurable sets of positive measure; for each $\Delta \in \mathcal{P}$, D_Δ is a union of elements of $\mathcal{P}$, and $T_\Delta \colon D_\Delta \to \Delta$ is one-one, onto, bimeasurable with T_Δ, T_Δ^{-1} both nonsingular: finally $\bigcup_{\Delta \in \mathcal{P}} D_\Delta = X$..

For convenience of notation we shall write $\Omega = (\chi, \mathcal{P}, \mathcal{T},)$ and refer to Ω as a *$\mathcal{T}$-triple*. Any $\mathcal{T}$-triple Ω determines a nonsingular many-to-one (in general) transformation $S_\Omega = S\colon X \to X$ by

(6.2) $$Sx = T_\Delta^{-1} x \qquad (x \in \Delta \in \mathcal{P}) \quad .$$

Given a $\mathcal{T}$-triple Ω, $n > 1$, and a sequence $\Delta = (\Delta_1, \ldots, \Delta_n) \in \mathcal{P}$, Δ is *consistent* if $\Delta_{j+1} \in D_{\Delta_j}$, $1 \leq j < n$. If Δ is consistent, then $T_\Delta = T_{\Delta_1} T_{\Delta_2} \ldots T_{\Delta_n}$ has a natural domain, $D_{\Delta_n} = D_\Delta$, and range $T_{\Delta_1} T_{\Delta_2} \ldots T_{\Delta_n} D_{\Delta_n} \subseteq \Delta_1$ which we also denote by Δ. Evidently $T_\Delta\colon D_\Delta \to \Delta$ is one-one, onto, bimeasurable with T_Δ, T_Δ^{-1} both nonsingular. Define $\mathcal{P}^{(n)} = \{\Delta = \Delta(\Delta_1, \ldots, \Delta_n) \mid (\Delta_1, \ldots, \Delta_n) \in \mathcal{P} \times \mathcal{P} \times \ldots \times \mathcal{P}$ consistent$\}$, and then set $\mathcal{T}^{(n)} = \{(T_\Delta, \Delta) \mid \Delta \in \mathcal{P}^{(n)}\}$. Then $\Omega^{(n)} = (\chi, \mathcal{P}^{(n)}, \mathcal{T}^{(n)})$ is again a $\mathcal{T}$-triple. The reader may check that

(6.3) $$S_{\Omega^{(n)}} = S_\Omega^n \qquad (n > 0) \quad .$$

If Ω is a $\mathcal{T}$-triple, there exists for each $(T_\Delta, D_\Delta) \in \mathcal{T}$ an element $J_\Delta \in L^1(\omega)$, supported on D_Δ, such that if $f \in L^1(\omega)$, then

$$\int_\Delta f(x)\,\omega(dx) = \int_{D_\Delta} f(T_\Delta x) J_\Delta(x)\,\omega(dx) \quad . \tag{6.4}$$

In terms of this we note that if $E \subseteq X$ is any Borel set and $n > 0$, then

$$S^{-n}E = \bigcup_{\Delta \in \mathcal{P}^{(n)}} T_\Delta(E \cap D_\Delta) \tag{6.5}$$

and

$$\omega(T_\Delta(E \cap D_\Delta)) = \int_{E \cap D_\Delta} J_\Delta(x)\,\omega(dx) \quad . \tag{6.6}$$

We shall be interested in the condition for $M < \infty$, $n > 0$, $\Delta \in \mathcal{P}^{(n)}$

$$M^{-1} J_\Delta(x) \leq J_\Delta(y) \leq M J_\Delta(x) \qquad (x, y \in D_\Delta) \tag{6.7}$$

which, when integrated over D_Δ relative to $\omega(dx)$ and then over $E \cap D_\Delta$ relative to $\omega(dy)$ yields, in view of (6.6),

$$\begin{aligned} M^{-1}\omega(\Delta)\,\omega(E \cap D_\Delta) &\leq \omega(T_\Delta(E \cap D_\Delta))\,\omega(D_\Delta) \\ &\leq M\omega(\Delta)\,\omega(E \cap D_\Delta) \quad . \end{aligned} \tag{6.8}$$

We use $\mathcal{P}^{(n)}(M)$, $n > 0$, $M < \infty$ to denote the set of $\Delta \in \mathcal{P}^{(n)}$ which satisfy (6.7). In our applications either $D_\Delta = X$ for all Δ or else $\mathcal{P}$ is finite, and in either case

$$\omega(D_\Delta) \geq \alpha > 0 \qquad (\Delta \in \mathcal{P}) \quad . \tag{6.9}$$

The same bound follows trivially for $\mathcal{P}^{(n)}$, $n > 0$.

6.10 Definition. A $\mathcal{T}$-triple Ω satisfies *Property* UCA (uniformly countably additive) if for every $\varepsilon > 0$ there exists $M = M(\varepsilon) < \infty$ such that

$$\sum_{\Delta \notin \mathcal{P}^{(n)}(M)} \omega(\Delta) < \varepsilon \qquad (n > 0) \quad . \tag{6.11}$$

6.12. Proposition. Let Ω be a $\mathcal{T}$-triple which satisfies (6.9) and Property UCA. There exists an S_Ω-invariant Borel probability measure μ such that $\mu < \omega$.

Proof: We shall prove that for every $\varepsilon > 0$ there exists $\delta > 0$ such that $\omega(E) < \delta$ implies $\omega(S^{-n}E) < 2\varepsilon$, all $n > 0$. To this end choose $M < \infty$ so that (6.11) holds for the given ε. By (6.5), the right hand inequality (6.8), and (6.11) we have for any $n > 0$

$$\begin{aligned} \omega(S^{-n}E) &\le \varepsilon + \sum_{\Delta \in \mathcal{P}^{(n)}(M)} \omega(T_\Delta(E \cap D_\Delta)) \\ &\le \varepsilon + \frac{M}{\alpha} \sum_{\Delta \in \mathcal{P}^{(n)}(M)} \omega(\Delta)\omega(E \cap D_\Delta) \\ &\le \varepsilon + \frac{M}{\alpha}\,\omega(E) \quad . \end{aligned}$$

Let $\delta < \frac{\varepsilon\alpha}{M}$. Then $\omega(E) < \delta$ implies $\omega(S^{-n}E) < 2\varepsilon$, $n > 0$. The criterion of Hajian-Kakutani [2] now implies the existence of μ.

6.13. Definition. A $\mathcal{T}$-triple Ω satisfies Property E if for all pairs $\Delta', \Delta'' \in \mathcal{P}$ there exist $n > 0$ and $\Delta = \Delta(\Delta_1, \ldots, \Delta_n) \in \mathcal{P}^{(n)}$ such that $\Delta_1 = \Delta'$ and $\Delta_n = \Delta''$.

6.14. Lemma. Assume Ω has property E, and suppose $E \in \mathcal{B}$ satisfies $S^{-1}E \subseteq E$ and $\omega(E) > 0$. Then $\omega(E \cap D_\Delta) > 0$ for all $\Delta \in \mathcal{P}$.

Proof: The sets D_Δ, $\Delta \in \mathcal{P}$, cover X, and therefore $\omega(E \cap D_{\Delta_0}) > 0$ for some $\Delta_0 \in \mathcal{P}$. Given another $\Delta^0 \in \mathcal{P}$, choose by Property E an $n > 0$ and $\Delta = \Delta(\Delta_1,\ldots,\Delta_n) \in \mathcal{P}^{(n)}$ such that $\Delta_1 = \Delta^0$. and $\Delta_n = \Delta_0$. Then $D_\Delta = D_{\Delta_0}$, and $E \cap \Delta^0 \supseteq E \cap \Delta \supseteq T_\Delta(E \cap D_{\Delta_0})$. Nonsingularity of T_Δ implies $\omega(E \cap \Delta^0) >$ As $\Delta^0 \in \mathcal{P}$ is arbitrary, and as each D_Δ, $\Delta \in \mathcal{P}$, is itself a union of elements of $\mathcal{P}$, the lemma follows.

An immediate corollary of Lemma 6.14 is

6.15. Lemma. Let Ω be a $\mathcal{T}$-triple such that either $D_\Delta =$ $\Delta \in \mathcal{P}$, or $\mathcal{P}$ is finite. In the latter case assume Ω satisfies Property E. If $E \in \mathcal{B}$ is such that $S_\Omega^{-1}E \subseteq E$ and $\omega(E) > 0$, there exists $\beta > 0$ such that

$$\omega(E \cap D_\Delta) \geq \beta \qquad (n > 0,\ \Delta \in \mathcal{P}^{(n)}) \quad . \tag{6.16}$$

Proof: Let $\beta = \min\limits_{\Delta \in \mathcal{P}} \omega(E \cap D_\Delta)$. β exists when $\mathcal{P}$ is finite or $D_\Delta = X$, $\Delta \in \mathcal{P}$. That $\beta > 0$ follows from Lemma 6.14; (6.16) follows from $\{D_\Delta | \Delta \in \mathcal{P}^{(n)}\} = \{D_\Delta | \Delta \in \mathcal{P}\}$.

Let Ω be a $\mathcal{T}$-triple, and define $\Delta^{(n)}(x)$, $x \in X$, $n > 0$ to be the element of $\mathcal{P}^{(n)}$ to which x belongs. We may and sometimes shall view $\Delta^{(n)}(x)$ as a consistent sequence,

$$\Delta^{(n)}(x) = \Delta(\Delta_1(x), \Delta_2(x), \ldots, \Delta_n(x)) \quad .$$

Let $\delta(\Delta^{(n)}(x)) = \delta_n(x)$ be the diameter of $\Delta^{(n)}(x)$ (relative to the given metric on X). Define $\Delta^\infty(x) = \bigcap\limits_{n=1}^{\infty} \Delta^{(n)}$ and $\mathcal{B}(\Omega) = \bigvee\limits_{n=1}^{\infty} \mathcal{P}^{(n)}$. The condition that $\mathcal{B}(\Omega) = \mathcal{B} \pmod{\omega}$ is equivalent to the condition $\Delta^{\infty}(x) = \{x\}$, a.e. $\omega(dx)$ in the presence of the condition

$$(*) \qquad \Delta^{\infty}(x) = \{x\} \text{ if and only if } \lim_{n} \delta_n(x) = 0 \quad .$$

(6.17) holds, for example, if $\Delta^{(n)}(x)$ is compact, $n > 0$.

6.17. Definition. A $\mathcal{T}$-triple Ω satisfies Property PF (for Perron-Frobenius) if for ω-a.e. x there exists $n > 0$ such that for any y $\Delta^{\infty}(y) = \{y\}$ as soon as the sequence $Sy, S^2y, \ldots$ visits $\Delta^{(n)}(x)$ infinitely often.

6.18. Proposition. Let Ω be a $\mathcal{T}$-triple such that either $\mathcal{P}$ is finite or else $D_{\Delta} = X$, $\Delta \in \mathcal{P}$, and in the former case assume Property E. If Ω satisfies Properties UCA, PF, and (*), and if μ is an absolutely continuous invariant probability measure for S_{Ω}, then μ is both ergodic and equivalent to ω. In particular, ω is ergodic for S_{Ω}, and μ is unique.

Proof: Let $E \in \mathcal{B}$ satisfy, $S^{-1}E \subseteq E$, and let $\mathcal{E} = \mathcal{E}_{\omega}$ be the conditional expectation operator

$$(6.19) \qquad \mathcal{E}_{\omega}(E|\mathcal{P}^{(n)})(x) = \frac{\omega(E \cap \Delta)}{\omega(\Delta)} \qquad (x \in \Delta \in \mathcal{P}^{(n)}) \quad .$$

As now $E \cap \Delta \supseteq T_{\Delta}(E \cap D_{\Delta})$ the left side of (6.8) implies for $x \in \Delta \in \mathcal{P}^{(n)}(M)$

$$(6.20) \qquad \begin{aligned} \mathcal{E}_{\omega}(E|\mathcal{P}^{(n)})(x) &\geq \frac{1}{M}\,\omega(E \cap D_{\Delta}) \\ &\geq \frac{\beta}{M} \quad . \end{aligned}$$

Now Property UCA implies the set of $x \in X$ such that $x \in \Delta^{(n)}(x) \in \mathcal{P}^{(n)}(M)$ infinitely often for some $M < \infty$ has ω-measure 1, and therefore by (6.20) and the martingale theorem

(6.21) $$\mathcal{E}_{\omega}(E|\mathcal{B}(\Omega))(x) > 0 \qquad (\omega\text{-a.e.} x \in X) \quad .$$

Hypothesis PF implies there is a partition, $\mathcal{P}^*$, of X (mod ω) such that if $\Delta^* \in \mathcal{P}^*$, then $\Delta \in \mathcal{P}^{(n)}$ for some $n > 0$; and if $S^k x \in \Delta$ for infinitely many $k > 0$, then $\Delta^\infty(x) = \{x\}$. Because $\mu < \omega$, $\mathcal{P}^*$ also partitions X (mod μ). By the Poincaré recurrence theorem μ-almost all $x \in X$ satisfy $S^k x \in \Delta \in$ for infinitely many k, where $x \in \Delta \in \mathcal{P}^*$, and therefore $\Delta^\infty(x) = \{x\}$ for μ-a.e.x. Property (*) implies then that the trace of $\mathcal{B}(\Omega)$ on the $\mathcal{B}(\Omega)$ set $\{x|\Delta^\infty(x) = \{x\}\}$ is, modulo μ, the trace of $\mathcal{B}$ on the same set. That is, in particular,

$$\mathcal{E}_{\omega}(E|\mathcal{B}(\Omega))(x) = \chi_E(x) \qquad (\mu\text{-a.e.}x) \quad .$$

Now (6.21) holds for ω-a.e.x and in particular for μ-a.e.x, and so $\mu(E) = 1$. We have proved that if $S^{-1}E \subseteq E$ and $\omega(E) > 0$, then $\mu(E) = 1$. Consider now the Lebesgue decomposit of ω with respect to μ, $\omega = \omega_a + \omega_s$. Then $\omega_a < \mu$, and ω_s is supported on a set $N \in \mathcal{B}$ with $S_\Omega^{-1}N \subseteq N$. As $\omega(N) = \omega_s(N)$ and $\mu(N) = 0$, it must be from the above that $\omega_s(N) = 0$. That is, $\omega = \omega_a$, and $\omega \sim \mu$. Ergodicity relative to μ (and ω) is now immediate from the above.

If Ω is a $\mathcal{T}$-triple, the tail σ-algebra, $\mathcal{B}_\infty(\Omega)$, is defined by

$$\mathcal{B}_\infty(\Omega) = \bigcap_{n=1}^{\infty} \bigcup_{k=n}^{\infty} \mathcal{P}^{(k)}$$

$$= \bigcap_{n=1}^{\infty} S_\Omega^{-n}\, \mathcal{B}(\Omega) \quad .$$

In the case that $\mathcal{P}$ is a generator ($\mathcal{B} = \mathcal{B}(\Omega)$), as in Proposi-

6.18, one has $\mathcal{B}_\infty(\Omega) = \mathcal{B}_\infty = \bigcap_{n=1}^{\infty} S_\Omega^{-n}\mathcal{B}$, the "usual" tail σ-algebra. This is the only case of interest to us.

We shall say a $\mathcal{T}$-triple Ω is *mixing* if for any $\Delta' \in \mathcal{P}$ there exists $n > 0$ such that

$$\bigcup_{\substack{\Delta\in\mathcal{P}^{(n)} \\ \Delta\subseteq\Delta'}} D_\Delta = X$$

We note that if this condition holds for an integer n it holds for all larger integers.

6.22. Lemma. Let Ω be a mixing $\mathcal{T}$-triple, and suppose $E \in \mathcal{B}_\infty$, $E = S^{-n}E_n$, $n > 0$, has $\omega(E) > 0$. Then for all $\Delta' \in \mathcal{P}$, $\omega(E \cap D_{\Delta'}) > 0$.

Proof: We shall prove $\omega(E_1 \cap D_{\Delta'}) > 0$ for all $\Delta' \in \mathcal{P}$. Replacing E by $S^{-1}E \in \mathcal{B}_\infty$ it will follow $\omega(E \cap D_{\Delta'}) > 0$. If $n > 0$ and $\Delta \in \mathcal{P}^{(n)}$, then $E \cap \Delta = T_\Delta(E_n \cap D_\Delta)$. If n is large the union of the sets D_Δ, $\Delta \subseteq \Delta'$, is X, and therefore at least one of them has $\omega(E_n \cap D_\Delta) > 0$ because $\omega(E_n) > 0$. It follows $\omega(E \cap \Delta') \geq \omega(E \cap \Delta) > 0$, and then $\omega(E_1 \cap D_\Delta) > 0$. The lemma is proved.

6.23. Proposition. Let Ω be a mixing $\mathcal{T}$-triple which satisfies the hypotheses of Proposition 6.18. Then $\mathcal{B}_\infty$ is trivial ($\equiv$ $(X,\mathcal{B},\mu,S_\Omega)$ is an exact endomorphism [12]).

Proof: Suppose $E \in \mathcal{B}_\infty$, $E = S^{-n}E_n$, $n > 0$, and let μ be the measure of Proposition 6.18, so that $\mu(E) = \mu(E_n)$, $n > 0$. Because $\mu \sim \omega$, $\mu(E) > 0$ implies there exists $\delta > 0$ such that $\mu(E_n) > \delta > 0$, all n. Using (6.8) we find for all

$n > 0$, $\Delta \in \mathcal{P}^{(n)}(M)$

$$\frac{\omega(E \cap \Delta)}{\omega(\Delta)} = \frac{\omega(T_\Delta(E_n \cap D_\Delta))}{\omega(\Delta)}$$

(6.24)

$$\geq \frac{1}{M}\,\omega(E_n \cap D_\Delta) \quad .$$

In case $D_\Delta = X$, $\Delta' \in P$, then because $D_\Delta = D_{\Delta'}$ for some $\Delta' \in P$, (6.24) and the Martingale theorem imply $\chi_E \geq \frac{\delta}{M} > 0$ ω-a.e and then $\omega(E) = 1$. In the remaining case P is finite, and we shall prove

(6.25) $$\inf_{\substack{n > 0 \\ \Delta \in P}} \omega(E_n \cap D_\Delta) > 0 \quad .$$

The argument then proceeds exactly as above. If (6.25) were not true, there would exist, because P is finite, a $\Delta \in P$ and a sequence $n_1 < n_2 < \dots$ such that $\omega(E_{n_k} \cap D_\Delta) \to 0$. Let f be a weak cluster point of $\{\chi_{E_{n_k}}\}$ in $L^2(\omega)$. Certainly f is B_∞ measurable; because $\omega(E_{n_k}) > \delta$ for all k, $\{f > 0\} = E^*$ is a B_∞ set of positive measure. Clearly $\omega(E^* \cap D_\Delta) = 0$, which contradicts Lemma (6.22). Thus, (6.25) must be true, and the proposition is proved.

We shall conclude this section with a discussion of conditions that a T-triple, Ω admit a σ-finite invariant measure μ, $\mu < \omega$.

To begin suppose $n > 0$, $\Delta_0 \in P^{(n)}$, and suppose there exists on Δ_0 an induced transformation (of S_Ω), $U: \Delta_0 \to \Delta_0$. That is, for ω-a.e. $x \in \Delta_0$, $Ux = S^q x$' where $q = q(x)$ is the least $q > 0$ such that $S^q x \in \Delta_0$. Let $q = q(x)$, and write $\Delta^{(q)}(x) = \Delta(\Delta_1(x),\dots,\Delta_q(x))$. If $\Delta_0 = \Delta(\Delta^1,\Delta^2,\dots,\Delta^n)$, then

$$\Delta^{(q+n)}(x) = \Delta(\Delta_1(x),\ldots,\Delta_q(x),\Delta^1,\ldots,\Delta^n)$$

that is, $\Delta_{q+j}(x) = \Delta^j$, $1 \leq j \leq n$. It is clear then that $q(\cdot)$ is constant on $\Delta^{(q+n)}(x)$, and moreover $U\Delta^{(q+n)}(x) = \Delta_0$. Therefore, if we define $\mathcal{P}(\Delta_0)$ to be the set of $\Delta^{(q+n)}(x)$, $q = q(x)$, $\mathcal{P}(\Delta_0)$ is a partition of Δ_0 into sets of positive measure. Next, define $\mathcal{T}(\Delta_0)$ to be the set of maps with domain Δ_0 defined by $T_\Delta|_{\Delta_0}$ where $\Delta = \Delta^{(q)}(x)$, $q = q(x)$, and $\Delta^{(q+n)}(x) \in \mathcal{P}(\Delta_0)$. Finally, let $\chi(\Delta_0)$ be the trace of χ on Δ_0 (normalizing ω in the process). Then $\Omega(\Delta_0) = (\chi(\Delta_0),\mathcal{P}(\Delta_0),\mathcal{T}(\Delta_0))$ is a $\mathcal{T}$-triple. We note that $D_\Delta = \Delta_0$, $\Delta \in \mathcal{P}(\Delta_0)$.

6.26 Definition. A $\mathcal{T}$-triple, Ω, satisfies *Property* LR if for ω-almost all x there exist $n > 0$ such that $\Delta^{(n)}(x)$ admits an S_Ω-induced transformation.

Remark. If, for example, S_Ω is known to be ergodic, then Property LR is automatic. In this situation one has that $\Delta^{(n)}(x)$ admits an induced transformation for all $n > 0$.

6.27 Definition. A $\mathcal{T}$-triple, Ω, satisfies Property LUCA if for a.e. $x \in X$ there exists $k > 0$ such that if $\Delta_0 = \Delta^{(k)}(x)$, then for some $M < \infty$ and all n $\mathcal{P}(\Delta_0)^{(n)}(M) = \mathcal{P}(\Delta_0)^{(n)}$.

Remark. The elements T_Δ, $\Delta \in \mathcal{P}(\Delta_0)^{(n)}$ are restrictions to Δ_0 of transformations with larger domains D_Δ. The Radon-Nikodym derivatives similarly are restrictions to Δ_0 of corresponding derivatives with domains D_Δ. In our applications $\Delta_0 \subseteq \mathcal{P}^{(n)}$ will be chosen to have compact closure in X, and the nature of the functions J_Δ, $\Delta \in \mathcal{P}^{(n)}$, $n > 0$ will be such that

$$\sup_{\substack{n>0 \\ \Delta\in\mathcal{P}^{(n)} \\ \Delta_0\subseteq D_\Delta}} \sup_{x,y\in\Delta_0} \frac{J_\Delta(x)}{J_\Delta(y)} < \infty \tag{6.28}$$

In this situation LUCA holds automatically.

We assume in what follows that S_Ω is ergodic, and consequently that Ω has property LR. Assume also Property LUCA, and choose $\Delta_0 \in \mathcal{P}^{(n)}$ for some n so that $\Omega(\Delta_0)$ satisfies Property UCA. By proposition 6.12 there exists a probability measure, μ_0, on Δ_0 such that $\mu_0 < \omega$ and μ_0 is $U = S_{\Omega(\Delta_0)}$ invariant.

If $x \in \Delta_0$, $E \in \mathcal{B}$, define $N_E(x)$ to be the number of integers j, $0 \leq j < q(x)$, such that $S_\Omega^j x \in E$, and define

$$\mu(E) = \int_{\Delta_0} N_E(x)\mu_0(dx) \quad .$$

It is readily verified that $\mu < \omega$ is S_Ω-invariant, and of course μ is σ-finite. As S_Ω is assumed to be ergodic, $\mu \sim \omega$. We have

6.29. Proposition. Let Ω be a $\mathcal{T}$-triple such that S_Ω is ergodic, and suppose Ω satisfies Property LUCA. Then S_Ω admits a σ-finite invariant Borel measure, μ, equivalent to ω.

With notations as above let ω_0 be ω normalized on Δ_0, so that $\mu_0 \sim \omega_0$. One sees readily that because $\mathcal{P}^{(n)}(\Delta_0)(M) = \mathcal{P}^{(n)}(\Delta_0)$, some $M < \infty$, $n > 0$, one has

$$(\alpha M)^{-1}\mu_0 \leq \omega_0 \leq \frac{M}{\alpha}\mu_0 \quad .$$

It follows therefore that μ is finite if and only if

$$(6.30) \qquad \int_{\Delta_0} q(x)\,\omega(dx) < \infty \quad .$$

As the integral (6.30) depends only upon Ω (and Δ_0), it is possible the question of finiteness of μ can be settled in certain specific instances.

In what follows we shall assume Ω is such that $\omega \times \omega$ is ergodic for $S \times S$ and that there exists $k > 0$, $\Delta_0 \in P^{(k)}$ such that (6.28) holds. Let $M_1 < \infty$ be the left side of (6.28), and assume

$$(6.31) \qquad M_2 = \sup_{x,y \in \Delta_0} \frac{J_{\Delta_0}(x)}{J_{\Delta_0}(y)} < \infty$$

We set $M = M_1 M_2$.

Suppose now $n > 0$, $\Delta = \Delta(\Delta_1,\ldots,\Delta_n) \in P^{(n)}$ are such that $\Delta_0 \subseteq D_\Delta$. If $\Delta_0 = \Delta(\Delta^1,\ldots,\Delta^k)$, then $\Delta^* = \Delta(\Delta_1,\ldots,\Delta_n,\Delta^1,\ldots,\Delta^k) \in P^{(n+k)}$, and by the chain rule

$$J_{\Delta^*}(x) = J_\Delta(T_{\Delta_0}x)J_{\Delta_0}(x) \qquad (x \in D_{\Delta^*}) \quad .$$

Therefore by (6.28) and (6.31), $\Delta^* \in P^{(n+k)}(M)$, $M = M_1 M_2$.

Suppose now $E \in B^\infty$, $E = S_\Omega^{-n} E_n, n > 0$. If Ω satisfies (*) and Property PF then because S_Ω is ergodic, we have from the Martingale Theorem

$$\lim_{n\to\infty} \mathcal{E}_\omega(E \mid P^{(n)})(x) = \chi_E(x) \qquad (\text{a.e.}\,\omega) \quad .$$

If Δ_0 is as above, there exist for a.e. $x,y \in X$ infinitely many integers n such that $(S^n x, S^n y) \in \Delta_0 \times \Delta_0$. If we let $\Delta' = \Delta^{(N+k)}(x)$, $\Delta'' = \Delta^{(n+k)}(y)$, then $D_{\Delta'} = D_{\Delta''}$, $E \cap \Delta' =$

$T_{\Delta'}(E_{n+k} \cap D_{\Delta'}),\ E \cap \Delta'' = T_{\Delta''}(E_{n+k} \cap \Delta'')$, and by (6.8)

$$\mathcal{E}_{\omega}(E|P^{(n+k)})(x) = \frac{\omega(E \cap \Delta')}{\omega(\Delta')} \leq \frac{M}{\omega(D_{\Delta'})}\,\omega(E_n \cap D_{\Delta'})$$

$$\leq M^2\,\frac{\omega(E \cap \Delta'')}{\omega(\Delta'')}$$

$$= M^2\mathcal{E}_{\omega}(E|P^{(n+k)})(y) \quad .$$

It follows that $\chi_E(x) \leq M^2\chi_E(y)$ for $\omega \times \omega$ a.e.x,y, and therefore $\omega(E) = 0$ or 1.

6.32 Proposition. Let Ω be a T-triple which satisfies Properties PF, (*), and the form (6.28), (6.30) of Property LUCA. If $S_\Omega \times S_\Omega$ is ergodic, then

$B_\infty = \bigcap_{n=1}^{\infty} S^{-n}B$ is trivial.

7. Partitions and $\mathcal{T}$-triples. In this section we fix $m > 1$ and set $X = \Delta_{m-1}$, the standard (m-1)-simplex. $\omega = \omega_{m-1} = \omega_{\Delta_{m-1}}$ is normalized Lebesgue measure on X. We shall observe that if $\mathcal{F}$ is a partition in the sense defined in Section 5, there is a naturally associated $\mathcal{T}$-triple, $\Omega(\mathcal{F})$.

If $A \in \mathcal{F}$, define $T_A: \Delta_{m-1} \to \Delta_{m-1}$ by

$$T_A \lambda = \frac{A\lambda}{|A\lambda|}$$

where, as before, $|\cdot|$ denotes sum-of-components. Then set $\Delta^A = T_A \Delta_{m-1}$, and let $\mathcal{P} = \mathcal{P}(\mathcal{F}) = \{\Delta^A | A \in \mathcal{F}\}$, $\mathcal{T} = \mathcal{T}(\mathcal{F}) = \{(T_A, \Delta_{m-1}) | A \in \mathcal{F}\}$. $\mathcal{P}(\mathcal{F})$ is a measurable partition (mod ω), and $\Omega(\mathcal{F}) = (\chi, \mathcal{P}((\mathcal{F}), \mathcal{T}(\mathcal{F}))$ is a $\mathcal{T}$-triple. We recall from [15] that the Radon-Nikodym derivative, J_A, is, because $|\det A| = 1$ given by

$$J_A(\lambda) = |A\lambda|^{-m} \quad . \tag{7.1}$$

If K is a compact or relatively compact subset of Δ_{m-1}, define $c(K)$ to be the supremum

$$c(K) = \sup_{\substack{\lambda, \lambda' \in K \\ 1 \leq i \leq m}} \left(\frac{\lambda_i}{\lambda'_i}\right)^m$$

We then have the bound, independent of $\mathcal{F}$,

$$\sup_{\substack{\lambda, \lambda' \in K \\ A \in \mathcal{F}}} \frac{J_A(\lambda)}{J_A(\lambda')} \leq c(K) < \infty \quad . \tag{7.2}$$

Let A be an $m \times m$ matrix all of whose entries are positive. We define a number, $\nu(A)$, by

$$\nu(A) = \max_{1 \leq i, j, j' \leq m} \frac{A_{ij}}{A_{ij'}} \quad . \tag{7.3}$$

If $r_j(A)$ is the jth column sum, $r_j(A) = \sum_{i=1}^{m} A_{ij}$, then (7.3) implies trivially

$$(7.4) \qquad r_j(A) \le \nu(A) r_{j'}(A) \qquad (1 \le j,\ j' \le m) \quad .$$

As $\min_j r_j(A) \le |A\lambda| \le \max_j r_j(A)$, we have in turn from (7.1),(7.4)

$$(7.5) \qquad \sup_{\lambda,\lambda' \in \Delta_{m-1}} \frac{J_A(\lambda)}{J_A(\lambda')} \le (\nu(A))^m \quad .$$

In general the partitions which arise in our study are partitions whose matrices *all* have 0 among their entries. In this situation $\nu(A)$, $A \in F$, is not defined. On the other hand there will exist for a.e. λ an $n > 0$ such that $A^{(n)}(\lambda)$ has all positive entries, and so $\nu(A^{(k)}(\lambda))$ is defined for all $k \ge n$.

If A has positive entries, and if B is a nonnegative, nonsingular matrix, then BA also has positive entries, and moreover

$$(7.6) \qquad \nu(BA) \le \nu(A) \quad .$$

We omit the simple proof.

7.7 Definition. Let F be a partition each element of which has positive entries. We define $\|F\| = \|F\|_\nu$ by

$$\|F\| = \sup_{A \in F} \nu(A)$$

and call F a *uniform partition* if $\|F\| \le \infty$.

As in Section 5, $F^{(n)}$, $n \ge 1$, denotes the set of n-fold products of elements of F. The importance of (7.6) lies in the fact $\|F^{(n)}\| \le \|F\|$, $n \ge 1$. If we set $M = \|F\|^m$, then (6.7) holds for this M and all $n > 0$, $A \in \mathcal{F}^{(n)}$, $\Delta = \Delta^A$.

7.8. Proposition. If $\mathcal{F}$ is a uniform partition, the associated $\mathcal{T}$-triple satisfies Property UCA.

7.9. Definition. A partition, $\mathcal{F}$, is a p-*partition* if for ω-a.e. $\lambda \in \Lambda_m$ there exists n such that $A^{(n)}(\lambda)$ has positive entries.

7.10. Proposition. If $\mathcal{F}$ is a p-partition, then $\mathcal{F}$ enjoys Property PF. In particular, if $\mathcal{F}$ is a uniform partition, $\mathcal{F}$ enjoys Property PF.

Proof. Let $\lambda \in \Delta_{m-1}$ be such that $B = A^{(n)}(\lambda)$ has positive entries. If $\lambda' \in \Delta_{m-1}$, then to say $S_\Omega^k \lambda' \in \Delta^B$ is to say $A_{k+1}(\lambda')A_{k+2}(\lambda')\ldots A_{k+n}(\lambda') = B$. Now in the notation of Section 6, $\Delta^\infty(\lambda') = \{\lambda'\}$ if and only if, in the notation of Section 5, $[\lambda']_{\mathcal{F}} = R^+\lambda'$. A sufficient condition for the latter is, by [15], that the matrix B occur as above for infinitely many k. The proposition is proved.

Remark. In the context of a partition, $\mathcal{F}$, it is immediate that $[\lambda]_{\mathcal{F}} = R^+\lambda$ only if diameter $(A^{(n)}(\lambda) \cap \Delta_{m-1}) \to 0$. That is, property (*) of Section 6 holds for $\Omega(\mathcal{F})$.

7.11. Proposition. Let $\mathcal{F}$ be a uniform partition. If $\Omega = \Omega(\mathcal{F})$, then $S = S_\Omega$ admits an invariant probability measure, μ, equivalent to Lebesgue measure. Moreover, $\mathcal{P}(\Omega)$ is a generator for S and, relative to μ, S is an exact endomorphism.

To say $\mathcal{P}(\Omega)$ is a generator is to say (in view of (*)) that $\Delta^\infty(\lambda) = \{\lambda\}$ ω-a.e., or what is the same $[\lambda]_{\mathcal{F}} = R^+\lambda$ a.e. We obtain then as a Corollary (or restatement of part) of Proposition 7.11.

7.12. Proposition. Every uniform partition is regular.

Remark. If $\lambda, \lambda' \in \Delta_{m-1}$, define

$$\delta(\lambda, \lambda') = \log \max_{1 \le i,j \le m} \frac{\lambda_i \lambda'_j}{\lambda'_i \lambda_j} .$$

(Recall the convention that Δ_{m-1} is the open (m-1)-simplex, i.e., $\lambda_i, \lambda'_j > 0$, all i,j.) The condition that $\|F\| < \infty$, i.e., that F be a uniform partition, translates in terms of the metric $\delta(\cdot,\cdot)$ into the condition that the elements of $P(F)$ have uniformly bounded diameters.

7.13. Definition. A partition, F', is a *subpartition* of a partition, F, if $F' \subseteq \bigcup_{n=1}^{\infty} F^{(n)}$; that is, if each $A \in F'$ is a product of elements of F.

If F' is a subpartition of F, then clearly $[\lambda]_{F'} = [\lambda]_F$ whenever these objects are defined. In particular, if F admits a regular subpartition, F itself must be regular. Proposition 7.12 yields.

7.14. Proposition. If a partition, F, admits a uniform subpartition, then F is regular.

Let F be a partition, and let $S_\Omega^{-1}E \subseteq E$, $\Omega = \Omega(F)$. We have seen that this means for all $n > 0$ and $A \in F^{(n)}$ that $\Delta^A \cap E \supseteq T_A E$; moreover, if this condition holds for all $A \in F$, then $S_\Omega^{-1}E \subseteq E$. Now if F' is a subpartition of F, each $A \in F'$ belongs to $F^{(n)}$ for some $n > 0$. Therefore, if $\Omega' = \Omega(F')$, it is immediate that $S_{\Omega'}^{-1}E \subseteq E$. If F' is *uniform*, then $S_{\Omega'}$ is ergodic (incompressible) and so $\omega(E) = 0$ or 1. We have

7.15. Proposition. If a partition, F, admits a uniform subpartition, and if $\Omega = \Omega(F)$, then S_Ω is ergodic relative to Lebesgue measure.

If $\mathcal{F}$ is a p-partition, then for ω-a.e. λ there exists $n > 0$ such that $\Delta^{(n)}(\lambda) = \Delta^{A^{(n)}(\lambda)}$ is relatively compact in Δ_{m-1}, and therefore (7.2) holds for $K = \text{closure}\,(\Delta^{(n)}(\lambda))$. Thus, if $\mathcal{F}$ is a p-partition, $\Omega(\mathcal{F})$ enjoys Property LUCA in the form (6.28), (6.30).

7.16. Proposition. Let $\mathcal{F}$ be a p-partition, and suppose for $\Omega = \Omega(\mathcal{F})$ that S_Ω is ergodic. For example, suppose that $\mathcal{F}$ admits a uniform subpartition. Then S_Ω admits a σ-finite invariant measure equivalent to Lebesgue measure.

Proof. Apply Proposition 6.29.

8. Caps. If $m > 1$, and if $\mathcal{F}$ is a partition (of Λ_m), we have associated to a.e. $\lambda \in \Lambda_m$ an expansion $\lambda \sim [A_1(\lambda), A_2(\lambda), \ldots]$. In terms of this expansion we define for every nonempty set, $\mathcal{G} \subseteq \mathcal{F}$, a set, $\Gamma = \Gamma(\mathcal{F}, \mathcal{G})$, by

$$\Gamma = \{\lambda \in \Delta_{m-1} | A_n(\lambda) \in \mathcal{G}, \text{ all } n \geq 1\} .$$

In terms of the sets $\mathcal{G}^{(n)}$, $n \geq 1$, of n-fold products of elements of $\mathcal{G}$, Γ has the definition

$$(8.1) \qquad \Gamma = \bigcap_{n \geq 1} \bigcup_{A \in \mathcal{G}^{(n)}} (A\Lambda_m \cap \Delta_{m-1}) .$$

Obviously Γ does not depend upon $\mathcal{F}$, and so we write $\Gamma = \Gamma(\mathcal{G})$ not $\Gamma(\mathcal{G}, \mathcal{F})$. Of course (8.1) reveals Γ as a generalized Cantor set or, more descriptively, a "projective Swiss cheese."

8.2. Proposition. Let $\mathcal{F}$ be a partition such that for the associated $\mathcal{T}$-triple, $\Omega(\mathcal{F})$, S_Ω is ergodic. Then for any proper subset $\mathcal{G} \subseteq \mathcal{F}$, $\omega(\Gamma(\mathcal{G})) = 0$.

Proof. Let $A \in \mathcal{G}^c$, $\Delta = \Delta^A$. Then as $\omega(\Delta^A) > 0$, ergodicity implies for ω-a.e. λ that $S_\Omega^n \lambda \in \Delta^A$ for infinitely many n. For any such n, $A_{n+1}(\lambda) = A$, and therefore $\omega(\Gamma(\mathcal{G})) = 0$, as claimed.

8.3. Remark. Let $n > 0$, $A_1, \ldots, A_n \in \mathcal{F}$, and set $A = A_1 \cdot\cdot$ Then $\omega(\Delta^A) > 0$, and the same argument shows that if S_Ω is ergodic, then for ω-a.e. λ, $A_{k+1}(\lambda) = A_1, \ldots, A_{k+n}(\lambda) = A_n$ holds for infinitely many k. Thus, if S_Ω is ergodic, almost every $\lambda \in \Delta_{m-1}$ (or Λ_m) is "normal" in the sense that any "finite $\mathcal{F}$-block" occurs infinitely often in the expansion of λ. Conversely, if $\mathcal{F}$ is a p-partition, and if a.e. $\lambda \in \Delta_{m-1}$ is normal in this sense, then we claim S_Ω is ergodic. For the hypothesis implies for any $n > 0$ and $A \in \mathcal{F}^{(n)}$ that for a.e. $\lambda \in \Delta_{m-1}$, $S^k\lambda \in \Delta^A$ infinitely often. We may suppose A has positive entries, and define $k = k(\lambda)$ small as possible with $S^k\lambda \in \Delta^A$. Then $A^{(k+n)}(\lambda) = A_1(\lambda)\ldots A_n(\lambda)A$, and $\nu(A^{(k+n)}(\lambda)) \le \nu(A)$. Clearly the set $\mathcal{F}' = \{A^{(k+n)}(\lambda) \mid k = k(\lambda), \lambda \in \Delta_{m-1}\}$ is a subpartition, and $\|\mathcal{F}'\| \le \nu(A) < \infty$. Now apply Proposition 7.15.

8.4. Proposition. Let $\mathcal{F}$ be a p-partition. If $\Omega = \Omega(\mathcal{F})$ is the associated $\mathcal{T}$-triple, then S_Ω is ergodic if and only if ω-a.e. λ has every finite $\mathcal{F}$-block occuring infinitely often in its "$\mathcal{F}$-expansion."

8.5. Definition. Let $\mathcal{F}$ be a partition. A set $\mathcal{H} \subseteq \mathcal{F}$ is a *cap* for $\mathcal{F}$ if $\omega(\Gamma(\mathcal{H}^c)) = 0$.

If $\mathcal{H}$ is a cap such that $\mathcal{H}^c = \mathcal{G} \neq \emptyset$, then $\Gamma(\mathcal{G})^c$ is the set

$$\Gamma(\mathcal{G})^c = \bigcup_{n=0}^{\infty} \bigcup_{A\in\mathcal{G}^{(n)}} \bigcup_{B\in\mathcal{H}} \Delta^{AB} \qquad . \tag{8.6}$$

Therefore, to say $\mathcal{H}$ is a cap is, because (8.6) is a disjoint union, to say

$$\mathcal{H}^* = \bigcup_{n=0}^{\infty} \mathcal{G}^{(n)}\mathcal{H} \tag{8.7}$$

is a partition (set $\mathcal{G}^{(0)} = \{I\}$).

Since $\|\mathcal{H}^*\|_\nu \le \|\mathcal{H}\|_\nu$ in (8.7), one procedure for constructing a uniform subpartition for $\mathcal{F}$ is to find a cap, $\mathcal{H}$, such that $\|\mathcal{H}\|_\nu < \infty$. However, as we have already mentioned, the partitions, $\mathcal{F}$, of immediate interest to this study have *no* elements with positive entries. Thus, we cannot procede so directly.

In what follows Ψ denotes a function with domain, $D(\Psi)$, a subset of the set of nonnegative matrices. If $\mathcal{C} \subseteq D(\Psi)$, define

$$\|\mathcal{C}\|_\Psi = \sup_{C\in\mathcal{C}} \Psi(C) \qquad .$$

8.8. Definition. Assume $\mathcal{F}$ is a given partition, $\mathcal{F}'$ is a subpartition, and $(\Psi, D(\Psi))$ is a function as above. A cap, $\mathcal{H}$, for $\mathcal{F}$ shall be called $(\mathcal{F}', \Psi)$-*stratified* if

a) There is a subset $\mathcal{H}_0 \subseteq \mathcal{H}$ and a biunique correspondence $H \Leftrightarrow H_{\alpha,j}$, $\alpha \in \mathcal{H}_0$, $j \geq 0$ between $\mathcal{H}$ and $\mathcal{H}_0 \times \mathbb{Z}^+$ such that $H_{\alpha,0} = \alpha$, $\alpha \in \mathcal{H}_0$.

and if there is a constant $K = K(\mathcal{H}) < \infty$ such that

b) If $B \in \mathcal{F}' \bigcup_{n=1}^{\infty} \mathcal{F}^{(n)}$, then $BH_{\alpha,j} \in D(\Psi)$, $\alpha \in \mathcal{H}_0$, $j \geq 0$,

and

$$\Psi(BH_{\alpha,j}) \leq K(j+1) \tag{8.9}$$

and

c) For all $\alpha \in \mathcal{H}_0$, $T > 0$, and B as in b)

$$\sum_{j \geq T} \omega(\Delta^{BH_{\alpha,j}}) \leq \frac{K}{T}\, \omega(\Delta^{B\alpha}) \quad . \tag{8.10}$$

8.11. Lemma. Let $\mathcal{F}$ be a partition and $\mathcal{F}'$ a subpartition of $\mathcal{F}$. If for a given function $(\Psi, D(\Psi))$ $\mathcal{F}$ admits an $(\mathcal{F}', \Psi)$-stratified cap, $\mathcal{H}$, then $\mathcal{F}$ admits a subpartition, $\mathcal{F}''$, such that $\|\mathcal{F}''\|_{\Psi} < \infty$.

Proof. Let $K = K(\mathcal{H})$ be as in the above definition, and choose any $T > 2K^2$ such that $\frac{T}{K} \in \mathbb{Z}$. Let $\mathcal{H}^* = \bigcup_{n=0}^{\infty} \mathcal{G}^{(n)}\mathcal{H}$ be as in (8.7), where $\mathcal{G} = \mathcal{H}^c$, $\mathcal{G}^{(0)} = \{I\}$. Define $\mathcal{F}_0 = \mathcal{F}'\mathcal{H}^*$. Each $A \in \mathcal{F}_0$ has the form $A = BCH_{\alpha,j}$, $\alpha \in \mathcal{H}_0$, $j \geq 0$, $C \in \bigcup_{n=0}^{\infty} \mathcal{G}$ $B \in \mathcal{F}'$, and we define $\mathcal{H}_0 \subseteq \mathcal{F}_0$ to be the set of such A for which $j \leq \frac{T}{K} - 1$ (an integer). For such A we have from (8.9) that $\Psi(A) \leq T$. From (8.10) we obtain

$$(8.12) \qquad \sum_{j \geq \frac{T}{K}} \omega(\Delta^{BCH_{\alpha},j}) \leq \frac{K^2}{T} \omega(\Delta^{BC\alpha}) \leq \tfrac{1}{2} \omega(\Delta^{BC\alpha}) \quad .$$

Now $\mathcal{F}'\mathcal{H}^*$ is itself a partition, and therefore from (8.12) if we let $\pounds_0$ be the complement of $\mathcal{H}_0$ in $\mathcal{F}_0$, we obtain

$$(8.13) \qquad \omega(\Delta^{\pounds_0}) \leq \tfrac{1}{2} \quad .$$

Suppose now that triples $(\mathcal{F}_i, \mathcal{K}_i, \pounds_i)$ have been constructed for some $n > 0$ and $0 \leq i < n$ with the properties

1) $\| \bigcup_{i=0}^{n-1} \mathcal{K}_i \|_{\Psi} \leq T$

2) $\pounds_{n-1} \cup \bigcup_{i=0}^{n-1} \mathcal{K}_i$ is a subpartition of $\mathcal{F}$

3) $\omega(\Delta^{\pounds_i}) \leq \frac{1}{2^{i+1}}$, $0 \leq i < n$

4) $\mathcal{F}_i = \mathcal{K}_i \cup \pounds_i$, $0 \leq i < n$.

Then define $\mathcal{F}_n = \pounds_{n-1}\mathcal{H}^*$, and divide $\mathcal{F}_n$ into sets $\mathcal{K}_n$, $\pounds_n$ in the same way $\mathcal{F}'\mathcal{H}^*$ was divided above. Properties 1), 2), and 4) are immediate either from the definitions or from the fact $\mathcal{H}^*$ is a partition. As for 3) we note that the matrices $BC\alpha$ which occur in (8.12) are restricted this time to satisfy

$B \in \pounds_{n-1}$, $C \in \bigcup_{\ell=0}^{\infty} \mathcal{G}^{(\ell)}$, $\alpha \in \mathcal{H}_0$. Therefore, the sum over all possibilities of the right hand side of (8.12) is dominated by $\frac{1}{2} \omega(\Delta^{\pounds_{n-1}})$. It follows then that

$$\omega(\Delta^{\pounds_n}) \leq \tfrac{1}{2}\,\omega(\Delta^{\pounds_{n-1}})$$

$$\leq \frac{1}{2^{n+1}}$$

the last inequality by the induction hypothesis. The construction of a sequence $(F_n, K_n, \pounds_n)$, $n \geq 0$, satisfying properties 1) - 4) above is now completed by induction. Setting $F'' = \bigcup_{n=0}^{\infty} K_n$, F'' is, by 2) and 3) a subpartition of F, and, by 1), $\|F''\|_\Psi < \infty$. The lemma is proved.

9. Two Swiss cheeses. Recall that if A is an $m \times m$ matrix, the matrix A^θ is defined by $A^\theta_{ij} = A_{m-i+1,m-j+1}$. In this section we shall have always $m = 4$.

Define 4×4 matrices A_ℓ, $\ell \geq 0$, by

$$(9.1) \qquad A_\ell = \begin{pmatrix} 1 & 1 & 1 & 1 \\ \ell & \ell+1 & 0 & 0 \\ 0 & 0 & 1 & 0 \\ 0 & 0 & 0 & 1 \end{pmatrix}$$

and then define $\mathcal{G}$ by

$$(9.2) \qquad \mathcal{G} = \{A \mid A = A_\ell \text{ or } A_\ell^\theta, \ \ell \geq 0\} \quad .$$

In the notation of Section 3 (Lemma 3.17) $\mathcal{G}$ is a subset of $\mathcal{F}_2(4)$. The main result of the section will be obtained from

9.3. Lemma. Let $\mathcal{G}$ be the set of 4×4 matrices defined by (9.1)-(9.2). Then $\omega(\Gamma(\mathcal{G})) = 0$.

Here first is a sketch of the proof of Lemma 9.3. Recalling the definition $J_A(\lambda) = |A\lambda|^{-4}$ from (7.1), define an operator, $\mathcal{L} = \mathcal{L}_{\mathcal{G}}$, on $f \in L^1(\Delta_3, \omega_3)$ by

$$\mathcal{L}f(\lambda) = \sum_{A \in \mathcal{G}} f(T_A\lambda) J_A(\lambda) \quad .$$

We have for any measurable set E that

$$\int_E \mathcal{L}f(\lambda)\omega(d\lambda) = \sum_{A \in \mathcal{G}} \int_E f(T_A\lambda) J_A(\lambda)\omega(d\lambda)$$

$$(9.4) \qquad = \sum_{A \in \mathcal{G}} \int_{T_A E} f(\lambda)\omega(d\lambda)$$

$$= \int_{\bigcup_{A \in \mathcal{G}} T_A E} f(\lambda)\omega(d\lambda) \quad .$$

Consider now the set $\Gamma = \Gamma(G)$ which by its definition (8.1) satisfies

$$\Gamma = \bigcup_{A \in G} T_A \Gamma \qquad .$$

If we take $E = \Gamma$ in (9.4), that equation then becomes

$$(9.5) \qquad \int_\Gamma \pounds f(\lambda)\,\omega(d\lambda) = \int_\Gamma f(\lambda)\,\omega(d\lambda) \qquad .$$

9.6. Lemma. With notations as above suppose that there exists $f \in L^1(\Delta_3, \omega_3)$ which satisfies for $\pounds = \pounds_G$

$$(9.7) \qquad \pounds f(\lambda) < f(\lambda) \qquad (\lambda \in \Delta_3) \qquad .$$

Then $\omega(\Gamma(G)) = 0$.

Proof. Immediate from (9.5).

The discussion which follows is directed toward proving

9.8. Lemma. Define $f(\lambda)$, $\lambda \in \Delta_3$, by

$$(9.9) \qquad f(\lambda) = \frac{1}{(\lambda_1+\lambda_2)(\lambda_1+\lambda_2+\lambda_3)(\lambda_2+\lambda_3+\lambda_4)(\lambda_3+\lambda_4)} \quad .$$

Then $f \in L^1(\Delta_3, \omega_3)$, and f satisfies (9.7).

We note that (9.9) is defined naturally on Λ_4 and is there homogeneous of degree -4. Our choice of f was motivated in part by this fact, in view of

9.10. Lemma. Let F be defined and homogeneous of degree -m on Λ_m, and let $f = F|_{\Delta_{m-1}}$. If A is a nonnegative m × m matrix with det A = ±1, then

$$f(T_A\lambda)J_A(\lambda) = F(A\lambda) \qquad (\lambda \in \Delta_{m-1}) \quad . \tag{9.11}$$

Proof. We have

$$f(T_A\lambda)J_A(\lambda) = f\left(\frac{A\lambda}{|A\lambda|}\right)|A\lambda|^{-m}$$

$$= F\left(\frac{A\lambda}{|A\lambda|}\right)|A\lambda|^{-m}$$

$$= F(A\lambda)$$

as claimed by (9.11). The lemma is proved.

It is convenient to recall here certain properties of the Digamma function, $\Psi(x) = \frac{\Gamma'(x)}{\Gamma(x)}$. The derivative of Ψ is given by

$$\Psi'(x) = \sum_{n=0}^{\infty} \frac{1}{(x+n)^2} \quad .$$

From this expression one obtains that for $0 < x < \infty$ a) $\Psi'(x)$ is monotone decreasing, and b) $\Psi'(x) > \frac{1}{x}$.

9.12. Lemma. Let $\Psi(x)$ be the Digamma function. If $a,b > 0$, then

$$\Psi(a+b) - \Psi(a) > \frac{b}{a+b} \quad . \tag{9.13}$$

Proof. By the mean value theorem the left hand side of (9.13) is $b\Psi'(x)$ for some x between a and $a+b$. By a) and b) above, this number exceeds the right hand side of (9.13).

If $\lambda \in \Delta_3$, define $u = u(\lambda) = \lambda_1+\lambda_2$ and $v = v(\lambda) = \lambda_3+\lambda_4$. Next, define $B_0(\lambda)$ by

$$(9.14) \qquad B_0(\lambda) = 1 - \frac{(1-\lambda_1)}{\lambda_3}\left(\Psi\left(\frac{1+\lambda_2+\lambda_3}{u}\right) - \Psi\left(\frac{1+\lambda_2}{u}\right)\right)$$

and let $B_0^\theta(\lambda) = B_0(\lambda_4,\lambda_3,\lambda_2,\lambda_1)$. Set up

$$(9.15) \qquad B(\lambda) = B_0(\lambda) + B_0^\theta(\lambda) \quad .$$

9.16. Lemma. The function $B(\lambda)$ defined by (9.15)-(9.14) satisfies

$$(9.17) \qquad 0 < B(\lambda) < 1 \qquad (\lambda \in \Delta_3) \quad .$$

Proof. Let $a = \frac{1+\lambda_2}{u}$, $b = \frac{\lambda_3}{u}$ in (9.13). We find

$$(9.18) \qquad B_0(\lambda) < 1 - \frac{1-\lambda_1}{\lambda_3} \cdot \frac{\lambda_3}{1+\lambda_2+\lambda_3} = 1 - \frac{1-\lambda_1}{1+\lambda_2+\lambda_3} \quad .$$

Similarly,

$$(9.19) \qquad B_0^\theta(\lambda) < 1 - \frac{1-\lambda_4}{1+\lambda_2+\lambda_3} \quad .$$

Because $|\lambda| = 1$, we have

$$1 - \frac{1-\lambda_1}{1+\lambda_2+\lambda_3} + 1 - \frac{1-\lambda_4}{1+\lambda_2+\lambda_3} = 1$$

and (9.17) follows from (9.18), (9.19) and (9.15). The lemma is proved.

9.20. Lemma Let F and $f = F|_{\Delta_3}$ be defined by (9.9), and let $\pounds = \pounds_G$ be as at the beginning of the section. Then

$$(9.21) \qquad \pounds f(\lambda) = B(\lambda)f(\lambda)$$

Where B is the function (9.17).

Proof. By Lemma 9.10

$$\mathcal{L}f(\lambda) = \sum_{A\in G} F(A\lambda)$$

or, by (9.2),

$$\mathcal{L}f(\lambda) = \sum_{\ell=0}^{\infty} F(A_\ell\lambda) + \sum_{\ell=0}^{\infty} F(A_\ell^\theta\lambda) \quad . \tag{9.22}$$

We shall prove that

$$\sum_{\ell=0}^{\infty} \frac{F(A_\ell\lambda)}{F(\lambda)} = B_0(\lambda) \qquad (\lambda \in \Delta_3) \quad . \tag{9.23}$$

Considerations of symmetry imply the same equality with A_ℓ replaced by A_ℓ^θ and B_0 replaced by B_0^θ. (This is because $F = F^\theta$.) Therefore (9.21) will follow from (9.23), (9.22) and (9.15).

The definition (9.1) of A_ℓ implies that if $\lambda \in \Delta_3$, then

$$A_\ell\lambda = (1, \ell u+\lambda_2, \lambda_3, \lambda_4)$$

where $u = u(\lambda) = \lambda_1+\lambda_2$. From (9.9) we have (recalling $v = v(\lambda) = \lambda_3+\lambda_4$)

$$vF(A_\ell\lambda) = \frac{1}{(1+\ell u+\lambda_2)(1+\ell u+\lambda_2+\lambda_3)(\ell u+\lambda_2+\lambda_3+\lambda_4)} \tag{9.24}$$

The following lemma is well known:

9.25. Lemma. Suppose $s \geq 2$ and $b_1,\ldots,b_s \in \mathcal{C}$ are given with $b_1,\ldots,b_s$ pairwise distinct and nonintegral. Then if $\alpha_1,\ldots,\alpha_s \in \mathcal{C}$ satisfy

$$\prod_{j=1}^{s} \frac{1}{(b_j+\ell)} = \sum_{j=1}^{s} \frac{\alpha_j}{\ell+b_j} \qquad (\ell \in \mathbb{Z}) \tag{9.26}$$

then

$$\sum_{\ell=0}^{\infty} \left(\prod_{j=1}^{s} \frac{1}{b_j+\ell} \right) = -\sum_{j=1}^{s} \alpha_j \Psi(b_j) \tag{9.27}$$

where $\Psi(\cdot)$ is the Digamma function.

Proof. We recall about the Digamma function that

$$\Psi(x) = -\gamma - \frac{1}{x} - \sum_{n=1}^{\infty} \left(\frac{1}{x+n} - \frac{1}{n}\right) \tag{9.28}$$

where γ is Euler's constant. Because $\sum_{j=1}^{s} \alpha_j = 0$ above, and because $\sum_{j=1}^{s} \frac{\alpha_j}{b_j} = \prod_{j=1}^{s} \frac{1}{b_j}$, (9.27) follows from (9.28) and (9.26). The lemma is proved.

Returning to (9.24), that equation implies

$$vF(A_\ell \lambda) = \frac{1}{u^2(1-\lambda_4)} \left\{ \left(\ell + \frac{\lambda_2+\lambda_3+\lambda_4}{u}\right)^{-1} - \left(\ell + \frac{1+\lambda_2+\lambda_3}{u}\right)^{-1} \right.$$
$$\left. - \frac{1}{u^2\lambda_3} \left\{ \left(\ell + \frac{1+\lambda_2}{u}\right)^{-1} - \left(\ell + \frac{1+\lambda_2+\lambda_3}{u}\right)^{-1} \right\} \right. .$$

Applying (9.27), we have

$$\sum_{\ell=0}^{\infty} vF(A_\ell \lambda) = \frac{1}{u^2(1-\lambda_4)} \left(\Psi\left(\frac{1+\lambda_2+\lambda_3}{u}\right) - \Psi\left(\frac{\lambda_2+\lambda_3+\lambda_4}{u}\right)\right)$$

$$- \frac{1}{u^2\lambda_3} \left(\Psi\left(\frac{1+\lambda_2+\lambda_3}{u}\right) - \Psi\left(\frac{1+\lambda_2}{u}\right)\right) \quad . \tag{9.29}$$

Another property of the Digamma function is that $\Psi(z+1) = \frac{1}{z} + \Psi(z)$ (which follows from $\Gamma(z+1) = z\Gamma(z)$). In particular,

$$\Psi\left(\frac{\lambda_2+\lambda_3+\lambda_4}{u}\right) = \frac{-u}{\lambda_2+\lambda_3+\lambda_4} + \Psi\left(\frac{1+\lambda_2}{u}\right) \quad . \tag{9.30}$$

Substituting (9.30) in (9.29), we find

$$\begin{aligned}\sum_{\ell=0}^{\infty} vF(A_\ell\lambda) &= \frac{1}{u(1-\lambda_1)(1-\lambda_4)} - \frac{1}{u\lambda_3(1-\lambda_4)}\Psi\left(\frac{1+\lambda_2+\lambda_3}{u}\right) \\ &+ \frac{1}{u\lambda_3(1-\lambda_4)}\Psi\left(\frac{1+\lambda_2}{u}\right)\end{aligned} \tag{9.31}$$

Now (9.23) follows directly from (9.31), upon noting

$$\frac{1}{F(\lambda)} = uv(1-\lambda_1)(1-\lambda_4) \quad .$$

This completes the proof of Lemma 9.20.

In what follows we adopt the convention that if $A, B, \ldots$ are nonnegative nonsingular $m \times m$ matrices, the column sums will be denoted by corresponding lower case letters, $a_j, b_j, \ldots$, $1 \leq j \leq m$. We recall that if ω is *normalized* Lebesgue measure on Δ_{m-1}, then if $\det A = \pm 1$,

$$\omega(\Delta^A) = \prod_{j=1}^{m} \frac{1}{a_j} \quad . \tag{9.32}$$

If G is the set (9.2), define ${}^G\Delta = (\Delta^G)^c$. The consequence of $\omega(\Gamma(G)) = 0$ corresponding to (8.7) for $\mathcal{H} = G^c$ is the relation

$$\Delta_3 = \bigcup_{n=0}^{\infty} \bigcup_{A \in G^{(n)}} T_A\, {}^G\Delta \qquad (\text{mod } \omega) \quad . \tag{9.33}$$

9.34. Lemma. Let B be a nonnegative 4×4 matrix with $\det B = \pm 1$. Then

$$\omega(T_B{}^G\Delta) = \frac{1}{b_2 b_3 (b_1+b_4)} \left(\frac{1}{b_1+b_3} + \frac{1}{b_2+b_4}\right) \quad . \tag{9.35}$$

Proof. We first compute $\omega(T_B\Delta^G)$ based upon symmetry and the fact

$$\sum_{\ell=0}^{\infty} \omega(T_B\Delta^{A_\ell}) = \sum_{\ell=0}^{\infty} \frac{1}{(b_1+\ell b_2)(b_1+(\ell+1)b_2)(b_1+b_3)(b_1+b_4)}$$

$$= \frac{1}{b_1 b_2 (b_1+b_3)(b_2+b_4)} \quad .$$

We find

$$\begin{aligned} \omega(T_B\Delta^G) &= \frac{1}{(b_1+b_4)} \left\{ \frac{1}{b_1 b_2 (b_1+b_3)} + \frac{1}{b_3 b_4 (b_2+b_4)} \right\} \\ &= \frac{1}{b_2 b_3 (b_1+b_4)} \left\{ \frac{b_3}{b_1 (b_1+b_3)} + \frac{b_2}{b_4 (b_2+b_4)} \right\} \\ &= \frac{1}{b_2 b_3 (b_1+b_4)} \left\{ \frac{1}{b_1} - \frac{1}{b_1+b_3} + \frac{1}{b_4} - \frac{1}{b_2+b_4} \right\} \\ &= \frac{1}{b_1 b_2 b_3 b_4} - \frac{1}{b_2 b_3 (b_1+b_4)} \left(\frac{1}{b_1+b_3} + \frac{1}{b_2+b_4}\right) \quad . \end{aligned} \tag{9.36}$$

Now (9.35) follows from (9.36) and (9.32). The lemma is proved.

We introduce a new collection of matrices, C_j and C_j^θ, $j \geq 1$, where

$$C_j = \begin{pmatrix} 1 & 1 & 1 & 1 \\ 0 & j & j-1 & 0 \\ 1 & 2 & 2 & 0 \\ 0 & 0 & 0 & 1 \end{pmatrix}$$

One has $\Delta^{C_j} \subseteq {}^G\Delta$, $\Delta^{C_j^\theta} \subseteq {}^G\Delta$, $j \geq 1$. This follows, for example, upon noting that in (3.23)

$$N\left(\begin{pmatrix} 1 & j \\ 0 & 1 \end{pmatrix}\right) = C_j \qquad (j \geq 1)$$

so that C_j, $C_j^\theta \in F_2(4)$, $j \geq 0$.

Let B be as in Lemma 9.34. A direct computation reveals

$$\omega(T_B \Delta^{C_j}) = \omega(\Delta^{BC_j}) = \frac{1}{(b_1+b_3)(b_1+jb_2+2b_3)(b_1+(j-1)b_2+2b_3)(b_1+b_4)}$$

and summing over j,

$$(9.37) \quad \sum_{j=1}^{\infty} \omega(\Delta^{BC_j}) = \frac{1}{b_2(b_1+2b_3)(b_1+b_3)(b_1+b_4)} .$$

In what follows B is a product of elements of $F_2(4)$, and therefore by Lemmas 3.27 and 3.31 we have the inequalities

$$(9.38) \quad \begin{array}{ll} b_2 \geq b_1 & b_3 \geq b_4 \\ b_1+b_3 \geq b_2 & b_2+b_4 \geq b_3 \end{array}$$

These imply
$$b_1+b_3 = \frac{b_1+b_3}{2} + \frac{b_1+b_3}{2} \geq \frac{b_2+b_4}{2}$$

We use this in dividing (9.37) by (9.35) to find

$$(9.39) \quad \frac{\omega\left(\bigcup_{j=1}^{\infty} T_B \Delta^{C_j}\right)}{\omega(T_B {}^G\Delta)} = \frac{b_2 b_3 (b_1+b_4)(b_1+b_3)(b_2+b_4)}{(b_1+b_2+b_3+b_4) b_2 (b_1+2b_3)(b_1+b_3)(b_1+b_4)}$$

$$= \frac{b_3(b_2+b_4)}{(b_1+2b_3)(b_1+b_2+b_3+b_4)}$$

$$\leq \tfrac{1}{2}\,\frac{1}{3/2} = \frac{1}{3} \quad .$$

Now set $\mathcal{C} = \{C \mid C = C_j \text{ or } C_j^\theta,\ j \geq 1\}$. Arguing by symmetry from (9.39), we obtain

$$\frac{\omega(T_B\Delta^{\mathcal{C}})}{\omega(T_B{}^{\mathcal{G}}\Delta)} \leq \frac{2}{3} \quad . \tag{9.40}$$

9.41. Lemma. With notations as above, let $\mathcal{G}_1 = \mathcal{G} \cup \mathcal{C}$. Then $\omega(\Gamma(\mathcal{G}_1)) = 0$.

Proof. We already know $\omega(\Gamma(\mathcal{G})) = 0$. Setting

$\mathcal{C}_1 = \bigcup_{n=0}^{\infty} \mathcal{G}^{(n)}\mathcal{C}$, it will therefore suffice to prove $\omega(\Gamma(\mathcal{C}_1)) =$

We use (9.33) and (9.40) to find that

$$\omega(\Delta^{\mathcal{C}_1^{(n+1)}}) = \sum_{B\in\mathcal{C}_1^{(n)}} \sum_{k=0}^{\infty} \sum_{C\in\mathcal{G}^{(k)}} \sum_{D\in\mathcal{C}} \omega(T_BT_C\Delta^D)$$

$$= \sum_{B\in\mathcal{C}_1^{(n)}} \sum_{k=0}^{\infty} \sum_{C\in\mathcal{G}^{(k)}} \omega(T_BT_C\Delta^{\mathcal{C}})$$

$$\leq \frac{2}{3} \sum_{B\in\mathcal{C}_1^{(n)}} \sum_{k=0}^{\infty} \sum_{C\in\mathcal{G}^{(k)}} \omega(T_BT_C{}^{\mathcal{G}}\Delta)$$

$$= \frac{2}{3} \sum_{B\in\mathcal{C}_1^{(n)}} \omega(T_B\Delta_3)$$

$$= \frac{2}{3}\,\omega(\Delta^{\mathcal{C}_1^{(n)}}) \quad .$$

10. Analysis of u_j, $1 \leq j \leq 3$. We have come finally to the proofs of the main results of this paper, Theorems 3.20 and 4.7. Both theorems will be consequences of a detaile study of u_2, which follows, together with the relationship between u_2 and u_1, u_3.

Let $G \subseteq SL(2,\mathbb{Z})$ be the set, defined in Section 3, in terms of which $F_2(4)$ is parametrized in (3.23). We define G_0 to be what remains of G after the matrices $g = \begin{pmatrix} \ell & 1 \\ -1 & 0 \end{pmatrix}$, $\ell \geq 1$, and $g = \begin{pmatrix} 1 & j \\ 0 & 1 \end{pmatrix}$, $j \geq 1$, have been discarded. We note that each $g \in G_0$ has *positive* entries. Let $\mathcal{H} \subseteq F_2(4)$ be the set corresponding to G_0 under the identification (3.23)

$$\mathcal{H} = \{A \mid A = N(g) \text{ or } N(g)^{\theta},\ g \in G_0\}$$

10.1. Lemma. With notations as above $\mathcal{H}$ is a cap for $F_2(4)$.

Proof. By definition, $\mathcal{H}^c = G_1$, where G_1 is defined in the statement of Lemma 9.41. That lemma asserts $\omega(\Gamma(G_1)) = 0$, which is to say $\mathcal{H}$ is a cap for $F_2(4)$, as claimed.

Define 2×2 matrices, σ and τ, by $\sigma = \begin{pmatrix} 1 & 1 \\ 0 & 1 \end{pmatrix}$ and $\tau = \begin{pmatrix} 1 & 0 \\ 1 & 1 \end{pmatrix}$. If $g \in SL(2,\mathbb{Z})$ has nonnegative entries, then as was noted in [15], there exist unique $n \geq 0$ and $a_0, a_1, \ldots, a_n \in$ $a_1, \ldots, a_n \geq 1$ such that

$$g = \sigma^{a_0} \tau^{a_1} \sigma^{a_2} \cdots \eta^{a_n} \tag{10.2}$$

where $\eta = \sigma$ or τ according to whether n is even or odd. When g has positive entries, one has $n \geq 1$; and if either $a_0 = 0$ or n is even, $n \geq 2$.

To distinguish the last exponent in (10.2) we define $\rho(g) = a_n$. If $g \in G_0$, $g = \begin{pmatrix} a & b \\ c & d \end{pmatrix}$ (all positive), the necessary and sufficient condition for $\rho(g) = 1$ is the set of inequali ties

$$\tfrac{1}{2}b \le a \le 2b$$

(10.3)

$$\tfrac{1}{2}d \le c \le 2d$$

Therefore, if $G_0' = \{g \in G_0 | \rho(g) = 1\}$, each element of G_0' satisfies (10.3).

Let $\mathcal{H}_0 \subseteq \mathcal{H}$ be the set corresponding to G_0' in (3.23),

$$\mathcal{H}_0 = \{A | A = N(g) \quad \text{or} \quad A = N(g)^\theta,\ g \in G_0'\} \quad .$$

If $g \in G_0'$ is as in (10.2), let $\alpha = N(g)$ (resp. $\alpha = N(g)^\theta) \in \mathcal{H}_0$, and define

(10.4) $$H_{\alpha j} = N(g\eta^j) \quad (\text{resp. } N(g\eta^j)^\theta) \qquad (j \ge 0)$$

Using (10.2) any $g \in G_0$ has the form $g = g'\eta^j$ ($j = \rho(g)-1$), and so (10.4) provides a biunique correspondence between $\mathcal{H}$ and $\mathcal{H}_0 \times \mathbb{Z}^+$. Of course, $H_{\alpha 0} = \alpha$, $\alpha \in \mathcal{H}_0$, by definition.

Let B be a nonnegative, nonsingular 4×4 matrix, and denote the columns and column sums of B by B_j and b_j, respectively, $1 \le j \le 4$. (As before, a corresponding lettering is adopted for matrices $A, C, \ldots$.)

If $g \in G_0$, $g = \binom{a\ b}{c\ d}$, the columns of $C = BN(g)$ are of the form

$$C_1 = B_1 + (a-1)B_2 + (c+1)B_3$$

$$C_2 = B_1 + (a+b-1)B_2 + (c+d+1)B_3$$

(10.5)

$$C_3 = B_1 + (b-1)B_2 + (d+1)B_3$$

$$C_4 = B_1 + B_4 \quad .$$

If inequalities between vectors are taken to mean componentwis inequalities, then from (10.5) we have

(10.6) $$C_2 \geq C_3 \quad .$$

If we assume now that $g \in G_0'$, so that the inequalities (10.3) hold, then for the column sums we have

(10.7)
$$\begin{aligned} c_2 &\geq a\ b_2 + c\ b_3 \\ &\geq \tfrac{1}{4}\ ((a+b)b_2 + (c+d)b_3) \quad . \end{aligned}$$

If in addition B is a product of elements of $F_2(4)$, then (3.32) tells us $b_2 \geq b_1$, and by (10.5) the last term in (10.7) exceeds $\frac{1}{8}\ c_2$. That is,

(10.8) $$c_2 \geq a\ b_2 + c\ b_3 \geq \frac{1}{8}\ c_2$$

By the same reasoning we have also

(10.9) $$c_2 \geq b\ b_2 + d\ b_3 \geq \frac{1}{8}\ c_2$$

Remark. Using the relation $B^\theta N(g) = (BN(g)^\theta)^\theta$, one could apply (10.6) and (10.8) - (10.9) to B^θ and $N(g)$ to obtain a set of inequalities for $BN(g)^\theta$. But the symmetry of our situation is obvious, and we shall not complicate the discussio by doing so.

Let $g \in G_0'$ be given by (10.2), and define $C^{(j)} = N(g\eta^j)$, $j \geq 0$. In the notation of (10.4), $C^{(j)} = H_{Cj}$. If n is even in (10.2), $C^{(j)}$ has columns

(10.10)
$$\begin{aligned} C_1^{(j)} &= C_1 \\ C_2^{(j)} &= C_2 + jaB_2 + jcB_3 \\ C_3^{(j)} &= C_3 + jaB_2 + jcB_3 \\ C_4^{(j)} &= C_4 \end{aligned}$$

When n is odd in (10.2), $jbB_2 + jdB_3$ is added to C_1 and C_2 while C_3 and C_4 are unchanged. The analysis below is otherwise similar, and so we shall treat only the case that n is even and $\eta = \sigma$ above. Using (10.10), (10.8) and (10.6), the column sums satisfy

$$c_2^{(j)} \geq \frac{8+j}{8} c_2$$

$$c_3^{(j)} \geq \frac{j}{8} c_2 \geq \frac{j}{8} c_3 \quad .$$

Therefore,

$$\begin{aligned} \omega(\Delta^{C^{(j)}}) &= \frac{1}{c_1^{(j)} c_2^{(j)} c_3^{(j)} c_4^{(j)}} \\ &\leq \frac{64}{(8+j)j} \frac{1}{c_1 c_2 c_3 c_4} \\ &= \frac{64}{(8+j)j} \omega(\Delta^C) \quad . \end{aligned}$$

Therefore if $T > 0$, one has

$$\sum_{j \geq T} \omega(\Delta^{C^{(j)}}) \leq \frac{64}{T} \omega(\Delta^C) \quad . \tag{10.11}$$

Therefore, the "stratification" $\mathcal{H} \cong \mathcal{H}_0 \times \mathbb{Z}^+$ satisfies the inequality (8.10) with $K = 64$.

If A is a nonnegative 4×4 matrix, define $\Psi_1(A)$ by (interpreting $\frac{0}{0} = 1$)

$$\Psi_1(A) = \max_{\substack{1 \leq j \leq 4 \\ 1 \leq j, j' \leq 3}} \frac{A_{ij}}{A_{ij'}} \quad . \tag{10.12}$$

In terms of Ψ_1, define $\Psi(A) = \min(\Psi_1(A), \Psi_1(A^\theta))$ and

$D(\Psi) = \{A \mid \Psi(A) < \infty\}$.

Remark. $\Psi_1(A^\theta)$ can be defined by (10.12) if $1 \leq j, j' \leq 3$ is replaced by $2 \leq j, j' \leq 4$. If $\nu(A)$ is defined by (7.3), then clearly $\max(\Psi_1(A), \Psi_1(A^\theta)) \leq \nu(A) \leq \Psi_1(A)\Psi_1(A^\theta)$.

10.13. Lemma. Let $g \in G'_0$ be as above with n even, and let B be a product of elements of $F_2(4)$ having positive entries and such that $B_{22}, B_{33} \geq 2$. Then

$$\Psi_1(BN(g\sigma^j)) \leq 8(j+1) \qquad (j \geq 0) \tag{10.14}$$

Proof. The assumption on B and the inequalities (3.28) (3.29) combine to imply

$$\left.\begin{aligned} B_{i1} + B_{i3} &\geq \tfrac{1}{2} B_{i2} \\ B_{i2} + B_{i4} &\geq \tfrac{1}{2} B_{i3} \end{aligned}\right\} \qquad (1 \leq i \leq 4) \tag{10.15}$$

Relations (10.15), (10.10) and (10.5) imply

$$\begin{aligned} C_1^{(j)} &= C_1 \\ &\geq (a - \tfrac{1}{2})B_2 + c\, B_3 \\ &\geq \tfrac{1}{2}(a\, B_2 + c\, B_3) \end{aligned} \tag{10.16}$$

Using the same inequalities with (10.3) yields

$$\begin{aligned} C_3 &= B_1 + (b-1)B_2 + (d+1)B_3 \\ &\geq \tfrac{1}{2} B_1 + (b-\tfrac{3}{4})B_2 + (d + \tfrac{1}{2})B_3 \\ &\geq \tfrac{1}{2} B_1 + \tfrac{a}{4} B_2 + \tfrac{c+1}{2} B_3 \end{aligned} \tag{10.17}$$

$$\geq \tfrac{1}{4}\,(B_1+(a-1)B_2 + (c+1)B_3)$$

$$= \tfrac{1}{4}\,C_1$$

As $C_3^{(j)} \geq C_3$, (10.17) implies

(10.18) $$C_3^{(j)} \geq \frac{1}{8}\,C_1^{(j)} \qquad (j \geq 0) \quad .$$

Because B is assumed to have positive entries, we have $B_2 \geq B_1$, $B_3 \geq B_4$ from (3.33). From (10.10), (10.5), (10.3) and (10.16) follow

$$C_2^{(j)} \leq (a+b)B_2 + (c+d+1)B_3 + jaB_2 + jcB_3$$

(10.19) $$\leq (3+j)aB_2 + (4+j)cB_3$$

$$\leq 2(4+j)C_1^{(j)} \leq 8(j+1)C_1^{(j)} \quad .$$

Collecting results (10.18), (10.19), and the obvious $C_2^{(j)} \geq C_3^{(j)}$ imply $\Psi_1(C^{(j)}) \leq 8(j+1)$, and (10.11) is established.

Remark. If B and g are as in the lemma, except that n is *odd*, then entirely analogous reasoning establishes

(10.14′) $$\Psi_1(BN(g\tau^j)) \leq 8(j+1) \quad .$$

Combining (10.14) - (10.14′) with the remark made earlier regarding $BN(g\eta^j)^\theta$, we have

10.20. Lemma. Let B be a product of elements of $\mathcal{F}_2(4)$ such that $B_{ij} \geq 2$ for all i,j. Then we have the inequalities

(10.21) $$\Psi(BH_{\alpha,j}) \leq 8(j+1) \qquad (\alpha \in \mathcal{H}_0,\ j \geq 0) \quad .$$

If $\lambda \in \Delta_3$ is such that (λ,θ_4) is a minimal interval exchange, and if $\lambda \sim [A_1(\lambda), A_2(\lambda), \ldots]$ is the expansion of λ relative to $\mathcal{F}_2(4)$, then arguing as in [15], if $A^{(n)}(\lambda) = A_1(\lambda)\ldots A_n(\lambda)$,

(10.22) $$\lim_{n\to\infty} A_{ij}^{(n)}(\lambda) = \infty \qquad (1 \leq i, j \leq 4)$$

This is a reflection of the fact each measure in $\vartheta(\lambda, \theta_4)$ is, because (λ, θ_4) is minimal, positive on nonempty open sets, in particular on I_j^λ, $1 \leq j \leq 4$. In the language of partitions (10.22) implies that $\mathcal{F}_2(4)$ admits a subpartition, $\mathcal{F}'$, such that each $B \in \mathcal{F}'$ satisfies the hypotheses of Lemma 10.20. From (10.21) and (10.11) we conclude

10.23. Lemma. With notations as above $\mathcal{F}_2(4)$ admits an $(\mathcal{F}', \Psi)$-stratified cap (namely $\mathcal{H}$).

We are now in a position to apply Lemma 8.11 to obtain

10.24. Lemma. $\mathcal{F}_2(4)$ admits a subpartition, $\mathcal{F}''$, such that

(10.25) $$\| \mathcal{F}'' \|_{\Psi} < \infty$$

where $\Psi(A) = \min(\Psi_1(A), \Psi_1(A^\theta))$ and Ψ_1 is defined by (10.12).

In what follows we reserve a special notation for two elements of $\mathcal{F}_2(4)$, γ and δ, defined by

$$\gamma = N\left(\begin{pmatrix} 1 & 1 \\ -1 & 0 \end{pmatrix}\right)^\theta = \begin{pmatrix} 1 & 0 & 0 & 0 \\ 0 & 1 & 0 & 0 \\ 0 & 0 & 1 & 0 \\ 1 & 1 & 1 & 1 \end{pmatrix}$$

and

$$\delta = N\left(\begin{pmatrix} 2 & 1 \\ -1 & 0 \end{pmatrix}\right)^\theta = \begin{pmatrix} 1 & 0 & 0 & 0 \\ 0 & 1 & 0 & 0 \\ 0 & 0 & 2 & 1 \\ 1 & 1 & 1 & 1 \end{pmatrix}$$

The reader may check that the inequalities

$$\Psi_1(A\gamma^{\ell}) \leq \Psi_1(A) \qquad (\ell \geq 0)$$

(10.26)

$$\nu(A\gamma^{\ell}\delta) \leq 2\Psi_1(A) \qquad (\ell \geq 0)$$

hold for 4×4 nonnegative matrix A, where ν is defined by (7.3). In what follows we set

$$X = (\Delta^{\gamma})^{c} \subseteq \Delta_3 \qquad .$$

10.27. Lemma. Let A be a product of elements of $F_2(4)$, and otherwise let the notations be as above. Then

$$\frac{\omega(T_A\Delta^{\delta})}{\omega(T_A X)} \geq \frac{1}{6}\,\frac{1}{(\Psi_1(A)+1)^3} \qquad . \tag{10.28}$$

Proof. As before a_j, $1 \leq j \leq 4$, are the column sums of A. From $\omega(\Delta^A) = \dfrac{1}{a_1a_2a_3a_4}$ and

$$\omega(\Delta^{A\gamma}) = \frac{1}{(a_1+a_4)(a_2+a_4)(a_3+a_4)a_4}$$

we obtain, after a simple calculation,

$$\omega(T_A X) = (1 - \frac{a_1}{a_1+a_4}\,\frac{a_2}{a_2+a_4}\,\frac{a_3}{a_3+a_4})\;\omega(\Delta^A) \qquad . \tag{10.29}$$

Also, we have

$$\omega(T_A\Delta^{\delta}) = \frac{a_1}{a_1+a_4}\,\frac{a_2}{a_2+a_4}\,\frac{a_3}{2a_3+a_4}\,\frac{a_4}{a_3+a_4}\;\omega(\Delta^A) \qquad . \tag{10.30}$$

Now define $0 < x_1, x_2, x_3 < 1$ by

$$x_1 = \frac{a_1}{a_1+a_4} \qquad x_2 = \frac{a_2}{a_2+a_4} \qquad x_3 = \frac{a_3}{a_3+a_4} \quad .$$

In terms of these quantities (10.29) and (10.30) combine to yield

$$\frac{\omega(T_A\Delta^\delta)}{\omega(T_A X)} = \frac{x_1 x_2 (1-x_3)}{1-x_1x_2x_3} \frac{a_3}{2a_3+a_4}$$

which, because $a_3 \geq a_4$, is

$$\geq \frac{1}{3} \frac{x_1x_2(1-x_3)}{1-x_1x_2x_3} \quad .$$

Now also, $1-x_1x_2x_3 \leq (1-x_1) + (1-x_2) + (1-x_3)$, and so we have

(10.31) $$\frac{\omega(T_A\Delta^\delta)}{\omega(T_A X)} \geq \frac{1}{3} \frac{x_1x_2(1-x_3)}{(1-x_1)+(1-x_2)+(1-x_3)} \quad .$$

Now set $T = \Psi_1(A) \quad (< \infty)$. Because $\frac{a_4}{a_1} \leq T\frac{a_4}{a_3}$, we have

(10.32) $$x_1 = \frac{1}{1+\frac{a_4}{a_1}} \geq \frac{1}{1+T\frac{a_4}{a_3}}$$

and

$$1-x_1 \le \frac{T\dfrac{a_4}{a_3}}{1+T\dfrac{a_4}{a_3}}$$

$$\le T\frac{a_4}{a_3+a_4} = T(1-x_3) \quad .$$

Similarly, one finds $1-x_2 \le T(1-x_3)$. Using the fact $a_3 \ge a_4$, (10.32) also implies $x_1 \ge \frac{1}{T+1}$, and similarly, $x_2 \ge \frac{1}{T+1}$. We now have from (10.31)

$$\frac{\omega(T_A\Delta^\delta)}{\omega(T_AX)} \ge \frac{1}{3}\frac{1}{(T+1)^2}\frac{1}{(2T+1)}$$

$$\ge \frac{1}{6}\frac{1}{(T+1)^3} \quad .$$

As $T = \Psi_1(A)$, (10.28) is established.

Lemmas 10.27 and 10.24 will be used now to construct a uniform subpartition of $\mathcal{F}_2(4)$. To begin we introduce two auxiliary partitions $\mathcal{A}$ and $\mathcal{A}^\theta$, by

$$\mathcal{A} = \bigcup_{\ell=0}^{\infty} \gamma^\ell\{\gamma\}^c$$

where $\{\gamma\}^c$ is relative to $\mathcal{F}_2(4)$, and

$$\mathcal{A}^\theta = \{A^\theta \mid A \in \mathcal{A}\} \quad .$$

Decompose $\mathcal{A}$ into subsets, $\mathcal{G}$ and $\mathcal{H}$, where

$$G = \bigcup_{\ell=0}^{\infty} \{\gamma^{\ell}\delta\}$$

$$\mathcal{H} = \bigcup_{\ell=0}^{\infty} \gamma^{\ell}\{\gamma,\delta\}^{c}$$

An analogous decomposition is defined for A^{θ}, $A^{\theta} = G^{\theta} \cup \mathcal{H}^{\theta}$.

In what follows $\mathcal{E}$ is a fixed subpartition of $F_2(4)$ such that $\|\mathcal{E}\|_{\Psi} < \infty$. $\mathcal{E}$ exists by Lemma 10.24, and we set $T = \|\mathcal{E}\|_{\Psi}$. By definition of Ψ, $\mathcal{E}$ divides into sets C and D such that $\|C\|_{\Psi_1} \le T$ and $\|D^{\theta}\|_{\Psi_1} \le T$. We write $C = C(\mathcal{E})$, $D = D(\mathcal{E})$, although a choice is involved in the constructions of C and D.

Suppose now that $\pounds$ is a partition or a subset of a partition and that $\|\pounds\|_{\Psi} \le T$. As above we associate $C(\pounds)$ and $D(\pounds)$ to $\pounds$. Now define new sets, $M_C(\pounds)$, $M_D(\pounds)$, $N_C(\pounds)$, $N_D(\pounds)$ by

$$M_C(\pounds) = C(\pounds)G$$

$$M_D(\pounds) = D(\pounds)G^{\theta}$$

$$N_C(\pounds) = C(\pounds)\mathcal{H}$$

$$N_D(\pounds) = D(\pounds)\mathcal{H}^{\theta}$$

By (10.26) we have

$$\|M_C(\pounds)\|_{\nu} \le 2T$$

(10.33)

$$\|M_D(\pounds)\|_{\nu} \le 2T$$

Setting $\beta = \frac{1}{6}\frac{1}{(T+1)^3}$, we also obtain from Lemma 10.27 that

$$\omega(\Delta^{N_C(\mathcal{L})}) \leq (1-\beta)\,\omega(\Delta^{C(\mathcal{L})})$$

$$\omega(\Delta^{N_D(\mathcal{L})}) \leq (1-\beta)\,\omega(\Delta^{D(\mathcal{L})})$$

Now write

$$F(\mathcal{L}) = M_C(\mathcal{L}) \cup M_D(\mathcal{L})$$

$$\mathcal{H}(\mathcal{L}) = N_C(\mathcal{L}) \cup N_D(\mathcal{L})$$

From the above discussion we have

$$\| F(\mathcal{L}) \|_\nu \leq 2T$$

(10.34)

$$\omega(\Delta^{\mathcal{H}(\mathcal{L})}) \leq (1-\beta)\,\omega(\Delta^{\mathcal{L}}) \quad .$$

10.35 Lemma. There exists a sequence of pairs $(F_n, \mathcal{H}_n)$ $n = 1,2,\ldots,$ such that

a) $F_1 \cup \ldots \cup F_n \,\dot{\cup}\, K_n$ is a subpartition of $F_2(4)$, $n \geq 1$.

b) $\| \bigcup_{j=1}^{n} F_j \|_\nu \leq 2T$, $n \geq 1$.

c) $\omega(\Delta^{K_n}) \leq (1-\beta)^n$, $n \geq 1$, where $\beta = \frac{1}{6}\,\frac{1}{(T+1)^3}$.

Proof. To begin, define $F_1 = F(\mathcal{E})$, $K_1 = K(\mathcal{E})$. Properties b) and c) are true by (10.34), while a) is obvious. Suppose now the sequence has been constructed for all integers $n \leq N$. Define $F_{N+1} = F(K_N)$, $K_{N+1} = K(K_N)$. That a) - c) remain true for $n = N+1$ follows from construction, the induction hypothesis, and (10.34).

If F_n, $n = 1,2,\ldots$ are as in the lemma just proved,

define $F = \bigcup_{n=1}^{\infty} F_n$. F is clearly a uniform subpartition of

$F_2(4)$, and we have proved

10.36 Lemma $F_2(4)$ admits a uniform subpartition, F.

Proposition 7.15 now implies that u_2 $(= S_{\Omega(F_2(4))})$ is

ergodic relative to ω, and then by Proposition 7.16 u_2 admit
a σ-finite invariant Borel measure equivalent to Lebesgue
measure.

By Proposition 8.6 almost every $\lambda \in \Delta_3$ is normal in the

sense that its expansion relative to $F_2(4)$
contains every finite block of elements infinitely often.
By Lemma 3.17 $F_1(4)$ is a subpartition of $F_2(4)$, and therefor
ω-a.e. $\lambda \in \Delta_3$ is normal relative to $F_1(4)$. Applying Propositi
8.4 once more, u_1 is ergodic, and finally by Proposition 7.16
u_1 admits a σ-finite invariant Borel measure equivalent to
ω.

Consider finally u_3. In terms of the Rauzy diagram
(4.6), u_2 is the induced transformation of u_3 on the set
$\Delta_3 \times \{\theta_4\} \subseteq \Delta_3 \times \mathcal{R}[\theta_4]$ (Proposition 4.11). As every u_3
orbits meets $\Delta_3 \times \mathcal{R}[\theta_4]$, ergodicity of u_2 implies u_3 is
ergodic. Proposition 6.29 now implies u_3 admits a σ-finite
invariant Borel measure equivalent to Lebesgue measure
($\omega \times$ counting measure) on $\Delta_3 \times \mathcal{R}[\theta_4]$. Theorems 3.20 and 4.7
are thereby proved.

Rice University
November, 1979

Note: (September, 1980) Conjecture 1.2 has been verified independently by H. Masur ("Interval exchange transformations and measured foliations") and the author "Gauss measures for transformations on the space of interval exchange maps." In the latter work one proves existence of invariant measure for the transformation u_3 of (4.5) for any m and Rauzy class. For example, in the case of u_2 (Section 3) the density of the invariant measure on Δ_3 is

$$f(\lambda) = \frac{1}{(\lambda_1+\lambda_2)}\frac{1}{(\lambda_2+\lambda_3)}\frac{1}{(\lambda_3+\lambda_4)}\left(\frac{1}{1-\lambda_1}+\frac{1}{1-\lambda_4}\right)$$

although we have no *direct* verification. In the case of u_3 one has:

Theorem. *Let* $m > 1$, *and let* $\mathcal{R} \subseteq \mathfrak{S}_m^0$ *be a fixed "Rauzy class." Then* u_3 *is conservative and ergodic on* $\Delta_{m-1} \times \mathcal{R}$ *and* u_3 *admits an absolutely continuous invariant measure,* μ, *unique in its measure class up to a scalar multiple. Moreover, the density of* μ *is the restriction to* $\Delta_{m-1} \times \mathcal{R}$ *of a function which is positive, rational, and homogeneous of degree*-m on $\Lambda_m \times \mathcal{R}$.

BIBLIOGRAPHY

[1] J. R. Blum and D. L. Hanson, "On invariant probability measures," Pac. J. Math., 10(1960), 1125-1129.

[2] A. Hajian and S. Kakutani, "Weakly wandering sets and invariant measures," Trans. Amer. Math. Soc., 110_1964), 136-151.

[3] A. B. Katok, "Interval exchange transformations and some special flows are not mixing," preprint, University of Maryland.

[4] _________, "Invariant measures of flows on oriented surfaces," Dokl. Nauk. SSR, 211(1973) = Sov. Math. Dokl. 14(1973), 1104-1108.

[5] _________, and A. M. Stepin, "Approximations in ergodic theory," Uspehi Mat. Nauk. 22(1967), no. 5 (137), 81-106 = Russian Math. Surveys 22(1967), no. 5, 77-102.

[6] M. Keane, "Interval exchange transformations," Math. Z. 141 (1975), 25-31.

[7] _________, "Non-ergodic interval exchange transformations," Israel J. Math. 26(1977), 188-196.

[8] _________, and G. Rauzy, "Unique Ergodicité des Echanges d'Intervalles," á paraitre, Math. Z.

[9] H. Keynes and D. Newton, "A minimal non-uniquely ergodic interval exchange transformation," Math Z., 148(1976), 101-105.

[10] G. Rauzy, "Echanges d'intervalles et Transformations Induites," Acta Arithmetica, 34(1979), 315-328.

[11] _________, "Une généralisation du développement en fraction continue," Séminaire de theorie des nombres-Année 1976-1977, Paris.

[12] V. A. Rohlin, "Exact endomorphisms of a Lebesgue space," Izvestia Acad. of Sciences, USSR, ser. Math. 1961, v 25., n. 4, 499-530; English translation, Amer. Math. Soc. Trans. (2) 39(1964), 1-36.

[13] _________, "On the fundamental ideas of measure theory," Mat. Sb. 25, 67(1949) 107-150; English translation Amer. Math. Soc. (1) 10(1962), 1-54.

[14] Ya. G. Sinai, *Introduction to Ergodic Theory*, Princeton Lecture Notes Series, Princeton University Press, 1977.

[15] W. A. Veech, "Interval exchange transformations," Journal D'Analyse Mathématique, 33(1978), 222-272.

[16] _________, "Quasiminimal invariants for foliations of orientable closed surfaces," preprint, Rice University.

[17] __________, "Strict ergodicity in zero dimensional dynamical systems and the Kronecker-Weyl theorem mod 2," Trans. Amer. Math. Soc. 140(1969),1-33.

[18] __________, "Topological dynamics," Bull. Amer. Math. Soc. 83(1977), 775-830.

[19] A. N. Zemlyakov and A. B. Katok, "Topological transitivity of billiards in polygons," Mat. Zametki 18(1975), 291-300. Translation: Math. Notes of the Academy of Sciences of the USSR, 18N2 (1975), 760-764.. Errata v20N6 (1976), 1051.

WHEN ALL POINTS ARE RECURRENT/GENERIC*

Yitzhak Katznelson
and
Benjamin Weiss

Abstract:

Several aspects of the title are explained. In particular an example is constructed in which there is a unique minimal set, all points are generic and there is a continuous arc of ergodic measures.

§1. Introduction

While preparing a talk on some recent results in recurrence in topological dynamics and ergodic theory the second author raised the naive question of what happens when one assumes that all points in a system are recurrent. It turns out that this sometimes implies that all orbit closures are minimal, but that in general all points can be recurrent without the minimality necessarily following. J. Auslander pointed out to us the old survey by Nemitsky [N] which has essentially the same example that is described below in §2, and also told us of the work of Y. N. Dowker and Lederer [DL]. They had investigated the question of what happens when one assumes that all points of a dynamical system are generic. It turns out that this property together with minimality implies unique ergodicity. Weakening minimality to the assumption that there be only one minimal set they were only able to show that if there is more than one ergodic measure there must be infinitely many, without settling the question as to whether or not this possibility can actually arise.

Motivated by the example in §2 we then constructed an example of a system with a unique minimal set and a whole arc of ergodic measures, where all points are generic for ergodic

---

*This research was supported in part by U.S.-ISRAEL BSF Grant No. 1927/79.

measures. This led us to reexamine the proof in [DL] and show that there cannot be an example with only a countable infinity of distinct ergodic measures. In §2 we discuss the recurrence questions and prove that for zero dimensional systems if all points are recurrent then all orbit closures are minimal. On the other hand we reproduce Bebutov's example of a topologically transitive system with all points recurrent that is not minimal. In §3 we refine the arguments of [DL] and show that if all points are generic and there is a unique minimal set then either there is only one ergodic measure or a continuum of ergodic measures. In §4 we construct an example that shows that the second alternative is not vacuous.

§2. Recurrent Points

If T is a homeomorphism of a compact metric space X we say that y is a recurrent point if for some $n_i \uparrow \infty$, $T^{n_i} y \to y$. The system (X,T) is minimal if the only closed T-invariant sets are ϕ and X. For $y \in Y$ the orbit closure of y is the closure of $\{T^n y: n \in \mathbb{Z}\}$ and is denoted by $ORC(y)$. Since orbit closures are closed invariant sets minimality is equivalent to the condition that all orbits are dense

If A is a finite set, say $A = \{1,2,\ldots,a\}$, $Y = A^{\mathbb{Z}}$ is given the product topology and $T: Y \to Y$ is the shift defined by

$$(Ty)(n) = y(n+1), \qquad n \in \mathbb{Z}$$

then any shift invariant subset is called a symbolic dynamical system. We say that a finite block $\beta \in A^k$ occurs in y at place n if $y(n)y(n+1)\cdots y(n+k-1) = \beta$ and β occurs in y if it occurs someplace.

For the shift it is straightforward to check that y is recurrent if any block that occurs in y occurs infinitely often to the right, while $(ORC(y),T)$ is minimal if and only if any block β that occurs in y, occurs at a set of n_i's that have bounded gaps. This last condition can also be expressed as follows: for any k there is some L, such that any block β of length k that occurs at all in y, also occurs in any L-block of y, i.e., $y(n+1)y(n+2)\cdots y(n+L)$.

Proposition 2.1: If every point of $(ORC(y),T)$ is recurrent then $(ORC(y),T)$ is minimal.

Proof: For $(ORC(y),T)$ to fail to be minimal, there must be some finite block β, that occurs in y, but whose occurences don't have bounded gaps. Since y is assumed recurrent, β occurs infinitely often to the right, and thus we have for some sequence $n_i \uparrow \infty$, occurring at n_i, but not at any j, $n_i < j \le n_k+i$. By passing to a subsequence we may suppose that $T^{n_i}y \to z$, and clearly β occurs in z at 0 but at no positive n, and thus $z \in ORC(y)$ is not recurrent. ∎

Essentially the same proof applies whenever (X,T) has a dense orbit and X is zero dimensional. To get an example where all points are recurrent but (X,T) is not minimal we will replace $A^{\mathbb{Z}}$ by $[0,1]^{\mathbb{Z}} = \Omega$. Denote the shift once again by T. We will construct an $\omega \in \Omega$ with the following properties:

(i) For some $p_i \uparrow \infty$, $\sup_n |\omega(n+p_i) - \omega(n)| \to 0$ as $i \to \infty$.

(ii) For any $k,\varepsilon > 0$ there is some L such that for all n, there is some $1 \le i \le L$ such that

$$1 - \varepsilon \le \omega(n+i+j), \qquad 1 \le j \le k.$$

The first property (i) implies that all points in $ORC(\omega)$ are recurrent, indeed even uniformly recurrent in the sense that for all $\varphi \in ORC(\omega)$, $T^{p_i}\varphi \to \varphi$. On the other hand (ii) implies that $\varphi \equiv 1 \in ORC(\omega)$, and thus $(ORC(\omega),T)$ is not minimal. In fact it is easy to see $\varphi \equiv 1$ is the unique minimal set in $ORC(\omega)$.

For fixed q, let $\omega_q(n)$ be defined by

$$\omega_q(i) = i/q, \qquad |i| \le q \quad \text{and for all} \quad i \quad \text{by}$$

$$\omega_q(i+2q) = \omega(i).$$

Clearly for any p, and q,

$$\sup_n |\omega_q(n+p) - \omega_q(n)| \le p/q.$$

Now it is clear that if p_i increase sufficiently rapidly so that $p_i/p_{i+1} \to 0$, $\omega(n) = \sup_i \omega_{p_i}(n)$ satisfies (i) and (ii).

§3. Generic Points

For the sake of completeness we will reproduce some of the arguments of [DL] and give a complete proof of the following stronger version of their theorem.

Theorem 3.1: If (X,T) is a compact dynamical system with a unique minimal set such that every point is generic for some measure then either (X,T) has a unique invariant measure or (X,T) has uncountably many distinct ergodic invariant measures.

Recall that $x \in X$ is generic for a measure μ if

$$(1) \qquad \lim_{N\to\infty} \frac{1}{N} \sum_{1}^{N} f(T^i x) = \int f \, d\mu \qquad \text{all} \quad f \in C(X)$$

where $C(X)$ denotes the space of continuous functions on X The individual ergodic theorem gives that if μ is an ergodic measure, μ-almost every $x \in X$ is generic for μ. For a preliminary orientation we take up a special case where the alternative can be pinpointed.

Proposition 3.2: If (X,T) is minimal and every point is generic for some measure then (X,T) has a unique invariant measure.

Proof: For each $f \in C(X)$, since (1) holds for all x (with $\mu = \mu_x$), denoting the limit by $\hat{f}(x)$ we have that $\hat{f}$ is everywhere the limit of a sequence of continuous functions. Thus by the theorem of Baire $\hat{f}$ has points of continuity. Let x_0 denote a point of continuity, and let x be arbitrary. Since $\hat{f}(x) = \hat{f}(T^j x)$ for all j, and $\{T^j x: j \geq 1\}$ is dense in X by the minimality, we get that $\hat{f}(x) = \hat{f}(x_0)$, thus $\hat{f}$ is a constant. Now, were there more than one invariant measure there would be at least two distinct ergodic measures, μ_1 and μ_2 and some continuous function f with

$$\hat{f}(x_1) = \int f \, d\mu_1 \neq \int f \, d\mu_2 = \hat{f}(x_2)$$

where x_i is generic for μ_i, $i = 1,2$, which cannot happen if $\hat{f}$ is a constant.

An important role will be played by the following elementary lemma which was abstracted by [DL] from the work of Kerekjarto:

Lemma K: Suppose that (X,T) is a compact dynamical system and that A is a closed invariant set. Then for any neighborhood U of A either

(i) there is a closed set $A \subsetneqq K \subset \overline{U}$ such that $TK \subset K$;

or

(ii) there is an open set $A \subset V \subset U$ such that $TB \supset V$.

Proof: Let $K_1 = \bigcap_1^\infty T^{-i}\overline{U}$. Clearly $TK_1 \subset K_1$ and $A \subset K_1$, while $TK_1 \subset \overline{U}$. If $A \neq K_1$ then also $A \neq TK_1$ and $K = TK_1$ satisfies (i). If $A = K_1$ then for some finite n,

$$\bigcap_1^n T^{-i}\overline{U} \subset \overline{U}$$

and

$$V = \bigcap_1^n T^{-i}U$$

satisfies (ii). ∎

The idea of the proof of the theorem is best explained by proving a special case, namely let's rule out the possibility that (X,T) satisfies the hypothesis of the theorem and has precisely <u>two</u> ergodic measures. Denote by $X_1 \subset X$ the unique minimal set and by μ_1 the unique measure supported on X_1. Denote by μ_0 the second ergodic measure, and then by restricting to the support of μ_0 which being invariant must contain X_1 we can assume that X itself is the support of μ_0. Now since μ_0 has generic points which are necessarily dense, the argument that proved Proposition 3.2 gives that

$$\hat{f}(x) = \int f \, d\mu_0$$

for all points of continuity of $\hat{f}$. From this it follows (via Baire's theorem) that if A denotes the closure of $\{x: x$ is generic for $\mu_1\}$ then A is closed and nowhere dense. Now in the basic lemma, applied to any open neighborhood U of A, $\overline{U} \neq X$ the second alternative is ruled out by the existence of an ergodic invariant measure μ_0 with global support - which leaves open only the first alternative, namely that there is some closed set $K \supsetneqq A$, with $TK \subset K$. Now if $x \in K \setminus A$ then x is generic, for a measure ν which is not μ_1, by the definition of A. Also ν is singular with respect to μ_0 since $\mu_0(K) = 0$. Thus the ergodic decomposition of ν involves other ergodic measures and (X,T) couldn't have had only two ergodic measures.

Now to prove the theorem, the argument above will be applied again and again to build a chain of ergodic measures μ_r, r a dyadic rational, such that if we denote by $\mathrm{Supp}(\lambda)$ the closed support of the measure λ,

(a) $\mathrm{Supp}(\mu_r) \subset \mathrm{Supp}(\mu_s)$ for $r > s$

(b) $\mu_s(\mathrm{Supp}(\mu_r)) = 0$ for $s < r$.

Next for any real $t \in (0,1)$, let μ_t be some weak limit of μ_{r_j}'s with $r_j \to t$. From (a) it follows that for $t > r$

(a)' $\mu_t(\mathrm{Supp}(\mu_r)) = 1$

while if we could conclude from (b) that for $t < r$,

(b)' $\mu_t(\mathrm{Supp}(\mu_r)) = 0$

we would be done with the proof. Because, since (a)' and (b)' involve only a countable number of sets (the $\mathrm{Supp}(\mu_r)$, r a dyadic rational), we can find for each t, an ergodic component $\tilde{\mu}_t$ of μ_t that also satisfies (a)' and (b)'. This clearly yields a continuum of distinct ergodic measures. To enable us to obtain (b)', which of course doesn't automatically follow from (b), the construction is complicated by introducing infinitely many continuous functions $f_{i,j}$ for each dyadic rational $i/2^j$, such that $0 \le f_{k,j} \le 1$, and $f_{i,j} \equiv 1$ on $\mathrm{Supp}(\mu_{i/2^j})$ while for all $k/2^\ell < i/2^j$ with $\ell > j$

$$\text{(b)''} \quad \int f_{i,j}\, d\mu_{k/2^\ell} \le \frac{1}{2^j} + \frac{1}{2^{j+1}} + \cdots + \frac{1}{2^\ell} .$$

Now it is easy to check that if μ_t is any weak limit of μ_{r_j} with r_j dyadic rationals converging to t and $r > t$ then (b)' holds. Having outlined the strategy let's first prove a lemma which will do most of the work.

Lemma 3.3: Suppose that (X,T) has a unique minimal set X_1, that all points of X are generic and that ν is an ergodic measure on X. If g_i are continuous functions from X to $[0,1]$, such that $g_i \equiv 1$ on X_1 for all $1 \le i \le \ell$, and

$$\int g_i d\nu \le 1 - c, \qquad c > 0, \qquad 1 \le i \le \ell$$

then for any $\varepsilon > 0$ there is another ergodic measure μ that satisfies:

(i) $\mathrm{Supp}(\mu) \subset \mathrm{Supp}(\nu)$

(ii) $\nu(\mathrm{Supp}(\mu)) = 0$

(iii) $\int g_i \, d\mu \leq \int g_i d\nu + \varepsilon$.

<u>Proof</u>: Restricting to $\mathrm{Supp}(\nu)$ we lose none of the assumptions so that we may assume $X = \mathrm{Supp}(\nu)$. Take $\varepsilon > 0$ and define

---

$$A = \{x \in X: \hat{g}_i(x) > \int g_i d\nu + \varepsilon \text{ for some } 1 \leq i \leq \ell\}$$

where as before $\hat{g}_i(x) = \lim_{N\to\infty} \frac{1}{N} \sum_{j=1}^{N} g_i(T^j x) = \int g_i d\mu_x$ where μ_x always denotes the measure that x is generic for. The argument above shows that if x is a point of continuity of $\hat{g}_i(x)$ then $\hat{g}_i(x) = \int g_i d\nu$, and since by Baire's theorem the points of continuity of $\hat{g}_i(x)$ form a dense G_δ it follows that A is nowhere dense. On the other hand $X_1 \subset A$, and thus A is a nonempty, closed invariant set.

Suppose first for simplicity that $\ell = 1$. Then we would let U be any proper neighborhood of A, and applying Lemma K we would conclude that there exists a closed set K for which $TK \subset K$ and $A \subsetneqq K$. Letting $x \in K \setminus A$ it follows that

(2) $$\int g_1 d\mu_x \leq \int g_1 d\mu + \varepsilon$$

and we conclude that the same inequality holds for some ergodic component μ of μ_x. This satisfies (i)-(iii). With more than one function g_i the proof just given breaks down since we can't conclude from (2) holding for $i = 1,\ldots,\ell$ that it also holds for some ergodic component of μ_x. To get around this difficulty we introduce an auxiliary function $0 \leq f \leq 1$, continuous, identically equal to one on A and such that $\int f \, d\nu \leq \frac{1}{2}$.

Arguing as before with f playing the role of g_1 we can find an ergodic measure such that

$$\int f\, d\mu \leq \frac{3}{4}$$

with $\nu(\mathrm{Supp}(\mu)) = 0$. Since f is identically equal to one on A, μ must have generic points outside of A, say $\mu_y = \mu$, $y \notin A$. Now (iii) follows from the definition of A, while (i) and (ii) are clear. □

To prove the theorem assume that there is more than one ergodic measure, set X_1 equal to the unique minimal set, μ_1 the measure supported on X_1, and restrict to the support of some ergodic measure μ_0, say $X_0 = \mathrm{Supp}(\mu_0)$. Since $\mu_0(X_1) = 0$, there is a continuous function $f_{0,1}$ that is $\equiv 1$ on X_1 and satisfies

$$\int f_{1,0}\, d\mu_0 \leq \frac{1}{2^2}\,.$$

Apply Lemma 3.3 to $\nu = \Delta_0$, $\ell = 1$, $g_1 = f_{1,0}$, $\varepsilon = 1/2^3$ and obtain an ergodic measure, which we denote by $\mu_{1/2}$ such that:

(i) $\mu_{1/2}(X_1) = 0$

(ii) $\int f_{1,0} d\mu_{1/2} \leq 1/2^2 + 1/2^3$

(iii) $\mu_0(\mathrm{Supp}(\mu_{1/2})) = 0.$

Set $X_{1/2} = \mathrm{Supp}(\mu_{1/2})$, and $E_1 = X_1$, $E_{1/2} = X_1 \cup X_{1/2}$.

We proceed to define $\mu_{1/4}$, $\mu_{3/4}$ as follows. Let $f_{2,1}$ be identically 1 on E_1 with

$$\int f_{2,1}\, d\mu_{1/2} \leq 1/2^3$$

$$\int f_{2,1}\, d\mu_0 \leq 1/2^3$$

and let $f_{1,1}$ be identically 1 on $E_{1/2}$ with

$$\int f_{1,1}\, d\mu_0 \leq 1/2^3.$$

For $\mu_{/34}$ apply Lemma 2 with $\nu = \mu_{1/2}$, $\ell = 2$, $g_1 = f_{1,0}$, $g_2 = f_{1,0}$, $\varepsilon = 1/2^4$ to get $\mu = \mu_{3/4}$ satisfying:

(i) $\mu_{/34}(X_1) = 0$

(ii) $\int f_{1,0}\, d\mu_{3/4} \leq 1/2^2 + 1/2^3 + 1/2^4$

$\int f_{2,1}\, d\mu_{3/4} \leq 1/2^3 + 1/2^4$

(iii) $\mathrm{Supp}(\mu_{3/4}) = X_{3/4} \subset X_{1/2}$.

For $\mu_{1/4}$ apply Lemma 2 with $\nu = \mu_0$, $\ell = 3$, $g_1 = f_{1,0}$, $g_2 = f_{2,1}$, $g_3 = f_{0,1}$, $\varepsilon = 1/2^4$ to get $\mu_{1/4}$ satisfying

(i) $\mu_{1/4}(E_{1/2}) = 0$

(ii) $\int f_{1,0}\, d\mu_{3/4} \leq 1/2^2 + 1/2^3 + 1/2^4$

$\int f_{2,1}\, d\mu_{3/4} \leq 1/2^2 + 1/2^4$

$\int f_{1,1}\, d\mu_{3/4} \leq 1/2^3 + 1/2^4$

(iii) $\mathrm{Supp}(\mu_{3/4}) = X_{3/4} \subset X_0$.

Set $E_{3/4} = X_{3/4} \cup X_1$, $E3/4 = E_{1/2} \cup X_{1/4}$. Continuing in this fashion, we get to each dyadic fraction $k/2^n$ an ergodic measure $\mu_{k/2^n}$, and a double sequence of continuous functions $f_{i,j}$ such that: $f_{i,j}$ is identically 1 on the support of $\mu_{k/2^n}$ with $i/2^j < k/2^n$; $\int f_{i,j}\, d\mu_{k/2^n} \leq 1/2^{j+2} + \cdots + 1/2^n$ if k is odd and $k/2^n < i/2^j$; and $\mu_{k/2^n}(\mathrm{Supp}(\mu_{k/2^{n'}})) = 1$, $k/2^n \geq k'/2^{n'}$.

For any $t \in [0,1]$ let μ_t be a weak limit of μ_{r_k} where r_k is some sequence of dyadic rationals converging to t. For any i,j with $1/2^j = r > t$, $\int f_{i,j} d\mu_t \le 1/2^{j+1}$ while $f_{i,j}$ is identically 1 on E_r. It follows that $\mu_t(E_r) = 0$. What is even more obvious from the construction is that $\mu_t(E_r) = 1$, $r < t$. For fixed t, there are a countable number of such conditions - and hence some ergodic $\tilde{\mu}_t$ in the ergodic decomposition of μ_t also satisfies the same conditions. It is manifest that for distinct t's, $\tilde{\mu}_t$ are unequal, hence we have proved the theorem. ∎

§4. The Example

The example mentioned in the introduction will be obtained as the orbit closure of a single element ψ in $\Omega = [0,1]^{\mathbb{Z}}$ with the shift transformation. To motivate the construction let's begin with some lemmas that will be used to prove that ψ has the desired properties.

Lemma 4.1: If for each N and $\varepsilon > 0$ there is an $L = L(N,\varepsilon)$ so that for all ℓ there is some $i \in (\ell,\ell+L)$ with

$$\psi(i+j) \ge 1 - \varepsilon, \qquad 0 \le j \le N \tag{1}$$

then $ORC(\psi)$ has a unique minimal set, the fixed point x_1 with $x_1(n) \equiv 1$ all $n \in \mathbb{Z}$.

Proof: By the nature of the hypothesis, any $x \in ORC(\psi)$ also satisfies the same hypothesis, so that each such x contains x_1 in its orbit closure. This implies that x_1 lies inside any closed invariant set in $ORC(\psi)$ whence the lemma. ∎

The next lemma is a criterion for ψ to be generic for an ergodic measure. Observe first that since one can approximate continuous functions on Ω uniformly by linear combinations of indicator functions of parallelipipeds of the form

$\tilde{J}(\ell) = \{x \in \Omega: x(\ell+j) \in J_j, \ 0 \le j \le k$ where the J_j's are intervals in $[0,1]$, and $J = J_0 \times J_1 \times \cdots \times J_{k-1}$, φ is generic for μ if an only if for all J and $\delta > 0$

$$\left| \frac{|\{0 \le i < N-k: \varphi(i+j) \in J_j \text{ all } 0 \le j < k\}|}{N} - \mu(\tilde{J}(0)) \right| < \delta \tag{2}$$

holds for all N sufficiently large. When (2) holds with a constant c replacing $\mu(\tilde{J}(0))$ we will say that the block $\varphi(0)\varphi(1)\cdots\varphi(N)$ is (J,δ,c)-<u>well-distributed</u>.

<u>Lemma 4.2</u>: If for all J and $\delta > 0$, there is some c and N_0 that for all $N \ge N_0$ the upper density of i's for which $\varphi(i)\cdots\varphi(i+N)$ is not (J,δ,c) - well-distributed is less than δ, then φ is generic for an ergodic measure.

<u>Proof</u>: Notice first that if N/N_1 is very small then if most of the N-blocks in $\varphi(i)\cdots\varphi(i+N_1)$ are (J,δ,c) - well-distributed so is the N_1-block $\varphi(i)\cdots\varphi(i+N_1)$ itself with a slightly larger δ. Using this observation one first checks that for fixed J and $\delta \downarrow 0$ the associated constants c tend to a limit and taking this limit to be $\mu(\tilde{J}(0))$ defines a measure μ which is invariant. It is also clear that φ is generic for μ and what remains to be seen is that μ is ergodic.

According to the mean ergodic theorem, if J is fixed, then letting f be the indicator function of $\tilde{J}(0)$, the averages

$$\frac{1}{N} \sum_{0}^{N-1} f(T^j \omega) \tag{3}$$

converge in k-measure to the projection of f on the space of invariant functions. For N sufficiently large, the fact that φ is generic for μ, together with our assumptions imply that but for a set of μ-measure at most δ the average in (3) is within 2δ of $\mu(\hat{J}(0))$. Since δ is arbitrary, it follows that the averages in (3) converge to a constant, i.e., the projection of f onto the space of invariant functions is a constant. Since this holds for all J's it follows that the entire space of invariant functions consists only of constants which gives the ergodicity of μ. □

The situation in which the above lemma will be applied is one where φ is extremely well approximated by periodic functions. This is formulated as the next lemma.

Lemma 4.3: If for every $\varepsilon > 0$, there is a periodic $\varphi_\varepsilon \in \Omega$ such that the upper density of i's for which $\varphi(i) \neq \varphi_\varepsilon(i)$ is at most ε then φ is generic for an ergodic measure.

Proof: Given $J = J_0 \times \cdots \times J_{k-1}$, $\delta > 0$ choose ε so that $\varepsilon < \frac{\delta^2}{100 \cdot k^4}$. Let φ_ε be periodic with period p and denoting by $I = \{i \geq 0: \varphi(i) \neq \varphi_\varepsilon(i)\}$ suppose that the upper density of I is at most ε. We may of course assume that k is small compared to p, say $k/p \leq \delta/10$. Since I has upper density at most ε, the upper density of j's for which

$$(4) \qquad \frac{|I \cap [pj,(p+1)j)|}{p} < \sqrt{\varepsilon}$$

has upper density at most $\sqrt{\varepsilon}$. Clearly φ_ε is generic for a measure μ_ε, and letting $c = \mu_\varepsilon(\tilde{J}(0))$, and $N_0 = p^2$ one readily checks that for all $N \gtrsim N_0$, those N-blocks of φ that are not (J,δ,c)-well-distributed have upper density at most δ, and thus the hypotheses of Lemma 2 are satisfied and the lemma is proved. ■

We can now give the construction of ψ. Let $\varphi_0(n)$ have period 2, with $\varphi(0) = 1$, $\varphi(1) = 0$. For any integer d, define now

$$\omega_d(i) = \begin{cases} 0 & |i| \le d^{d+1} \\ \frac{1}{d} & d^{d+1} < |i| \le d^{d+1} + d^d \\ \frac{2}{d} & d^{d+1} + d^d < |i| \le d^{d+1} + d^d + d^{d-1} \\ \vdots & \\ 1 & d^{d+1} + \cdots + d^2 < |i| \le d^{d+1} + \cdots + d^2 + d \end{cases}$$

by periodicity for all larger i.

Take a sequence of d_k's tending to infinity sufficiently rapidly, and satisfying the condition that

$$P_k = 2\left(d_k^{d_k+1} + \cdots + d_k\right),$$

the period of ω_{d_k}, divides P_{k+1} for all k, define

$$\varphi_k = \omega_{d_k}$$

and set

$$\psi_k = \max(\varphi_0, \varphi_1, \ldots, \varphi_n)$$

$$\chi_n = \sup(\varphi_{n+1}, \varphi_{n+2}, \ldots)$$

$$\psi = \lim_{n\to\infty} \psi_n = \sup(\varphi_0, \varphi_1, \varphi_2, \ldots).$$

By our arithmetic assumptions on the d_k's the period of ψ_n is also p_n. In order to be able to use Lemma 3 we assume that

(5) $$\sum_1^{\infty} p_n/p_{n+1} < +\infty .$$

This gives that ψ itself is generic for an ergodic measure μ_0. The fact that all points in $ORC(\psi) = X$ are generic will come from the very slow rise of the φ_k's, which readily gives the following.

Lemma 4.4: If $d_n \uparrow \infty$ sufficiently rapidly and $\delta_n = 3/d_n$ then for any finite interval $I \subset \mathbb{Z}$:

$$\frac{|\{i \in I: X_n(i) > (\min_{j \in I} X_n(j) + \delta_n\}|}{I} \leq \delta_n .$$

Suppose now that $x \in X$ and that $\inf_n x(n) = 0$. Denote the set of such x by X_0. We claim that all $x \in X_0$ are generic for μ_0. To see this fix some $\varepsilon > 0$, and choose some n_0 with $x(n_0) < \varepsilon$. For any $N > n_0$ there is some ℓ_N so that

$$|x(n_0+i) - \psi(\ell_N+i)| < \varepsilon, \qquad 0 \leq i < N. \tag{6}$$

If m_0 is chosen so that $\delta_{m_0} < \varepsilon$, it follows from Lemma 4 that

$$\frac{\{0 \leq i < N: |x(n_0+i) - \psi_{m_0}(\ell_N+i)| < 3\varepsilon\}}{N} < \varepsilon. \tag{7}$$

Here N can be as large as we please with m_0 fixed, and so arguments similar to those used above give that x is generic for the same measure that ψ is generic for, i.e., μ_0.

Denote by X_t the set $x \in X$ for which $\inf_n x(n) = t$. All $x \in X$ will be generic for the ergodic measure μ_t which is obtained as follows. Define

$$\psi_t = \max(\psi, t).$$

Clearly, $\psi_t \in X$, and ψ_t is generic for an ergodic measure μ_t and the same kinds of arguments as above show that x is generic for μ_t. We summarize this discussion in:

Example: The system (X,T) has the following properties:

(i) The closed invariant sets E_t, $0 \le t \le 1$ satisfy $E_s \supset E_t$ for $s < t$, $E_0 = X$, $E_1 = \{x_1\}$, and these are the only closed invariant sets in X. In particular the fixed point x_1 is the unique minimal set.

(ii) All points in $X_t = E_t \setminus \bigcup_{s>t} E_s$ are generic for an ergodic measure μ_t whose support is E_t.

(iii) The map $t \to \mu_t$ is continuous in the ω^*-topology.

We can say a little more about the points in X. Namely, if $x \in X$ is not the fixed point then it isn't hard to see that if ℓ_n is a sequence of translates so that $T^{\ell_n}\psi \to x$ then the ℓ_n's have to converge module p_k for all k to a fixed integer a_k. These are consistent of course, i.e., $a_{k+1} \equiv a_n \pmod{p_k}$. Since the ψ_k's are increasing

$$\bar{x} = \lim T^{a_k} \psi_k$$

exists and it's not hard to see that $\bar{x} \le x$. Indeed x is determined by $\bar{x}$ and the level t, which can still be any number in the interval $(\inf_n \bar{x}(n), 1)$.

References

[DL] Yael Naim Dowker and G. Lederer, On Ergodic Measures, Proc. Amer. Math. Soc. 15 (1964), 65-69.

[N] V. V. Nemitsky, Topological Problems of the Theory of Dynamical Systems, Usp. Math. Nauk (1949), 91-153. English translation in Translations of A.M.S Series One Vol. 5, Stability and Dynamic Systems, pp. 414-497.

SINGULARITIES IN CLASSICAL MECHANICAL SYSTEMS

Robert L. Devaney *

Singularities in the equations of motion of a classical mechanical system usually play a dominant role in the global phase portrait of the system. By a singularity we mean a point or set of points where the system is undefined, as in the case of a collision between two or more of the particles in the n-body problem. Such singularities often lead to a complicated global orbit structure. Not only do certain solutions tend to run off the phase space, but also nearby solutions tend to behave in an erratic or unpredictable manner. Numerical studies of such systems are often inconclusive because of this erratic behavior. And power series or other analytic techniques often yield only a very local description of solutions near the singularity, one which gives no hint of the global complexity of the system.

Our goal in these notes is to use geometric techniques to gain a complete picture of the local behavior of the solutions near a singularity. Our main tool is a device due to McGehee by which we "blow up" the singular set and replace it with an invariant boundary. The dynamical system extends smoothly (after a scaling of time) over this boundary, and so we get a new flow on an augmented phase space. It turns out that this

* Research partially supported by NSF Grant MCS 79-00430.

new flow restricted to the boundary is extremely simple to understand — usually it is a gradient-like, Morse-Smale flow. So on the boundary of one of the most complicated types of dynamical systems — Hamiltonian systems — we find one of the simple types of systems. It is this fact that enables us to readily understand the behavior of solutions near the singularity.

It also turns out that this geometric approach to singularities gives us at least a partial grasp on the global orbit structure of systems with singularities. Using symbolic dynamics, one can often associate to each trajectory of the system a doubly infinite sequence of integers in a natural way which adequately describes the qualitative behavior of solutions. This association is often onto the space of all possible such sequences, and so we get a rough idea of the possible complexity of the set of all solution curves in the system.

We will illustrate this idea with two important classical mechanical systems: the anisotropic Kepler problem and the isosceles three body problem in the plane. In fact, examples such as these play an important role throughout these notes. In each of the three major parts, we work mainly with two or three specific examples, leaving the formulation of general theorems to others.

These notes are divided into three parts. In the first we deal with McGehee's technique of blowing up the singularity. We illustrate this technique by describing the flow on the invariant boundary in the case of Newtonian and non-Newtonian central force problems, as well as in the examples mentioned above. Here we prove that the extended flow is generically Morse-Smale.

In the second part, we switch our emphasis to mappings with singularities, rather than vector fields. We introduce symbolic dynamics via the classical Smale horseshoe mapping, and show how the shift automorphism arises in the well-known Hénon mapping. The baker transformation and another mapping associated to the restricted three body problem provide important examples of mappings with singularities, and we show how symbolic dynamics enters in these cases.

The final part combines the techniques of the previous two to describe at least partially the global orbit structure of the anisotropic Kepler problem and the isosceles three-body problem. The key idea here is the reduction of each differential equation to a Poincaré mapping with singularities on a surface of section. The symbolic dynamics of part two as well as the McGehee method of part one then give an adequate description of this mapping, and thereby the associated phase portrait.

These notes grew out of a series of lectures delivered by the author at the University of Maryland in the spring of 1980. It is a pleasure to thank A. Katok and N. Markley for arranging my visit, and M. Brin, W. Neumann, and M. Paul for many helpful discussions and comments.

Part One: Singularities in Classical Mechanical Systems.

Our goal in this section is to study some of the properties of solutions of Newton's equations

$$Mq'' = -\nabla V(q).$$

Here q is a point in Q, an open subset of $\mathbb{R}^n$, and the potential energy function $V: Q \to \mathbb{R}$ is sufficiently smooth. M, the mass matrix, is a diagonal matrix with positive entries $m_1, \ldots m_n$.

Newton's equations arise in many different areas of mechanics, and we will study in detail several such examples. Most important are:

1. The Kepler or Newtonian central force problem. Here $M = I$ and $V(q) = -1/|q|$.
2. Other central force problems. In this case, $M = I$, but $V(q) = -1/k|q|^k$. All of these systems are integrable systems and their phase portraits are fairly well understood.
3. The anisotropic Kepler problem. Here we keep the Kepler potential $V(q) = -1/|q|$, but make the mass matrix anisotropic, i.e.,

$$M = \begin{pmatrix} \mu & 0 \\ 0 & 1 \end{pmatrix}$$

with $\mu > 1$.
4. Special cases of the three body problem.

A. The collinear (rectilinear) problem with potentia

$$V(q_1,q_2,q_3) = - \sum_{i<j\leq 3} m_i m_j/|q_i - q_j|.$$

B. The isosceles problem with mass matrix

$$\begin{pmatrix} 1 & 0 \\ 0 & 1+\varepsilon/\varepsilon \end{pmatrix}$$

and potential $V(q_1,q_2) = -1/8q_1 - 2\varepsilon/|q|$.

These last two problems are non-integrable and their dynamics are considerably more complicated than the first two. Note that, for each of these examples, V is real analytic and homogeneous of degree $-k$, i.e., $V(\lambda q) = \lambda^{-k}V(q)$. We will in general restrict attention to such potentials, though much that follows can be extended to other classes of potentials.

We also note that, for each of the examples above, the potential suffers a singularity at the origin, and hence the differential equation is undefined there. Our first goal will be to understand how this affects the local behavior of solutions of the system. In particular, for each of these examples, we will examine in this section how the singularity affects orbits which pass close to it. Later, we will describe how the presence of a singularity affects the global phase portrait of the system.

§1.1 Hamiltonian Systems.

Before considering specific examples, we deal with Newton's equations in a general setting. Assume that the potential energy function V is real analytic, homogeneous of degree $-k$, and has an isolated singularity at the origin. Newton's equation

$$Mq'' = -\nabla V(q) \tag{1}$$

is more conveniently written as a first order system by introducing the momentum vector $p = Mq'$. Then (1) may be written

$$\begin{aligned} q' &= M^{-1}p \\ p' &= -\nabla V(q). \end{aligned} \tag{2}$$

Let $Q = \mathbb{R}^n - \{0\}$. Q is called the <u>configuration space</u> of the system. (2) is a first order system of differential equations or a vector field on $Q \times \mathbb{R}^n$, the <u>phase space</u> of the system.

This system may be written in Hamiltonian form by introducing the <u>Hamiltonian</u> or <u>total energy function</u>

$$H(q,p) = \tfrac{1}{2}p^tM^{-1}p + V(q). \tag{3}$$

Then (2) assumes the form

$$\begin{aligned} q' &= \frac{\partial H}{\partial p} \\ p' &= -\frac{\partial H}{\partial q}. \end{aligned} \tag{4}$$

Systems which can be written in this form are called <u>Hamiltonian systems</u>, and they enjoy many special properties. We refer the reader to [AM, Ar] for detailed studies of Hamiltonian systems; in this section we will simply recall a few of their basic properties which we will use over and over again.

We remark that our Hamiltonian function is especially simple; one can generate more complicated Hamiltonian systems by taking an arbitrary function H on $\mathbb{R}^{2n}$ or any other symplectic manifold.

The term $\tfrac{1}{2}p^tM^{-1}p$ is called the <u>kinetic energy</u> of the system. Since M is positive definite, the kinetic energy is a positive definite quadratic form in the momentum variables.

It is well known that H is a first integral or constant of the motion for (4). Indeed,

$$\frac{dH}{dt} = \sum_{i=1}^{n} \frac{\partial H}{\partial q_i}\frac{dp_i}{dt} + \frac{\partial H}{\partial p_i}\frac{dq_i}{dt}$$

which vanishes identically by (4). Hence we can reduce the dimension of the system by one by considering (4) as a vector field on the surface of constant energy (or energy level) $H^{-1}(e)$. When e is a regular value of H, this is a smooth submanifold of $Q \times \mathbb{R}^n$.

Since the kinetic energy is non-negative, it follows from (3) that, on $H^{-1}(e)$, we have $V(q) \leq e$. The corresponding region in Q is called the <u>Hill's region</u> for energy level e. If we take a solution curve $(q(t),p(t))$ of the system on $H^{-1}(e)$, then $q(t)$ must lie for all time in the corresponding Hill's region.

We can now describe $H^{-1}(e)$ topologically as follows. Over each point q^* in the interior of the Hill's region, we have an ellipsoid in the momentum coordinates defined by

$$\tfrac{1}{2}\, p^t M^{-1} p = e - V(q^*) > 0. \tag{5}$$

On the boundary $V(q) = e$, the ellipsoid degenerates to a point. So $H^{-1}(e)$ is a "pinched" sphere bundle over the corresponding Hill's region. We visualize $H^{-1}(e)$ as a union of configurations q^* with attached momentum vectors p^* satisfying (5). See Fig. 1.

In many of our examples, the Hill's region is a two dimensional disk minus a point. So in this case, $H^{-1}(e)$ is

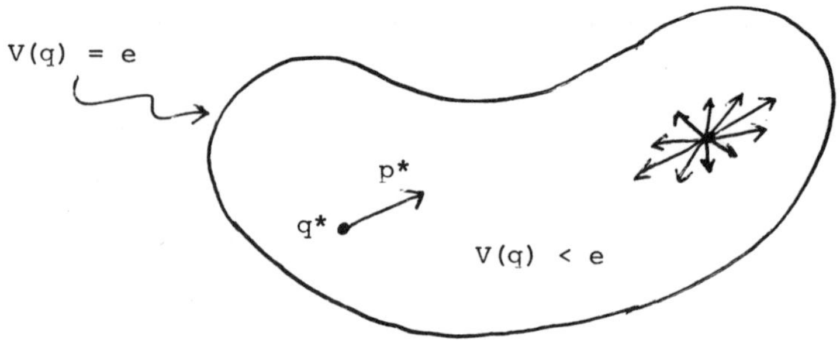

Fig. 1. The Hill's region.

topologically an open solid torus, as one checks easily using the description above.

Note that the projection of a solution curve to the Hill's region is a curve $q(t)$ which may cross itself; at such crossings one has of course different values for the momentum coordinates. Also, solution curves may meet the boundary $V(q) = e$. This boundary is called the <u>zero velocity set</u>, since here the momentum coordinates are all 0. In this case a solution curve $q(t)$ must fall back upon itself and retrace its path in the opposite direction. This follows immediately from the fact that if $(q(t),p(t))$ is a solution of (4), then so to is $(q(-t),-p(-t))$. Systems with this property are called <u>reversible systems</u>.

In particular, we note that if a solution curve has two distinct points of intersection with the zero velocity set, then the corresponding solution curve in phase space is necessarily periodic. These solutions are called <u>symmetric periodic solutions</u>. They are often the easiest types of closed orbits to find in a reversible system. See Fig. 2.

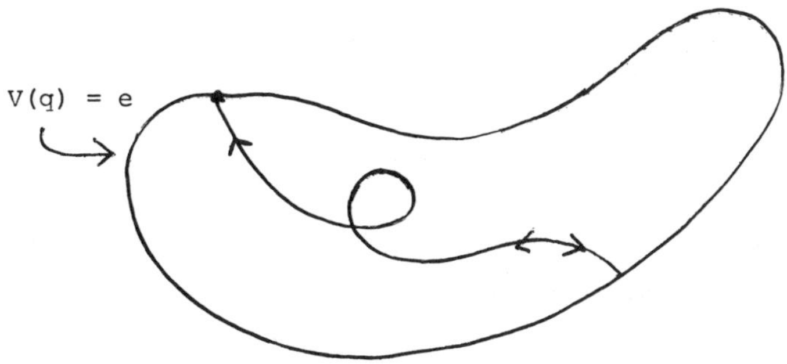

Fig. 2. A symmetric periodic solution in the Hill's region.

§1.2 McGehee Coordinates.

In this section we introduce a remarkable change of coordinates due to R. McGehee [McG 1]. Recall that V is homogeneous of degree $-k$ with an isolated singularity at the origin in configuration space. Since $-\nabla V(q)$ is not defined at $q = 0$, we have a "hole" in phase space at $\{0\} \times \mathbb{R}^n$ where the vector field is undefined. Certain orbits of the system reach this singularity in finite time (the collision orbits), while others begin at the singularity (the ejection orbits). Moreover, the local behavior of the system near $\{0\} \times \mathbb{R}^n$ is usually quite complicated. However, the McGehee coordinates allow us to read off the behavior of these solutions with relative ease.

We introduce these coordinates via a series of three coordinate changes. The first step is to introduce "polar" coordinates generated by the moment of inertia of the system. Let

$$r = (q^t Mq)^{\frac{1}{2}}$$
$$s = r^{-1}q. \tag{1}$$

The quantity r^2 is called the <u>moment of inertia</u> of the system and s is the <u>configuration</u>. Since $s^t Ms = 1$, we think of s as a point in the unit sphere S in $\mathbb{R}^n$ in the metric induced by M, i.e., under the inner product $v_1{}^t Mv_2$. Also define

$$y = s^t p$$
$$x = M^{-1}p - ys. \tag{2}$$

Then y is the radial component of velocity $q' = M^{-1}p$ and x is the component tangent to S. Indeed, we have $s^t Mx = 0$, so we think of the pair (s,x) as defining a point in the tangent bundle TS of S.

In these coordinates, the original system becomes

$$\begin{aligned} r' &= y \\ y' &= r^{-1}(x^t Mx) + r^{-k-1}k\ V(s) \\ s' &= r^{-1}x \\ x' &= -r^{-1}(yx + (x^t Mx)s) - r^{-k-1}(M^{-1}\nabla V(s) + k\ V(s)\ s). \end{aligned} \tag{3}$$

Here we have used the fact that ∇V is homogeneous of degree $-k-1$ so that $\nabla V(rs) = r^{-k-1}\ V(s)$. Also, Euler's formula gives

$$s^t \nabla V(s) = -k\ V(s).$$

The system (3) is a real analytic vector field on the manifold $(0,\infty) \times \mathbb{R} \times TS$, and the energy relation gives

$$\tfrac{1}{2}(x^t Mx + y^2) + r^{-k}V(s) = e \tag{4}$$

The system (3) is no longer Hamiltonian, but (4) defines a codimension one invariant set which we continue to call the energy level.

Note that the singularity set for (3) now corresponds to $r = 0$. As $r(t) \to 0$, we in general have $x^t Mx + y^2 \to \infty$, so the next step is to scale down these components of velocity. Introduce

$$\begin{aligned} u &= r^{k/2}x \\ v &= r^{k/2}y \end{aligned} \tag{5}$$

The system (3) becomes

$$\begin{aligned} r' &= r^{-k/2}v \\ v' &= r^{-k/2-1}(u^t Mu + (k/2)v^2 + k\,V(s)) \\ s' &= r^{-k/2-1}\,u \\ u' &= r^{-k/2-1}((k/2-1)vu - (u^t Mu)s - kV(s)\,s - \\ &\qquad -M^{-1}\nabla V(s)) \end{aligned} \tag{6}$$

This system still has singularities at $r = 0$, but now they can be removed by a change of time scale. Introduce a new time variable τ via

$$\frac{dt}{d\tau} = r^{k/2+1} \tag{7}$$

Then (6) becomes

$$\begin{aligned} \dot{r} &= rv \\ \dot{v} &= u^t Mu + (k/2)v^2 + kV(s) \\ \dot{s} &= u \end{aligned} \tag{8}$$

$$\dot{u} = (k/2 - 1)vu - (u^t Mu)s - kV(s)s - M^{-1}\nabla V(s)$$

where the dot indicates differentiation with respect to τ. The energy relation goes over to

$$r^k e = \tfrac{1}{2}(u^t Mu + v^2) + V(s). \tag{9}$$

We will study this system in detail in §4, but for now we list some of the most important observations.

1. This is an analytic vector field on $[0,\infty) \times \mathbb{R} \times TS$; that is, the singularities at $r = 0$ have been removed.
2. In their place, the vector field has been extended analytically to the boundary $r = 0$, and this boundary is invariant under the flow, since $\dot{r} = 0$ when $r = 0$.
3. The energy relation also extends to the boundary, giving

$$\tfrac{1}{2}(u^t Mu + v^2) + V(s) = 0 \tag{10}$$

when $r = 0$. In effect, we have glued a boundary onto each energy level, and this boundary is also invariant under the flow. We denote by Λ the boundary of $H^{-1}(e)$ determined by (10); Λ is called the collision manifold of the system.
4. Λ is independent of e; so each energy surface has the same boundary in $r = 0$.
5. Orbits which previously reached $r = 0$ in finite time are now asymptotic to Λ. And orbits which previously passed close to the singularity now spend

a long time near Λ. Thus the flow on Λ determines the local behavior of solutions near the singularity. In the following sections we will show that in fact the flow on Λ is extremely simple - generically it turns out to be a gradient-like, Morse-Smale flow.

All of the examples we will discuss are systems with two degrees of freedom, i.e., $n = 2$. In this case, the spherical variables (s,u) may be replaced by the more usual polar coordinate θ (in the M-metric) and the component of velocity in the θ direction, which we also call u. The system (8) becomes in this simple case

$$\begin{aligned} \dot{r} &= rv \\ \dot{v} &= u^2 + (k/2)v^2 + kV(\theta) \\ \dot{\theta} &= u \\ \dot{u} &= (k/2 - 1)vu - V'(\theta) \end{aligned} \tag{11}$$

with the energy relation

$$\tfrac{1}{2}(u^2 + v^2) + V(\theta) = r^k e. \tag{12}$$

We call the variables (r,v,θ,u) <u>McGehee coordinates</u>.

§1.3 The Kepler Problem and Other Central Forces.

Before discussing the general properties of the flow on Λ, we pause to describe two simple examples, the Kepler problem and other (non-Newtonian) central force problems. We will concentrate on how the flow on Λ describes the qualitative behavior of orbits which pass close to the singularity.

For the Kepler problem, the potential energy is simply

$$V(q) = -1/|q| \tag{1}$$

and $M = I$. For simplicity, we will restrict to negative energ levels $e < 0$. It follows from (1) that the Hill's regions for negative energy are disks of radius $-1/e$ with the origin removed. Restricted to S, we have $V(\theta) = -1$, so that the Kepler problem in McGehee coordinates is given by

$$\begin{aligned} \dot{r} &= rv \\ \dot{v} &= \tfrac{1}{2}v^2 + u^2 - 1 \\ \dot{\theta} &= u \\ \dot{u} &= -\tfrac{1}{2}vu \end{aligned} \tag{2}$$

with the energy relation

$$\tfrac{1}{2}(u^2 + v^2) - 1 = re. \tag{3}$$

When $r = 0$, the energy relation shows that Λ is a two dimen- sional torus in $r = 0$ defined by

$$\begin{aligned} &\tfrac{1}{2}(u^2 + v^2) = 1 \\ &\theta \quad \text{arbitrary} \end{aligned} \tag{4}$$

The vector field on Λ is then given by

$$\begin{aligned} \dot{v} &= \tfrac{1}{2}u^2 \\ \dot{\theta} &= u \\ \dot{u} &= -\tfrac{1}{2}vu \end{aligned} \tag{5}$$

where we have used the energy relation to simplify $\dot{v}$. This system is easy to understand. Note first that (5) has rest points iff $u = 0$, $v = \pm\sqrt{2}$, θ arbitrary. That is, there are two circles of equilibrium points on Λ. All other solution curves move from the lower circle $v = -\sqrt{2}$ to the upper circle $v = +\sqrt{2}$, since $\dot{v} > 0$ when $u \neq 0$. In fact, there is a unique unstable manifold associated to the rest point at $\theta = \theta^*$, $v = -\sqrt{2}$, and this unstable manifold has the property that it also forms the stable manifold of the rest point directly "above" it, i.e., at $\theta = \theta^*$, $v = +\sqrt{2}$. To see this, we introduce the angular variable ψ via

$$u = \sqrt{2}\cos\psi$$
$$v = \sqrt{2}\sin\psi.$$

The system on Λ becomes

$$\dot{\psi} = (1/\sqrt{2})\cos\psi$$
$$\dot{\theta} = \sqrt{2}\cos\psi$$

so that we have

$$\frac{d\psi}{d\theta} = \frac{1}{2}.$$

The solutions of this vector field are sketched in Fig. 1.

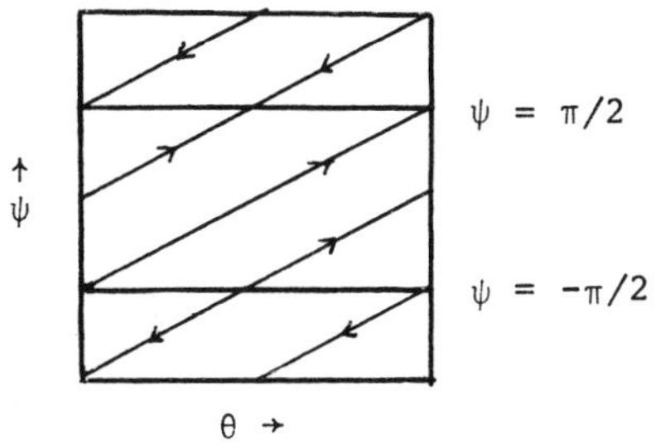

Fig. 1. The flow on Λ for the Kepler problem.

Away from Λ, we also wish to describe the set of orbits which begin and end in collision. For this, we fix $\theta = \theta^*$ and $u = 0$. Then $\dot{\theta} = \dot{u} = 0$, so that the corresponding r,v-plane is invariant under the flow. Restricted to this plane, the system (2) becomes

$$\dot{r} = rv$$
$$\dot{v} = \tfrac{1}{2}v^2 - 1 = re.$$

This system is easily solved and the phase portrait is sketched in Fig. 2.

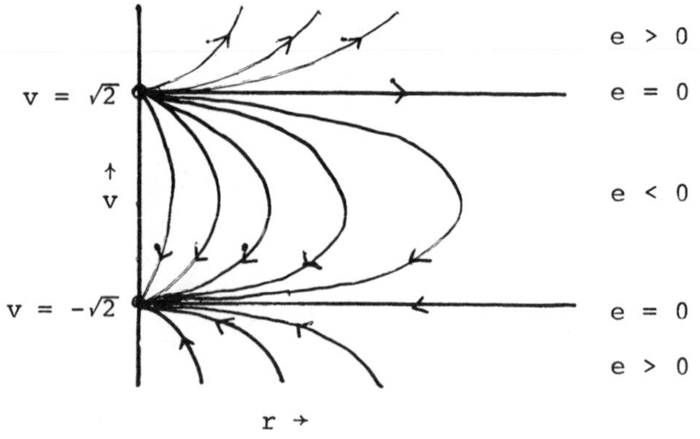

Fig. 2. The flow on an invariant r,v-plane.

For each fixed negative energy level, there is a unique solution which begins and ends at collision with $r = 0$, and which satisfies $\theta = \theta^*$ and $u = 0$ for all time. Such orbits are called <u>homothetic orbits</u>. In configuration space, they simply traverse the ray $\theta = \theta^*$ until meeting the zero velocity curve, whereupon they fall back to collision. See Fig. 3.

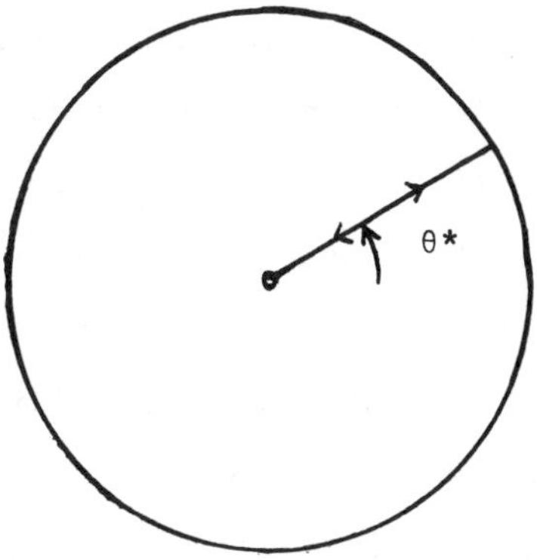

Fig. 3 Homothetic orbits in configuration space.

One may check using the normal hyperbolicity of the circles of equilibria that the homothetic orbits are the only collision and/or ejection orbits for the system. Hence the flow on and near Λ may be pictured as in Fig. 4.

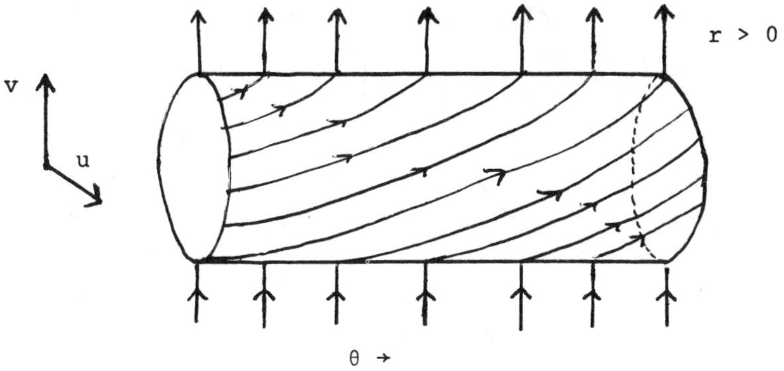

Fig. 4 The flow on and near the collision manifold in the Kepler problem.

We can now read off the local behavior of solutions which pass near the origin. Such an orbit follows a collision orbit until it comes close to Λ. Then it has two choices depending upon which branch of the unstable manifold in Λ it then follows. Each branch, however, forces θ to change by 2π before yielding escape from a neighborhood of collision near an ejection orbit. In one case, θ increases by 2π before leaving a neighborhood of the origin. That is, such solutions circle the origin in a counter-clockwise direction before escaping. In the other case, θ decreases by 2π, forcing solutions to circle the origin clockwise before escaping. See Fig. 5.

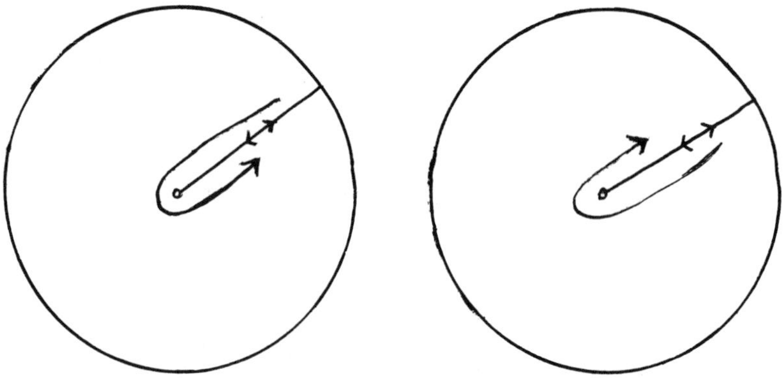

Fig. 5 Local behavior of solutions of the Kepler problem which pass close to collision.

Actually, it is well known that these near-collision orbits lie on ellipses, so the typical near-collision orbit looks like Fig.6.

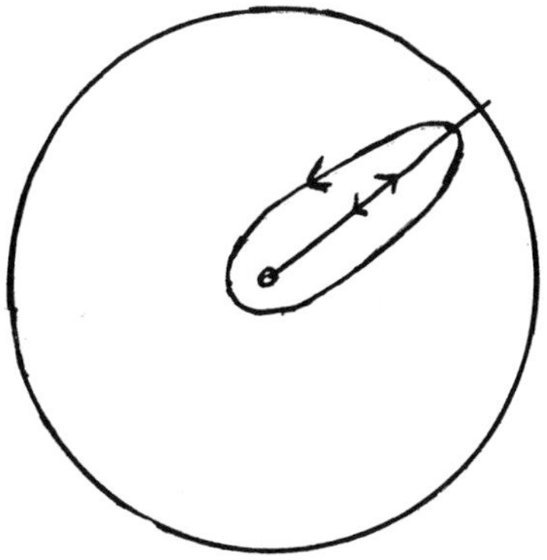

Fig. 6 Near-collision orbits in the Kepler problem.

The same techniques as above also yield a description of collision for other central force potentials, specifically, for those of the form

$$V(q) = -1/k|q|^k \qquad k > 0$$

In McGehee coordinates, the system on Λ becomes

$$\dot{v} = (1 - k/2)u^2$$
$$\dot{\theta} = u$$
$$\dot{u} = (k/2 - 1)vu$$

with $\frac{1}{2}(u^2 + v^2) = 1/k$. One checks easily that the circles $u = 0$, $v = \pm\sqrt{2/k}$ are again circles of rest points, and that there is a cylinder of homothetic orbits both coming into and leaving collision. As before, when $e < 0$, these orbits match up exactly, yielding collision/ejection orbits.

The major difference between these systems and the Kepler problem is the non-stationary orbits on Λ. Introducing the angular variable ψ via

$$u = \sqrt{2/k}\ \cos\ \psi$$

$$v = \sqrt{2/k}\ \sin\ \psi$$

one finds

$$\frac{d\psi}{d\theta} = 1 - k/2.$$

The resulting phase portrait is different depending on whether $k < 2$, $k = 2$, or $k > 2$. In the first case, $\dot{v} > 0$, so all orbits still travel from the lower circle of rest points to the upper circle, whereas when $k > 2$, exactly the opposite is true. For $k = 2$, $\dot{v} = 0$, so all orbits on Λ except the rest points are periodic. Fig. 7 gives a catalogue of these different cases.

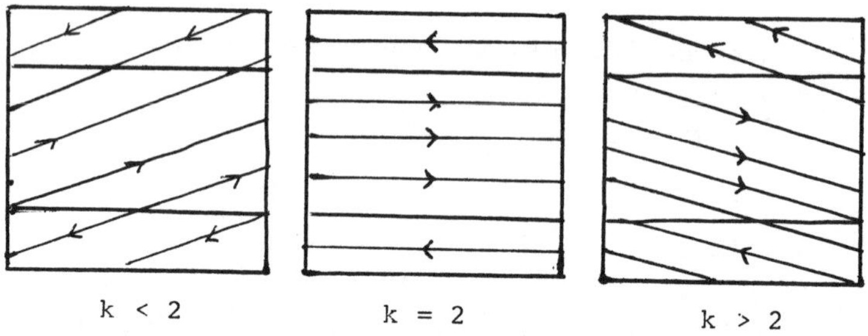

Fig. 7 The flow on the collision manifold for the non-Newtonian central force problems.

When $k > 2$, the lower circle of rest points is now an attractor for the flow off Λ. Hence a large open set of orbits reach collision in this case. Similarly, a large open set of orbits also begins at collision. We remark that this does not contradict the volume preserving property of the original system, since as the orbits tend to $r = 0$, the velocities become unbounded.

When $k = 2$, we have a transitional case: open sets of orbits still tend to collision.

For $k < 2$, only the homothetic orbits reach $r = 0$, but the local behavior changes dramatically with k. The unstable manifolds at $\theta = \theta^*$, $v = -\sqrt{2/k}$ in Λ do not necessarily join up with the stable manifolds at $\theta = \theta^*$, $v = \sqrt{2/k}$ as was the case in the Kepler problem. Only when we have

$$\frac{d\psi}{d\theta} = 1 - k/2 = 1/2n \qquad n = 1,2,3...$$

do we have this property. That is, if $k = 2 - 1/n$ for a positive integer n, then each branch of the unstable manifold makes n circuits of Λ before rejoining the upper circle of rest points at the same θ-value. When

$$\frac{d\psi}{d\theta} = 1 - k/2 = 1/(2n+1) \qquad n = 0,1,2...$$

or equivalently when $k = 2 - 2/(2n+1)$, the unstable manifolds leaving $\theta = \theta^*$ join up with the stable manifolds at $\theta = \theta^*+\pi$ after making $n+\frac{1}{2}$ circuits. See Fig. 8.

In all other cases, the two branches of the unstable manifolds reach distinct equilibrium points, as in Fig. 9.

The local behavior of solutions may then be described as follows. For $k < 2$, a near-collision orbit makes n revolu-

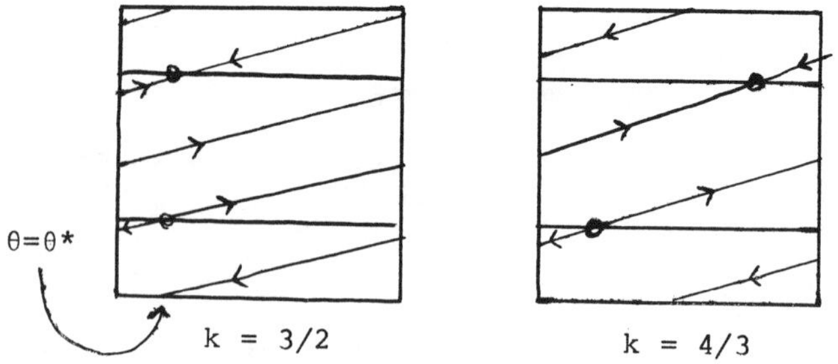

Fig. 8 The flow on Λ in two special cases. The stable and unstable manifolds at $\theta = \theta^*$ match up when $k = 3/2$, while the unstable manifold at θ^* equals the stable manifold at $\theta^* + \pi$ when $k = 4/3$.

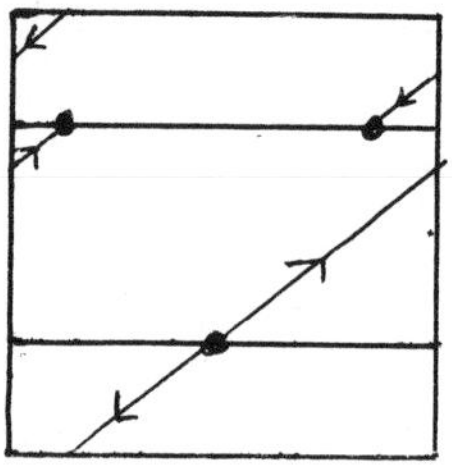

Fig. 9 In this case, the branches of the unstable manifold at θ^* tend to distinct rest points.

tions about $r = 0$ before exiting at an angle which depends on k. In the two special cases depicted in Fig. 8, the orbit either exits in the direction in which it approached collision $(k = 3/2)$ or else in exactly the opposite direction $(k = 4/3)$. See Fig. 10.

If k is not of one of these two forms, then nearby initial conditions will lead to quite different behavior near collision. This is the basic idea behind Easton's notion of topological regularization [E]. In this case we cannot join orbits coming to collision with orbits leaving collision in a meaningful way so to make the resulting flow continuous. See Fig. 11.

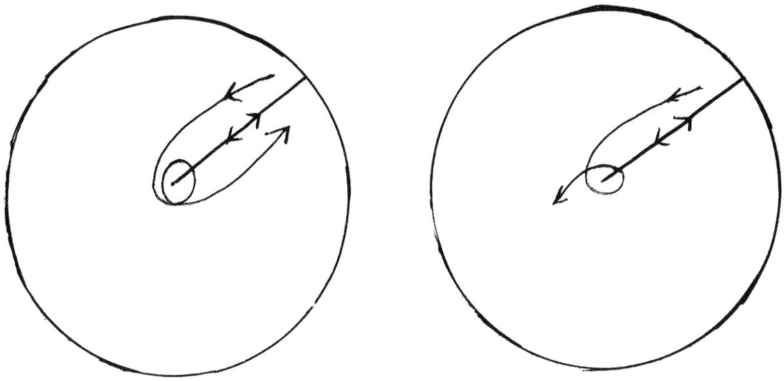

k = 3/2 k = 4/3

Fig. 10 Behavior of solutions near a homothetic orbit.

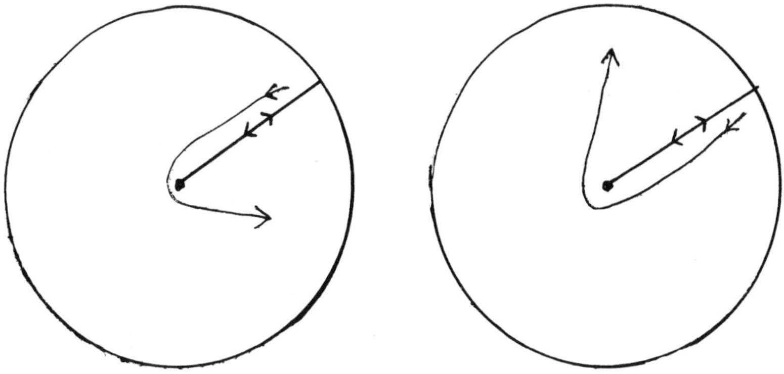

Fig. 11 Behavior of near-collision orbits in the case illustrated in Fig. 9. Note that nearby initial conditions lead to quite different exiting behavior.

§1.4 The Flow on the Collision Manifold

In this section we return to the general setting to describe the behavior of the flow on the collision manifold. Recall that, from §2, in McGehee coordinates, Newton's equations assume the form

$$\begin{aligned}
\dot{r} &= rv \\
\dot{v} &= (k/2)v^2 + u^tMu + kV(s) \\
\dot{s} &= u \\
\dot{u} &= (k/2 - 1)vu - (u^tMu)s - kV(s)s - M^{-1}\nabla V(s)
\end{aligned} \tag{1}$$

with the energy relation

$$r^k e = \tfrac{1}{2}(u^tMu + v^2) + V(s). \tag{2}$$

When $r = 0$, the collision manifold Λ is defined by

$$0 = \tfrac{1}{2}(u^tMu + v^2) + V(s) \tag{3}$$

Recall that V is real analytic, and that $V(s)$ is the restriction of V to the unit sphere S in configuration space.

Proposition 1. *If* 0 *is a regular value of* $V(s)$, *then* Λ *is a compact smooth submanifold of the boundary* $r = 0$.

Proof. Let $g(v,s,u) = \tfrac{1}{2}(u^tMu + v^2) + V(s)$ and note that Λ is $g^{-1}(0)$. Λ is clearly closed and bounded. Suppose that 0 is a critical value of g. Then, since $\partial g/\partial u = 0$ and $\partial g/\partial v = 0$, it follows that both $u = 0$ and $v = 0$. Hence $V(s) = 0$ also. But then $dV(s) = 0$, contradicting our assumption that 0 is a regular value of $V(s)$. qed

We remark that, in specific examples, one can compute $dV(s)$ explicitly, so that the hypotheses of this proposition are easy to verify in practice. We will call a singularity <u>non-degenerate</u> if $V(s)$ has 0 as a regular value. Clearly, an open and dense set of homogeneous potentials of degree $-k$ have non-degenerate singularities.

We now turn our attention to the flow on Λ in the non-degenerate case. When $r = 0$, (1) reads

$$\begin{aligned} \dot{v} &= (1 - k/2)u^t Mu \\ \dot{s} &= u \\ \dot{u} &= (k/2 - 1)vu - (u^t Mu)s - kV(s)s - M^{-1}\nabla V(s) \end{aligned} \tag{4}$$

where we have used the energy relation (3).

We remark that the expression $kV(s)s + M^{-1}\nabla V(s)$ is the gradient vector field associated to the restriction of V to S in the metric induced by M. Indeed, in this metric,

$$\langle kV(s)s + M^{-1}\nabla V(s), s\rangle = kV(s)s^t Ms + \nabla V(s)^t s = 0$$

by Euler's formula. Furthermore, for any vector w tangent to S at s, we have

$$\langle kV(s)s + M^{-1}\nabla V(s), w\rangle = \nabla V(s)^t w = dV(s)(w)$$

since $s^t Mw = 0$. This proves that

$$kV(s)s + M^{-1}\nabla V(s) = \text{grad}_S(V) \tag{5}$$

where $\text{grad}_S(V)$ denotes the gradient vector field of the restriction of V to S at s.

Now the system (4) has rest points whenever

$$\begin{cases} u = 0 \\ \text{grad}_S(V) = 0 \\ v = \pm\sqrt{-2V(s)} \end{cases} \tag{6}$$

Hence there are two rest points for the flow corresponding to each point s where $\text{grad}_S(V)$ vanishes. Such points are called <u>central configurations</u>. Now $\text{grad}_S(V) = 0$ iff $dV(s) = 0$, so rest points for the flow are in two-to-one correspondence with critical points of the restriction of V to S. We summarize this in a proposition.

Proposition 2. *The flow on Λ has a rest point at (v_0, s_0, u_0) iff s_0 is a critical point of the restriction of V to S, $u_0 = 0$, and $v_0 = \pm\sqrt{-2V(s_0)}$.*

Recall that a vector field is called gradient-like with respect to a function h if h increases along all non-equilibrium point orbits. For the flow on Λ we have:

Proposition 3. *If $k < 2$ (resp. $k > 2$), then the flow on Λ is gradient-like with respect to the v-coordinate (resp. $-v$).*

Proof. We have from (4) that

$$\dot{v} = (1 - k/2)u^t Mu$$

so that $\dot{v} \neq 0$ when $u \neq 0$. If $u = 0$, we have $\dot{u} = -\text{grad}_S(V)$

so that $u(\tau)$ is not identically 0 unless $\mathrm{grad}_s(V) = 0$. Consequently, $v(\tau)$ either increases or decreases, depending on whether $k < 2$ or $k > 2$. qed

Gradient-like vector fields on compact manifolds are extremely simple dynamical systems: all non-equilibrium point orbits both begin and end at rest points and there is no non-trivial recurrence.

For the flow on Λ we can in fact say more. Recall that $V(s)$ is called a Morse function if all critical points of $V(s)$ are non-degenerate, i.e., $d^2V(s)$ is a non-degenerate bilinear form at each critical point. On the other hand, a rest point p for the flow on Λ is hyperbolic if all of the eigenvalues of the linearization of the system at p have non-zero real parts. These two notions are linked by the following proposition.

Proposition 4. *Suppose $k \neq 2$. If $V(s)$ is a Morse function, then all of the rest points for the flow on Λ are hyperbolic.*

Proof. From (1), one computes that the linearization of the flow at a rest point with $r = 0$, $u = 0$, $dV(s_0) = 0$, and $v_0 = \pm\sqrt{-2V(s_0)}$ is given by the matrix

$$A = \begin{pmatrix} v_0 & 0 & 0 & 0 \\ 0 & kv_0 & 0 & 0 \\ 0 & 0 & 0 & I \\ 0 & 0 & -B & (k/2 - 1)v_0 I \end{pmatrix} \qquad (7)$$

Here I is the $(n-1) \times (n-1)$ identity matrix and B represents the linearization of the flow of $\text{grad}_s(V)$ at s_0. Restricting to an energy surface, it follows easily that the eigenvalues of the linearization of the flow are v_0 together with the eigenvalues of the submatrix

$$A' = \begin{pmatrix} 0 & I \\ -B & (k/2 - 1)v_0 I \end{pmatrix}$$

These eigenvalues are supplied by the following lemma.

Lemma 5. <u>Suppose</u> λ <u>is an eigenvalue of</u> B. <u>Then</u>

$$\zeta^{\pm} = \tfrac{1}{2}(\mu \pm \sqrt{\mu^2 - 4\lambda})$$

<u>is an eigenvalue of</u> A', <u>where</u> $\mu = (k/2 - 1)v_0$.

Proof. Suppose x is an eigenvector of B associated to the eigenvalue λ. Then the vector $(x, \alpha x)$ is an eigenvector for A' iff

$$\alpha^2 - \mu\alpha + \lambda = 0$$

iff $\alpha = \zeta^{\pm}$. This completes the proof of the lemma.

If $V(s)$ is a Morse function and $k \neq 2$, then all of th eigenvalues of B are real and non-zero, and so all of the eigenvalues of A' have non-zero real part. This follows sinc in the non-degenerate case, $v_0 \neq 0$. qed

Hence for each rest point p in Λ one has a stable

and unstable manifold denoted by $W^s(p)$ and $W^u(p)$. One can in fact relate the dimensions of these manifolds to the index of the corresponding critical point and sgn v_0 using Lemma 5. See [De 1] for details. Also, one can prove that, generically, the stable and unstable manifolds of all equilibria meet transversely in Λ. By generically we mean within the set of potentials that are homogeneous of degree $-k$. We again refer to [De 1] for the proof.

Flows on Λ having the above properties are called Morse-Smale flows. More precisely, a flow on a manifold is Morse-Smale if

1. The non-wandering set consists of a finite number of hyperbolic rest points and closed orbits.
2. All stable and unstable manifolds meet transversely.

In our case, the flow on Λ is gradient-like, so that there are no closed orbits on Λ, only rest points. In such a case we call the singularity a Morse-Smale singularity. The central force problems of the previous section do not have Morse-Smale singularities, since the rest points in Λ are not hyperbolic. However, the examples which follow are Morse-Smale and, as we have shown above, are more typical.

To summarize this section, we have shown that

Theorem 6. *An open and dense set of potentials which are homogeneous of degree $-k$ and which have isolated singularities at the origin have only non-degenerate, Morse-Smale singularities.*

§1.5 Collision Orbits.

We now discuss what happens near the collision manifold in the case of a Morse-Smale singularity. Let C denote the set of solutions which end in collision with the singularity, and let E denote the set of ejection orbits, i.e., those orbits which begin at the singularity.

Proposition 1. <u>Suppose there is a Morse-Smale singularity at the origin. Then both C and E consist of a finite union of submanifolds, one for each central configuration.</u>

Proof. Let s_0 be a central configuration, i.e., a critical point of $V(s)$. So the point $p = (v_0, s_0, 0)$ is a hyperbolic rest point for the flow on Λ. We claim that p is a hyperbolic rest point for the entire flow on an energy surface. Using the results of the previous section and particularly the linearization matrix (7), one checks easily that v_0 is an eigenvalue for the linearization whose associated eigenvector is transverse to Λ. Now $v_0 = \pm\sqrt{-2V(s_0)}$, so that p is a hyperbolic rest point in $H^{-1}(e)$, not just on the boundary Λ, since $V(s_0) \neq 0$ in the non-degenerate case. If $v_0 < 0$, we thus have a stable manifold at p in $H^{-1}(e) - \Lambda$ which is one dimension larger than the dimension of the stable manifold at p in Λ. These are the collision orbits which asymptotically attain the configuration s_0. When $v_0 > 0$, we similarly find a submanifold of ejection orbits in $H^{-1}(e) - \Lambda$ given by the corresponding unstable manifold. Since the flow on Λ is gradient-like, it follows easily that in fact all collision and ejection orbits lie in such a stable or unstable manifold. And since there are only a finite number of critical points for $V(s)$, the result follows. qed

We remark that the above proof shows that corresponding to any central configuration s_0, there are two invariant manifolds. One consists of collision orbits asymptotic to the rest point $(v_0, s_0, 0)$ with $v_0 < 0$, and the other consists of ejection orbits emanating from $(-v_0, s_0, 0)$.

As in the central force potentials, there are special collision and ejection orbits which are associated to each central configuration. These are the <u>homothetic orbits</u> given as follows. From equation (1) of §1.4, it follows that if s_0 is a central configuration and $u = 0$, then $\dot{s} = 0$ and $\dot{u} = 0$. Hence the r,v-plane defined by $s = s_0$, $u = 0$ is invariant. As in the Kepler problem, the vector field on this plane is given by

$$\dot{r} = rv$$
$$\dot{v} = (k/2)v^2 + kV(s_0) = r^k e.$$

The phase portrait in the r,v-plane is sketched in Fig. 1. In particular, there is a unique homothetic orbit in each negative energy level which begins and ends in collision and which projects to configuration space along the ray $s = s_0$. These homothetic orbits can therefore be interpreted as heteroclinic orbits connecting distinct rest points. As such, they play a key role in the global dynamics of the examples discussed below.

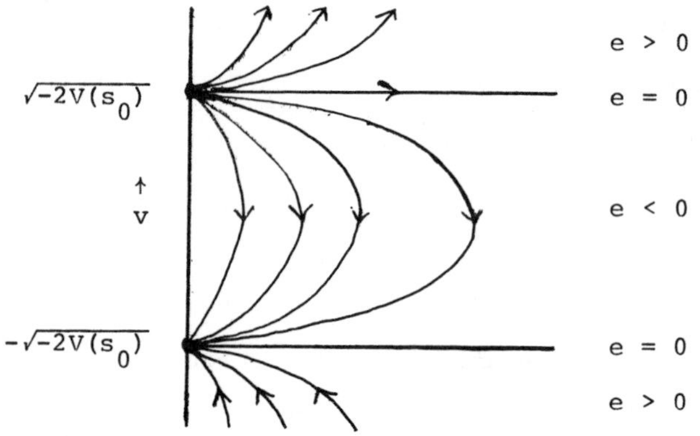

Fig. 1 The flow on an invariant r,v-plane. Each orbit is a homothetic orbit.

§1.6 The Anisotropic Kepler Problem

Our goal in this section is to describe the dynamics of a classical mechanical system which is much more complicated than the previous examples. This is the anisotropic Kepler problem introduced by Gutzwiller in [Gu 1] to model certain quantum mechanical systems. The system is actually a one parameter family of Hamiltonian systems, depending analytically on a parameter $\mu \geq 1$. When $\mu = 1$, we have the usual Kepler problem discussed in §1.3. As μ increases, the system becomes more and more complicated until, for $\mu > 9/8$, numerical evidence indicates that the system is ergodic. We will give partial results in this direction in the third part of these notes.

For $\mu \geq 1$, the anisotropic Kepler problem is described by the equations

$$\ddot{q}_1 = -\mu q_1/|q|^3$$
$$\ddot{q}_2 = -q_2/|q|^3$$

This system differs from the central force problems in that the force vector is directed more toward the q_2-axis, forcing the solution curves to cross that axis more often than the q_1-axis. Introducing the mass matrix

$$M^{-1} = \begin{pmatrix} \mu & 0 \\ 0 & 1 \end{pmatrix}$$

and the potential energy $V(q) = -1/|q|$, one may write this system in Hamiltonian form

$$\begin{aligned} \dot{q} &= M^{-1}p \\ \dot{p} &= -\nabla V(q) \end{aligned} \qquad (1)$$

where the momentum vector p is defined by the first equation. The total energy function is

$$H(q,p) = \tfrac{1}{2}p^t M^{-1} p + V(q).$$

For negative energy, the Hill's region is the disk given by $|q| \leq -1/e$.

We remark that an equivalent system results if one uses the Hamiltonian

$$H'(q,p) = \tfrac{1}{2}|p|^2 - \frac{1}{\sqrt{\mu q_1^2 + q_2^2}} .$$

Here the potential rather than the kinetic energy is anisotropic.

In McGehee coordinates, (1) goes over to

$$\begin{aligned}
\dot{r} &= rv \\
\dot{v} &= u^2 + \tfrac{1}{2}v^2 + V(\theta) \\
\dot{\theta} &= u \\
\dot{u} &= -\tfrac{1}{2}vu - V'(\theta)
\end{aligned} \tag{2}$$

where the potential $V(\theta)$ is given by

$$V(\theta) = \frac{-1}{\sqrt{\mu\cos^2\theta + \sin^2\theta}} \tag{3}$$

$V(\theta)$ has two non-degenerate maxima at $\theta = 0$ and $\theta = \pi$, and two non-degenerate minima at $\theta = \pi/2$ and $3\pi/2$. The graph of $V(\theta)$ is given in Fig. 1.

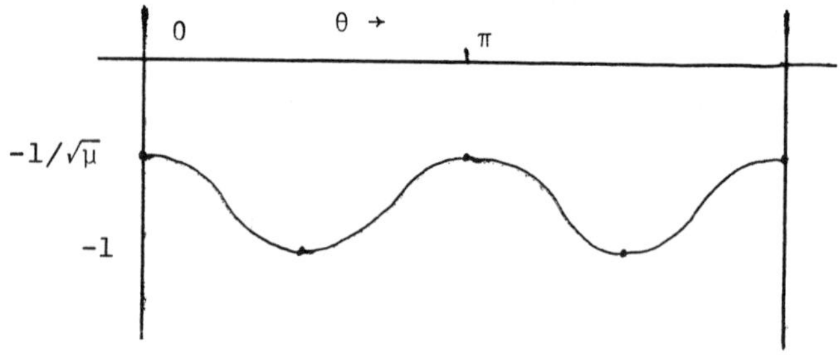

Fig. 1 The graph of $V(\theta)$.

The collision manifold Λ is determined from the energy relation

$$re = \tfrac{1}{2}(u^2 + v^2) + V(\theta) \tag{4}$$

and is sketched in Fig. 2. When $r = 0$, this equation deter-

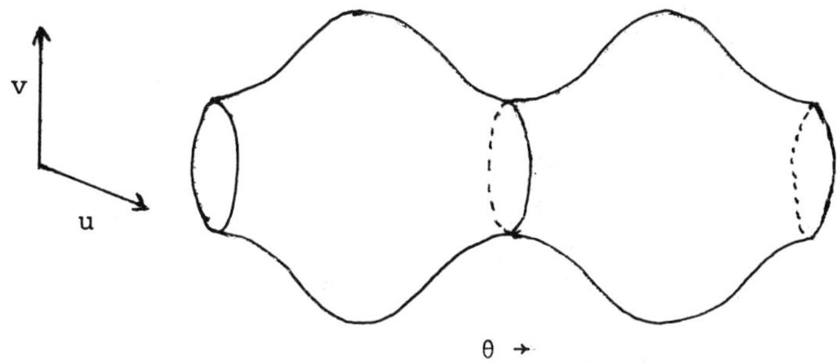

Fig. 2 The collision manifold for the anisotropic Kepler problem.

mines a torus.

From the results of §4, there are 8 rest points for the flow on Λ: two sinks, two sources, and four saddles. The eigenvalues of the linearization at a critical point of the form $(v_0, \theta_0, 0)$ are given by

$$-\tfrac{1}{4}v_0 \pm \tfrac{1}{2}\sqrt{\tfrac{1}{4}v_0^{\,2} - 4V''(\theta_0)}$$

where $v_0 = \pm\sqrt{-2V(\theta_0)}$. Hence the sinks and sources occur when $V''(\theta_0) > 0$, i.e., at $\theta_0 = \pi/2$ and $3\pi/2$. At these points we have $V(\theta_0) = -1$ and $V''(\theta_0) = (\mu-1)/2$, so that the radical in the above expression reduces to $\sqrt{(9-8\mu)}$. Hence the eigenvalues at the sinks and sources are complex when $\mu > 9/8$. This implies that orbits spiral into and away from the sinks and sources, and, as we show below, this causes nearby solutions to behave quite erratically.

The ultimate behavior of the stable and unstable manifolds of the saddles has been studied by Gutzwiller [Gu 1,2]. Briefly, for all $\mu > 1$, he finds that each branch of one of these invariant manifolds is asymptotic to a distinct sink or source, after

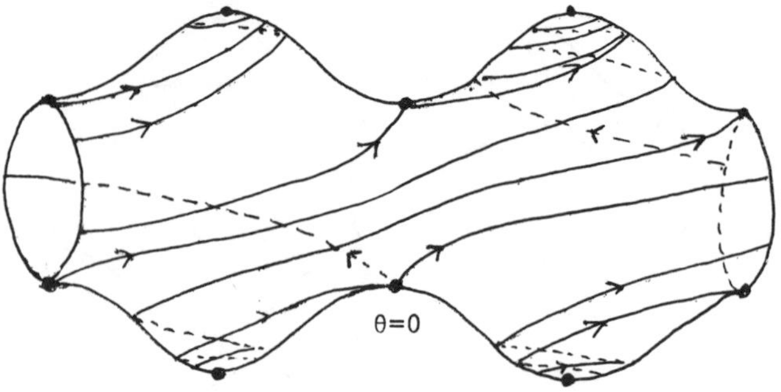

Fig. 3 The stable and unstable manifolds of the saddles. The branches of the invariant manifolds on the backside of Λ can be obtained by the reflection $v \to -v$, $u \to -u$.

θ increases or decreases by $3\pi/2$ or $\pi/2$. That is, the two branches of the stable manifold of any saddle emanate from different sources, while the two branches of the unstable manifold die in different sinks.

From §5, we also have homothetic orbits traveling along each ray $\theta = 0, \pi/2, \pi$, and $3\pi/2$. For negative energy, these solutions are bounded and connect distinct rest points in Λ. For $\mu > 9/8$, solutions near these orbits behave quite erratically. Consider an orbit near the collision orbit on $\theta = 0$. The collision orbits along $\theta = \pi$ are handled completely analogously. After passing close to the rest point in Λ, the orbit follows one of the two branches of the unstable manifold emanating from the rest point. At this rest point, $v_0 < 0$, so from Fig. 3, along one of these branches, θ decreases by $3\pi/2$, while along the other, θ increases by $3\pi/2$. Thus we have two distinct types of qualitative behavior for nearby solutions, depending on which branch the solution follows.

Now after passing close to collision, each solution leaves a neighborhood of the origin spiralling about the corresponding ray $\theta = \pm\pi/2$, at least when $\mu > 9/8$. See Fig. 4.

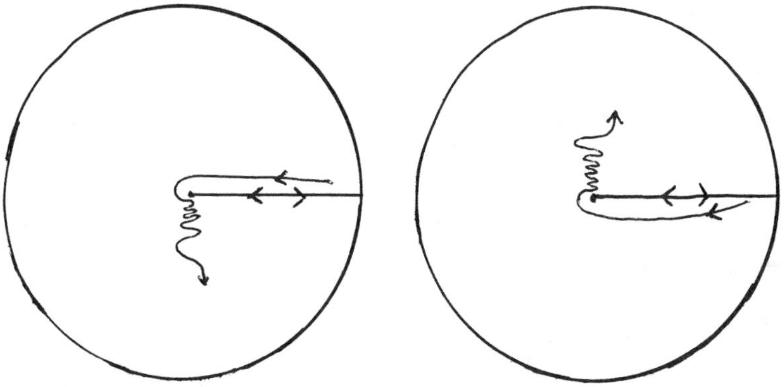

Fig. 4 The qualitative behavior of solutions near the homothetic orbit along $\theta = 0$.

Thus the flow on the collision manifold completely determines the qualitative behavior of orbits close to the collision orbit on $\theta = 0$. One easily makes a similar analysis for the other homothetic orbits.

We also wish to point out how the flow on the collision manifold enables us to find infinitely many distinct collision/ejection and periodic solutions when $\mu > 9/8$. Consider the piece of Λ near the rest points at $\theta_0 = 0$ and $\pi/2$ with $v_0 > 0$. Erect an annular surface of section A for the flow defined by $v = v^*$ with $\sqrt{-2V(0)} < v^* < \sqrt{-2V(\pi/2)}$. For r sufficiently small, A is transverse to the flow.

Now one branch of the unstable manifold at $\theta = 0$ in Λ crosses A at a point, say p^*. From the results of §1.5 the entire unstable manifold is two dimensional, and it follow that it must meet A transversely (at least near p^*) in a smooth curve γ tending to p^*. Only p^* lies in Λ, howeve

Now the homothetic orbit leaving $\theta = \pi/2$, $v_0 > 0$ reaches the zero velocity set Z at a point which we call q^* Consider a sufficiently small neighborhood N of q^* in Z. Under backwards time, the orbit through each point in N passes close to the rest point and then moves on to meet A. One checks easily that the trace of these points of intersection with A form a spiral winding down to the circular intersection of A with Λ. In particular, this spiral meets γ at infinitely many distinct points, say p_i, $i = 1,2,\ldots$ The orbit through each p_i is an ejection orbit which also meets the oval of zero velocity. By symmetry, such an orbit must fall back upon itself and is therefore also a collision orbit at $\theta = 0$. See Fig. 5.

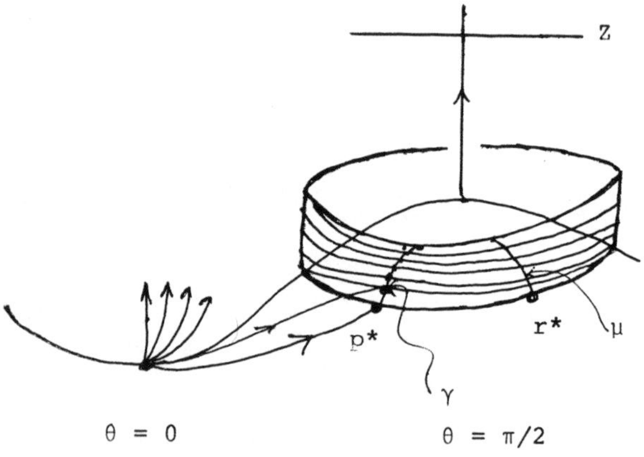

Fig. 5 The proof of Proposition 1.

We have thus proven:

Proposition 1. <u>There exist infinitely many distinct collision/ejection orbits in the anisotropic Kepler problem on each negative energy level when</u> $\mu > 9/8$.

In Fig. 6 we have sketched several of these collision/ejection orbits. Note that they are characterized by a rapid oscillation about the q_2-axis.

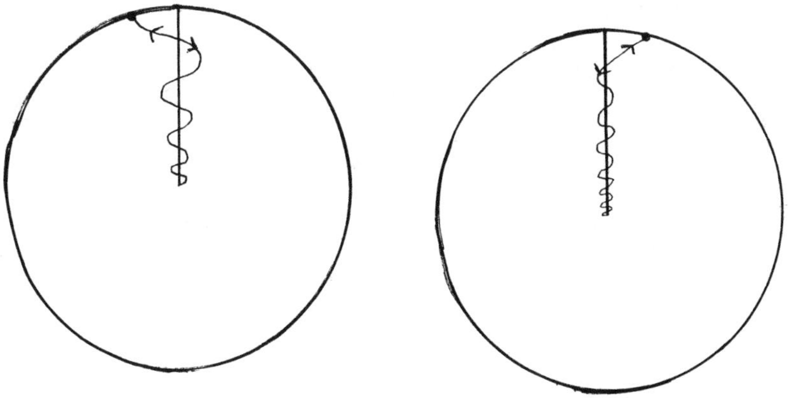

Fig. 6 Collision/ejection orbits in the anisotropic Kepler problem.

We conclude this section by proving the existence of infinitely many symmetric closed orbits for the anisotropic problem, again when $\mu > 9/8$. The homothetic orbit along $\theta = \pi$ meets the zero velocity set at a point s^*; let P be a small neighborhood of s^* in Z. It is known that Z is transverse to the stable manifold containing s^*. See [De 3].

This time we follow P forward in time under the flow. The orbit of each point in P comes close to the saddle in Λ at $\theta = \pi$ and then follows one of the branches of the unstable manifold. One of these branches meets the annular surface of section at a point r^*, and one may check easily that there is an arc of points μ in A limiting at r^* which comes from P. See Fig. 5. Now recall that there is a spiralling curve in A which comes from the oval of zero velocity. As before, this spiral intersects μ in infinitely many points. These points lie on orbits which meet the zero velocity set at two distinct points, and by our remarks in §1.1, therefore lie on symmetric closed orbits. We have therefore established:

Proposition 2. <u>There exist infinitely many distinct symmetric closed orbits in each negative energy level when</u> $\mu > 9/8$.

Some of these solutions are sketched in Fig. 7.

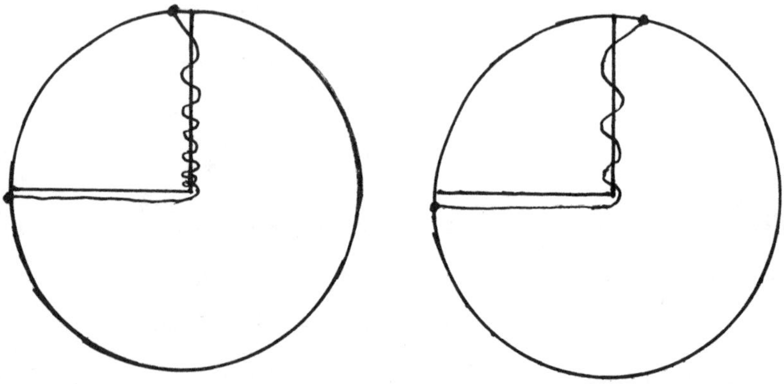

Fig. 7 Symmetric periodic solutions on a negative energy level in the anisotropic Kepler problem.

Again we note that these orbits are characterized by a rapid oscillation about the q_2-axis, followed by at most one crossing of the q_1-axis. The number of oscillations about the q_2-axis may be arbitrarily large. Hence these periodic solutions are quite different from the periodic solutions which occur on negative energy levels for the Kepler problem. Compare Fig. 6 in §1.3. Also, the existence of such orbits suggests the possibility of codifying closed orbits of the anisotropic problem by listing the number of crossings of the q_2-axis in between successive crossings of the q_1-axis. This is an idea of Gutzwiller [Gu 2] which motivates the introduction of symbolic dynamics into the problem and which we will treat in much more detail in part 3.

§1.7 The Planar Isosceles Three Body Problem.

In this section, we discuss another example of a non-integrable classical mechanical system, a special case of the three body problem. Our aim here is to show how the techniques of the previous sections allow us to recover classical results of Euler, Lagrange, Siegel, and others relatively easily, as well as to describe the local behavior of near-collision orbits relatively simply.

This problem is remarkably similar to the anisotropic Kepler problem in certain respects, as we will show later. One major difference is that the singularity at the origin is not isolated as in the previous examples. Presumably, our remedy for this situation extends to other examples with non-isolated singularities.

To state the problem, we take three point masses $m_1 = m$ and $m_3 = \varepsilon$. Without loss of generality, we may assume that $m_1 = m_2 = 1$. Suppose m_1 and m_2 are positioned symmetricall with respect to the y-axis, and with symmetric initial velocities. And suppose also that the third mass lies on the y-axis with initial velocity parallel to the axis. See Fig. 1.

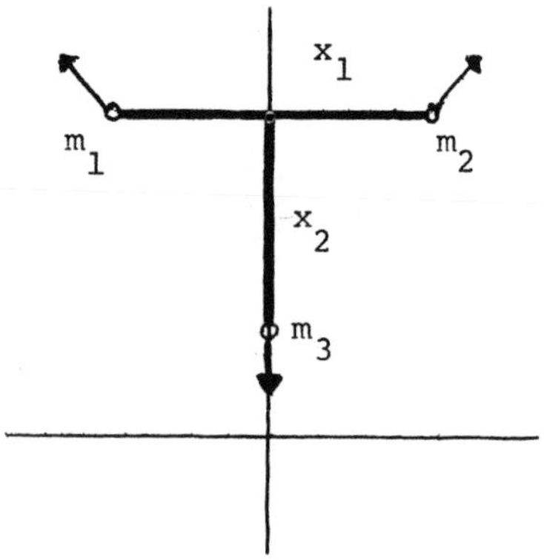

Fig. 1 Jacobi coordinates in the isosceles problem.

If q_i denotes the position of the ith mass, then the differential equation may be written

$$m_i q_i'' = -\nabla_i V$$

where

$$V(q_1,q_2,q_3) = -\sum_{k<j} m_i m_j / |q_i - q_j| \; .$$

We may make several standard reductions to reduce the number of degrees of freedom in the problem to two. First, by symmetry, if we know $q_1(t)$, then $q_2(t)$ is determined, since the particles always remain in a (possibly degenerate) isosceles triangle configuration. Secondly, we may fix the center of mass at the origin since

$$\sum_i m_i q_i'' = 0$$

and the equations of motion remain unchanged. Hence the positions of all three particles may be determined from Jacobi coordinates x_1, x_2 where $x_1 \geq 0$ denotes the distance between m_1 and m_2, and x_2 the signed distance between m_3 and the center of mass of m_1 and m_2. We take $x_2 > 0$ when m_3 lies above the other masses. See Fig. 1.

In these coordinates, the differential equation is given by

$$\begin{aligned} x_1'' &= -2/x_1 - 8\varepsilon x_1/(x_1^2 + 4x_2^2)^{3/2} \\ x_2'' &= -8(2+\varepsilon)/(x_1^2 + 4x_2^2)^{3/2} \quad . \end{aligned}$$

We refer to Pollard's book [Po] for the derivation of these equations. Note that these equations are singular when $x_1 = 0$, which corresponds to a double collision between m_1 and m_2, and also when $x_1 = x_2 = 0$, which corresponds to a triple collision or total collapse of the system. Hence this system differs from the previous examples in that the singularity is not isolated.

We first remove the singularity at the origin. Toward that end, we introduce the momenta

$$p_1 = \tfrac{1}{2}x_1'$$
$$p_2 = (2\varepsilon/2+\varepsilon)x_2'$$

Then the system may be written in Hamiltonian form with the Hamiltonian

$$H = \tfrac{1}{2}p^t M^{-1} p + V(x)$$

where

$$M^{-1} = \begin{pmatrix} 2 & 0 \\ 0 & (2+\varepsilon)/2\varepsilon \end{pmatrix}$$

and

$$V(x_1,x_2) = -\frac{1}{x_1} - \frac{4\varepsilon}{\sqrt{x_1^2 + 4x_2^2}} \qquad (1)$$

The Hill's region for negative energy is sketched in Fig. 2.

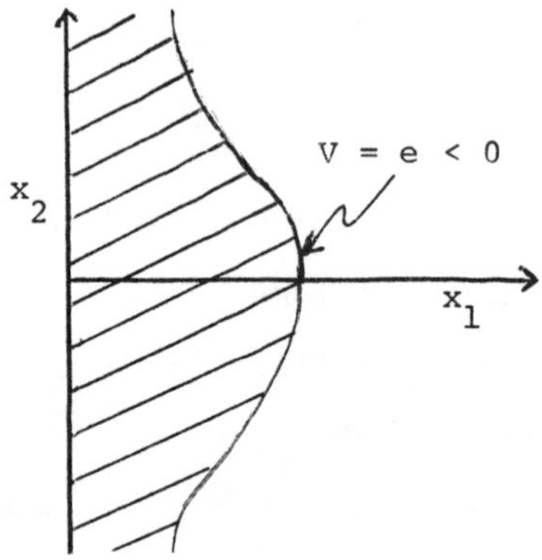

Fig. 2 The Hill's region for negative energy in the isosceles problem.

In McGehee coordinates the equations become

$$\begin{aligned} \dot{r} &= rv \\ \dot{v} &= \tfrac{1}{2}v^2 + u^2 + V(\theta) \\ \dot{\theta} &= u \\ \dot{u} &= -\tfrac{1}{2}vu - V'(\theta) \end{aligned} \qquad (2)$$

where

$$V(\theta) = -\frac{1}{\sqrt{2}\cos\theta} - \frac{4\varepsilon^{3/2}}{\sqrt{2\varepsilon + 4\sin^2\theta}} \qquad (3)$$

is defined for $-\pi/2 < \theta < \pi/2$. The graph of $V(\theta)$ is given in Fig. 3.

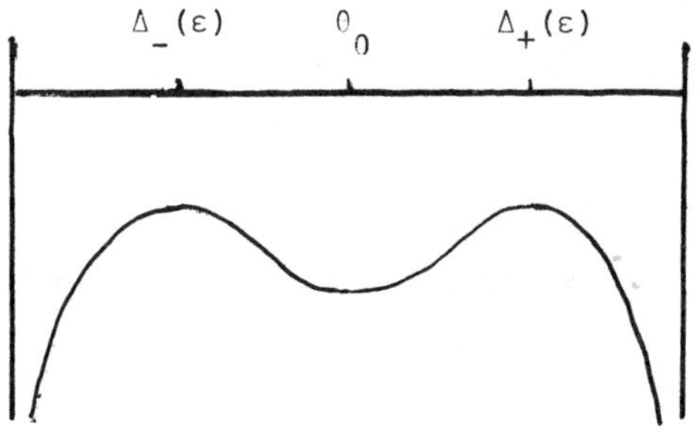

Fig. 3 The graph of $V(\theta)$ in the isosceles problem.

There are three non-degenerate critical points: a local minimum at $\theta = 0$, and two maxima at $\Delta_{\pm} = \arcsin(\sqrt{3\varepsilon/(2+4\varepsilon)})$

We will interpret the flow on the triple collision manifold later. First, however, we will eliminate the singularities in the above equations when $\theta = \pm\pi/2$. This will be achieved by simultaneously regularizing each double collision via a procedure due to Sundman [Su]. This necessitates a further change of variables. We remark that McGehee carried out a similar procedure for the collinear three body problem [McG 1]

Let $W(\theta) = -\cos\theta\, V(\theta)$. One checks easily that $W(\theta)$ is a positive, real analytic function on $[-\pi/2,\pi/2]$. Now introduce a new tangential component of velocity

$$w = u\cos\theta/\sqrt{W(\theta)}$$

as well as a further change of time scale

$$\frac{dt}{ds} = \frac{\cos\theta}{\sqrt{W(\theta)}}\ .$$

The differential equation becomes

$$\begin{aligned}
\dot r &= \frac{\cos\theta}{\sqrt{W(\theta)}}\, rv \\
\dot v &= \sqrt{W(\theta)}\ \left[1 - \frac{\cos\theta}{2W(\theta)}(v^2 - 4re)\right] \\
\dot\theta &= w \\
\dot w &= \sin\theta\left[-1 + \frac{\cos\theta}{W(\theta)}(v^2 - 2re)\right] - \frac{vw\cos\theta}{2\sqrt{W(\theta)}} + \\
&\qquad + \frac{W'(\theta)}{W(\theta)}\,(\cos\theta - w^2/2)
\end{aligned} \tag{4}$$

with energy relation

$$w^2/2\cos\theta - 1 = \frac{\cos\theta}{W(\theta)}\,(re - v^2/2)\ . \tag{5}$$

Here the dots indicate differentiation with respect to s. This gives a real analytic vector field on $[0,\infty) \times \mathbb{R} \times [-\pi/2,\pi/2] \times \mathbb{R}$. Note that the vector field does not vanish when $\theta = \pm\pi/2$. Hence we have extended solution curves through double collision via an "elastic bounce".

The triple collision manifold is given by

$$w^2/2 + v^2\cos^2\theta/W(\theta) = \cos\theta \ . \qquad (6)$$

When $\theta = \pm\pi/2$, this gives $w = 0$, v arbitrary, i.e., a straight line. For all other values of θ, this equation determines an ellipse in the corresponding v,w-plane. The triple collision manifold can then be sketched as in Fig. 4.

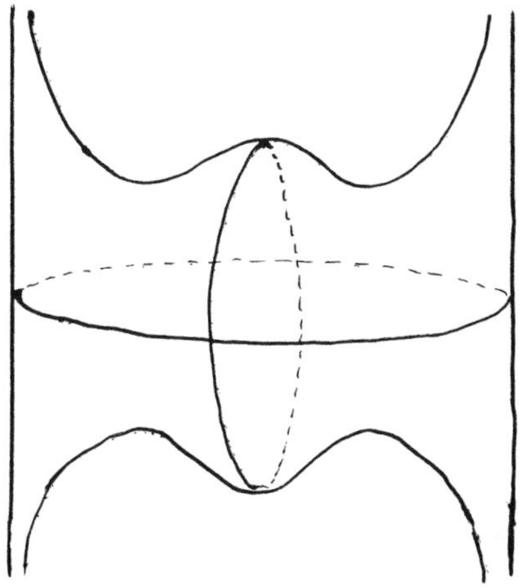

Fig. 4 The triple collision manifold in the isosceles problem after regularization of double collisions.

Note that, in this case, the triple collision manifold is noncompact; topologically, it is a sphere with four points removed.

Using the results of §1.4, one checks easily that there are four saddle points for the flow corresponding to the critical points at $\theta = \Delta_\pm$, and a sink and source at $\theta = 0$. Moreover, a simple computation shows that the eigenvalues at the sink and source are complex when $\varepsilon < 55/4$. See Fig. 5.

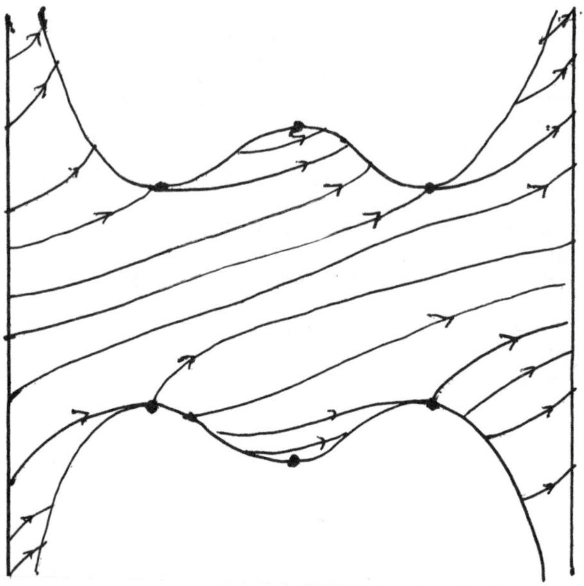

Fig. 5 The flow on the triple collision manifold.

The rest point structure allows us to prove several classical theorems quite easily:

Theorem 1 (Siegel) <u>The set of solutions of the isosceles three body problem which begin or end in triple collision is given by a union of lower dimensional submanifolds.</u>

Proof. There are two dimensional stable manifolds in each energy level asymptotic to the saddle points with $v < 0$, and a one dimensional stable manifold at the source in Λ. All of these orbits are collision orbits. We claim that, in fact, these are the only orbits which end in triple collision. For any orbit which ends in triple collision satisfies $r \to 0$.

The ω-limit set of such an orbit must then lie in $r = 0$, and one checks easily that it must be a compact set in Λ. Since the flow on Λ is gradient-like, it follows that the ω-limit set must be a rest point.

The proof for the ejection orbits is essentially the same. qed

The relatively simple structure of the flow on the triple collision manifold also allows us to interpret what happens to the three masses as they approach triple collision.

Theorem 2 (Sundman) <u>As they approach triple collision, the three masses tend to either a straight line or else to the vertices of an equilateral triangle.</u>

Proof. First suppose the orbit is asymptotic to the rest point at $\theta = 0$. In terms of the original Jacobi coordinates, x_1, x_2, it follows that

$$x_2/x_1 = (2+\varepsilon)\sin\theta/4\varepsilon\cos\theta \to 0$$

so that the particles tend toward a horizontal line. On the other hand, if the orbit is asymptotic to one of the rest points with $\theta = \Delta_\pm$, then we find

$$x_2^2/x_1^2 = (2+\varepsilon)\sin^2\theta/4\varepsilon\cos^2\theta \to 3/4$$

so that the particles tend to form an equilateral triangle in the limit. qed

In the classical literature on the subject, these asymptotic configurations are called central configurations, motivating our more general use of the term.

We remark that, since the corresponding invariant manifolds are one-dimensional, there is a unique orbit tending to and away from the rest point at $\theta = 0$ (the straight line configuration). This must then be the homothetic solution discusse in §1.5. Since $\theta = 0$ along this solution, it follows that x_2 is identically equal to zero along the solution also. Hence the third mass must remain fixed at the center of mass of the binary system. This special solution was first found by Euler [Eu] in the more general setting of the full three body probler

There are also a pair of homothetic solutions satisfying $\theta = \Delta_\pm$. For these special solutions, the particles remain for all time at the vertices of an equilateral triangle. Lagrange [La] was the first to find these solutions. We have sketched the projections of these solutions to configuration space in Fig. 6 for negative energy.

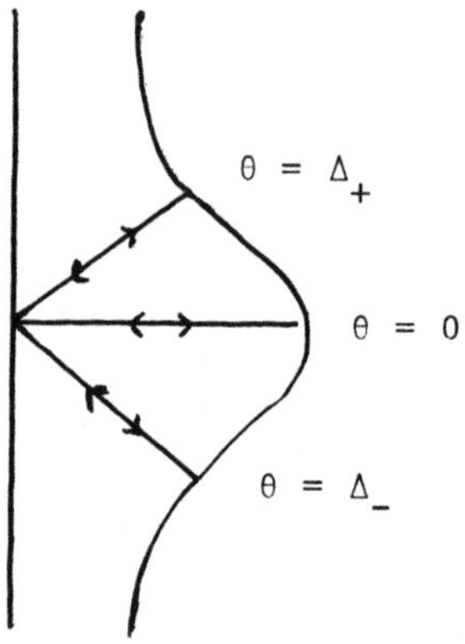

Fig. 6 The projection to the Hill's region of the special homothetic orbits discovered by Euler and Lagrange.

To study the behavior of solutions passing close to triple collision, we need some information about the behavior of the stable and unstable manifolds of the saddle points in Λ. This has been studied by Simo [Sim]. It turns out that there are many possibilities, depending on ε. We will discuss only the case of $0 < \varepsilon < \varepsilon_0$, for some small but positive ε_0 determined by Simo. For this case, the invariant manifolds are depicted in Fig. 7.

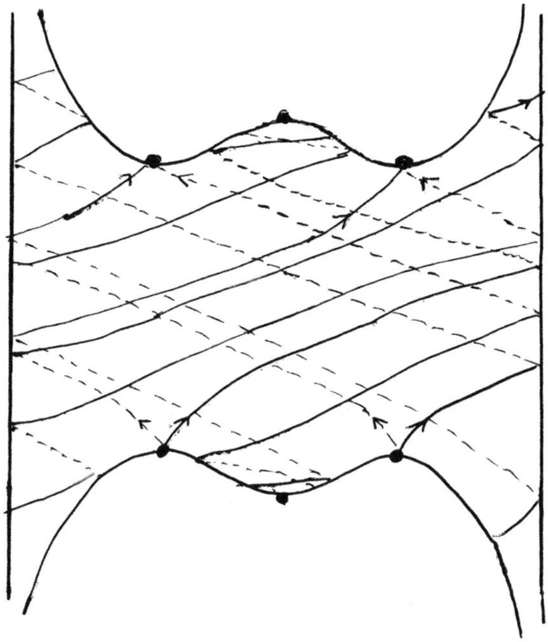

Fig. 7 The behavior of the stable and unstable manifolds of the saddle points in the isosceles problem.

In Fig. 7, note that one branch of each unstable manifold dies in the sink, while the other branch runs up the "arm" of the triple collision manifold. The behavior of the stable manifolds at the upper saddle points can be obtained via the reflection $v \to -v$, $\theta \to \theta$, $w \to -w$ which maps the unstable manifolds (resp. stable manifolds) of the rest points with $v < 0$ to the stable (resp. unstable) manifolds of the rest points with $v > 0$. Fig. 7 displays most of this information.

When ε is sufficiently small, the behavior of a solution curve which passes close to triple collision is easily decided. Consider a solution curve near an orbit which reaches triple collision at $\theta = \Delta_+$. If this orbit is close enough to the triple collision orbit, there are precisely two possibilities for its eventual behavior: either the orbit follows the branch of the unstable manifold which dies in the sink, or else it follows the branch which runs up the arm with $\theta = -\pi/2$ In the first case the orbit leaves a neighborhood of $r = 0$ near the line $\theta = 0$. Note that because of the spiralling near the sink, the orbit oscillates rapidly about $\theta = 0$, i.e., the third mass is left oscillating about the center of mass of the binary system, which in turn separates roughly along the x-axis

In the second case, one finds dramatically different behavior. One can prove that this branch of the unstable manifold meets the line $\theta = -\pi/2$ infinitely often as it runs up the arm of the collision manifold. One may also check that nearby solutions cross $\theta = -\pi/2$ an arbitrarily large number of times. Each crossing corresponds to a binary collision between m_1 and m_2.

Finally note that orbits following this branch of the unstable manifold exit from a neighborhood of triple collision with an arbitrarily large value for v, the radial component of velocity. This means that m_3 is separating from the binary system with arbitrarily large velocity. The third mass travels down the y-axis, while the binary pair moves up the axis in a tight oscillation featuring many double collisions. One may check that this large value of v leads to escape orbits, i.e., with $m_3 \to -\infty$, $m_1, m_2 \to +\infty$. The structure of the set of escape orbits is not as yet known.

We remark that the fact that the particles may leave a neighborhood of triple collision with arbitrarily large velocity was first noticed in the collinear three body problem by McGehee [McG 1]. Later, McGehee and Mather exploited this fact to construct a solution of the collinear four body problem which becomes unbounded in finite time [MM]. Their solution, however, admits infinitely many binary collisions, which they regularize as we did above. This then leaves open the question of the existence of a non-collision singularity in the n-body problem. Is there a solution which becomes unbounded in finite time and which does not experience any collisions at all?

Note that the rapid oscillation of solutions leaving a neighborhood of triple collision is reminiscent of the behavior of non-collision orbits in the anisotropic Kepler problem. We will exploit this idea in part three of these notes.

Finally, in Fig. 8 we have displayed the two different types of local behavior of solutions which are close to the homothetic solution along $\theta = \Delta_+$.

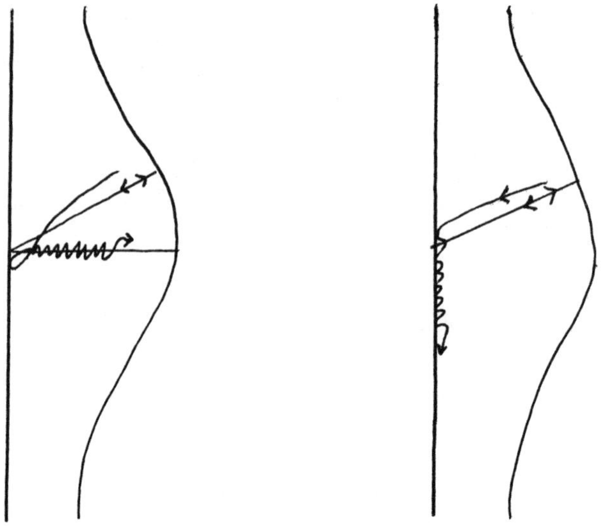

Fig. 8 Local behavior of near-collision orbits in the isoscele
problem.

Part Two: Symbolic Dynamics and Mappings with Singularities.

In this section, we move away from the setting of vector fields and classical mechanics to discuss mappings which exhibit random or stochastic behavior. In particular we will describe the shift automorphism and show how it enters into the dynamics associated to mappings. Later, in part three, we will return to the examples of part one and apply the methods of this section to at least partially describe the global dynamics associated with these systems.

We begin with a description of the shift automorphism and the "Smale horseshoe" mapping due to Conley. For proofs, we will refer to Moser's book [Mo]. Conley's axioms are then applied to an interesting example, the Hénon mapping. Later we describe mappings which possess singularities, notably the baker transformation and a nonlinear version of this mapping which is associated with the restricted three body problem.

Mappings which admit singularities are important for our purposes since, in part three, we will show that both the isosceles problem and the anisotropic Kepler problem can be reduced to Poincaré mappings on two dimensional spaces which admit singularities. The singularities will correspond to points on collision or ejection orbits. We will also present a model mapping for the anisotropic problem which is a type of "infinite" baker transformation.

§2.1 The Smale Horseshoe.

In this section we review the construction of the now classical Smale horseshoe mapping. Historically, this was one of the first mappings which had infinitely many periodic points and was also structurally stable. This mapping admits a complicated Cantor set as a closed, invariant set on which the mapping is equivalent to the well known shift automorphism. The presentation below is due mainly to Conley. For proofs we will refer to Moser's book [Mo]. See also Smale's original paper [Sm] and Nitecki's book [Ni].

We first recall the construction of the shift automorphism. Let A be a finite or infinite set of positive integers. The elements of A are called symbols and A itself is called the alphabet. Let $\Sigma = \Sigma_A$ denote the set of all doubly infinite sequences of the form

$$(s) = (\ldots s_{-2}\ s_{-1}\ s_0; s_1\ s_2 \ldots)$$

with $s_j \in A$. We define the shift automorphism $\sigma\colon \Sigma \to \Sigma$ by $(\sigma(s))_j = s_{j-1}$. That is,

$$\sigma(\ldots s_{-1}\ s_0; s_1\ s_2 \ldots) = (\ldots s_{-2}\ s_{-1}; s_0\ s_1\ s_2 \ldots).$$

To topologize Σ we take as a neighborhood basis of $(s^*) = (\ldots s^*_{-1}\ s^*_0;\ s^*_1\ s^*_2 \ldots)$ all "cylinder" sets of the form

$$U_j = \{(s) \mid s_i = s^*_i \text{ for } |i| \leq j\}.$$

Note that for any $\alpha_1, \alpha_2, \ldots \alpha_k \in A$ and any set of distinct indices $i_1, \ldots i_k$, the requirement that $s_{i_j} = \alpha_j$ for all $j \leq k$

defines an open subset of Σ. Also, in this topology, two sequences (s) and (t) are "close" if $s_i = t_i$ for $|i| \leq k$.

For finite alphabets, it is easy to see that Σ is a Cantor set. When A consists of infinitely many symbols, Σ turns out to be non-compact. Later we will describe and use several different compactifications of Σ.

In the topology above, the shift automorphism is clearly a homeomorphism. Moreover, periodic points for σ are given by repeating sequences. Thus periodic points for σ are dense in Σ. Also, σ admits a dense orbit. For example, if A consists of the two symbols 1 and 2, then the following sequence represents a dense orbit in Σ:

$$(s) = (\ldots s_{-1}\ s_0;\ 1\ 2\ \underbrace{11\ 12\ 21\ 22}_{\text{all 2 blocks}}\ \underbrace{111\ 112\ \ldots}_{\text{all 3 blocks}})$$

We simply enumerate all possible n blocks of symbols in some order, then all n+1 blocks, etc.

We now describe how the shift automorphism arises as an invariant subsystem of a smooth mapping of the plane. Let $F\colon \mathbb{R}^2 \to \mathbb{R}^2$ be given by

$$x_1 = F_1(x_0, y_0)$$
$$y_1 = F_2(x_0, y_0).$$

For simplicity, we will concentrate on the restriction of F to the square S in the plane given by $0 \leq |x|, |y| \leq 1$. We define a horizontal strip in S to be the region bounded by the non-intersecting graphs of two smooth functions

h_1, h_2: $[-1,1] \to [-1,1]$ which satisfy $|h_j'(x)| < 1$ for $j = 1,2$. Vertical strips are defined analogously. See Fig. 1

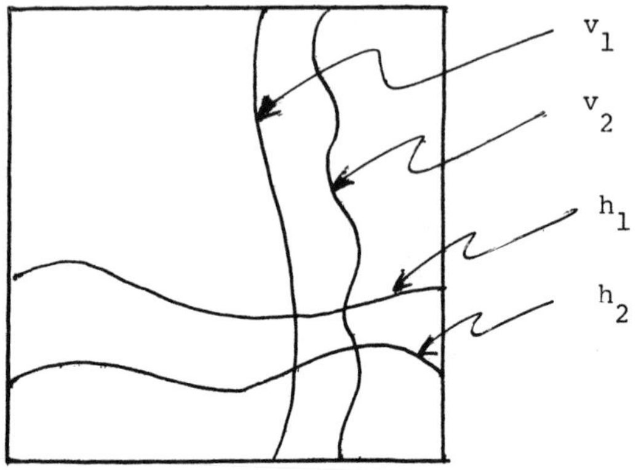

Fig. 1 Vertical and horizontal strips in s.

We now list two axioms which will guarantee that F possesses an invariant set in S on which F is equivalent to σ.

<u>Axiom 1</u>: There exist a finite or infinite number of horizontal and vertical strips H_i and V_i with $i \in A$. The mappin F takes V_i diffeomorphically onto H_i, with horizontal boundaries mapped to horizontal boundaries and vertical boundaries mapped to vertical boundaries.

We call this axiom the "strip" condition. Note that we need only verify this condition for F, not for its powers. Nevertheless, we will eventually be able to conclude something about F and all of its powers.

The next condition is the hyperbolicity condition. Let (ξ_0,η_0) be a tangent vector to S, and denote its image under dF by (ξ_1,η_1) and under dF^{-1} by (ξ_{-1},η_{-1}). Consider the sector bundles in the tangent bundle defined by

$$S_p^+ = \{(\xi_0,\eta_0) \in T_p\mathbb{R}^2 \mid |\eta_0| \le |\xi_0|\}$$

$$S_p^- = \{(\xi_0,\eta_0) \in T_p\mathbb{R}^2 \mid |\eta_0| \ge |\xi_0|\} .$$

Then the second axiom for F is:

<u>Axiom 2</u>: For some $\lambda > 1$, and for any $p \in \cup V_i$, we have

i. $dF(S_p^+) \subset S_{F(p)}^+$

ii. For $(\xi_0,\eta_0) \in S_p^+$, $|\xi_1| \ge \lambda|\eta_0|$.

Also, for $p \in \cup H_i$, we have

iii. $dF^{-1}(S_p^-) \subset S_{F^{-1}(p)}^-$

iv. For $(\xi_0,\eta_0) \in S_p^-$, $|\eta_{-1}| \ge \lambda|\eta_0|$.

S^+ is called the bundle of unstable vectors and S^- is the bundle of stable vectors. According to Axiom 2, dF stretches vectors in S^+ while dF^{-1} stretches vectors in S^-.

If F satisfies Axioms 1 and 2, then we have the following theorem due to Smale:

Theorem 1. <u>Suppose</u> F <u>is a smooth mapping</u> $F: \cup V_i \to \cup H_i$ <u>satisfying Axioms 1 and 2. Then there exists a subset</u> Λ <u>homeomorphic to</u> $\Sigma = \Sigma_A$ <u>on which</u> F <u>is topologically conjugate to the shift automorphism on</u> Σ. <u>More precisely, there is</u>

a homeomorphism $s\colon \Lambda \to \Sigma$ *such that the following diagram* commutes

$$\begin{array}{ccc} \Lambda & \xrightarrow{F} & \Lambda \\ {\scriptstyle s}\downarrow & & \downarrow{\scriptstyle s} \\ \Sigma & \xrightarrow{\sigma} & \Sigma \end{array}$$

For the proof we refer to Theorem 3.1 in Moser's book [Mo, pg. 72].

We remark that this theorem says that the restriction of F to Λ behaves dynamically exactly as the shift automorphism does on Σ. That is, F-periodic points are dense in Λ, F has a dense orbit in Λ, etc.

In the case of a finite alphabet or in the case of one of the compactifications discussed in later sections for the infinite alphabet case, one can also assert that Λ is a hyperbolic set. We will discuss this in more detail below. We simply remark here that the existence of hyperbolic sets like Λ for a diffeomorphism is the cornerstone for Smale's definition of Axiom A diffeomorphisms. This important class of mappings have been studied exhaustively in recent years. See [Bo] for more details.

One can generalize the notion of the full shift automorphism somewhat. This leads to the notion of subshifts of finite type. These are restrictions of the full shift to certain invariant subsets of Σ_A. For the definition, let B be a $k \times k$ matrix of 0's and 1's where k = card A. Define the subset $\Sigma_B \subset \Sigma_A$ by $(s) \in \Sigma_B$ iff $B_{s_i s_{i+1}} = 1$ for all i.

That is, Σ_B consists of all sequences which are allowable in the following sense: β can follow α in such a sequence iff the (α,β)-entry of B is 1. So the matrix with all entries 1 corresponds to the full shift, while any other matrix yields a σ-invariant subset Σ_B of Σ_A, and the induced shift automorphism is a subshift of finite type. See [Fr] for more details.

We conclude this section by briefly describing the conjugacy s in Theorem 1 above. We will use this construction over and over again in the sequel.

Define inductively

$$V_{s_0 s_{-1} \dots s_{-n}} = \text{closure}(V_{s_0} \cap F^{-1}(V_{s_{-1}}) \cap \dots \cap F^{-n}(V_{s_{-n}}))$$
$$= \text{closure}(V_{s_0} \cap F^{-1}(V_{s_{-1} \dots s_{-n}}))$$

It is easy to check that the $V_{s_0 \dots s_{-n}}$ form a nested sequence of vertical substrips in V_{s_0} as $n \to \infty$.

Similarly define

$$H_{s_1 \dots s_n} = \text{closure}(F(V_{s_1}) \cap \dots \cap F^n(V_{s_n}))$$
$$= \text{closure}(F(V_{s_1}) \cap F(V_{s_2 \dots s_n}))$$
$$= \text{closure}(H_{s_1} \cap F(H_{s_2}) \cap \dots \cap F^{n-1}(H_{s_n})) .$$

Again the $H_{s_1 \dots s_n}$ form a nested sequence of horizontal strips in H_{s_1}. The main part of the proof of the above Theorem consists of showing that

$$\bigcap_{n=1}^{\infty} V_{s_0 \dots s_{-n}} \cap H_{s_1 \dots s_n}$$

consists of a unique point in S. Then the homeomorphism s assigns the sequence $(\ldots s_{-1}\, s_0;\, s_1\, s_2 \ldots)$ to this point.

§2.2 The Hénon Mapping.

Axioms 1 and 2 in the previous section are often easy to verify in practice. In this section we give a rather simple example which illustrates this: we find a shift automorphism embedded in the dynamics of a quadratic diffeomorphism of the plane. The example is a two parameter family of mappings recently introduced by Hénon [H1] as a simple example of a mapping which seems to possess a "strange attractor". We will not discuss this particular phenomenon; rather, we will consider parameter values far away from those considered by Hénon. For our parameter values, the mapping of Hénon will be a concrete example of the Smale horseshoe.

The original Smale horseshoe mapping may be defined geometrically as follows. The mapping G is pictured in Fig. 1. $G|V_i$ is linear and hyperbolic and maps V_i onto H_i for $i = 1,2$. Axioms 1 and 2 are easy to verify, so there exists a subset Λ of the square on which G is conjugate to the two-shift.

The mapping of Hénon is geometrically similar and is given analytically by $(x_1, y_1) = F(x_0, y_0)$ where

$$x_1 = A + By - x^2$$
$$y_1 = x$$

where $A \in \mathbb{R}$ and $|B| < 1$ are two parameters. Actually, Hénon considers a slightly different form of this mapping given by

$$X_1 = 1 + Y - AX^2$$
$$Y_1 = BX$$

but these two forms are related for $A \neq 0$ by the linear conjugacy

$$x = X/A \qquad y = BY/A \, .$$

F has constant Jacobian determinant $-B$, and its global inverse is given by

$$x_{-1} = y$$
$$y_{-1} = (x - A + y^2)/B \, .$$

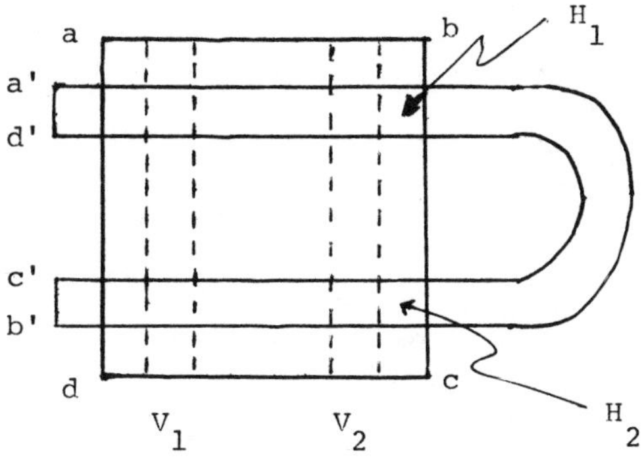

Fig. 1 The geometric description of the Smale horseshoe mapping.

For simplicity we will consider only the orientation reversing case with $0 < B \leq 1$. The orientation preserving case is handled in a similar fashion.

We assume throughout that

$$A > 3(1 + B)^2 \tag{1}$$

Actually, slightly lower A-values work as well, but this estimate is sufficient for our purposes. Let

$$R = R(A) = \tfrac{1}{2}(1 + B + \sqrt{(1 + B)^2 + 4A}\ ) \tag{2}$$

So R is positive for $A > 0$ and is the larger root of

$$R^2 - (1+B)R - A = 0 \ . \tag{3}$$

Note that

$$A - BR > R \tag{4}$$

for $A > 2(1+B)^2$.

Consider now the square S centered at the origin and having sides of length $2R$. We claim that F maps S into $\mathbb{R}^2$ as in Fig. 2. Indeed, the verticals $x = \pm R$ are mapped by F to the horizontals $y = \pm R$, while the fact that $F(a) = d$ and $F(b) = a$ follows immediately from (3). The fact that $F(c)$ and $F(d)$ lie to the left of S follows from the fact that $A - BR - R^2 < A + BR - R^2$. Finally, note that F maps $y = \pm R$ to parabolas opening to the left, with vertices on the x-axis. Since $F(0,R) = (A + BR, 0)$ and $F(0,-R) = (A - BR,0)$ it follows from (4) that both parabolas cut completely across S as in Fig. 2.

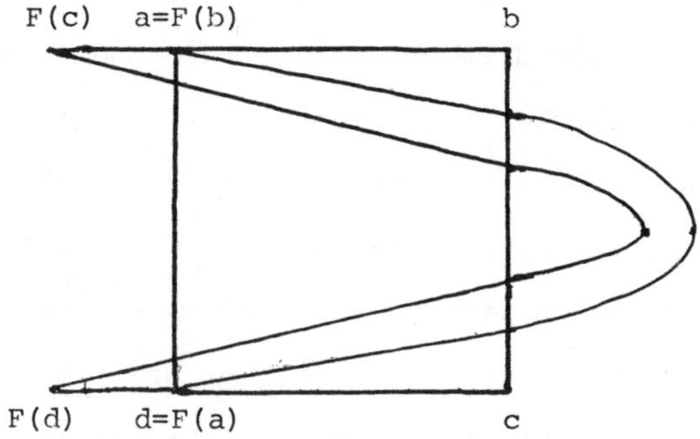

Fig. 2 Geometric description of F.

Similarly, F^{-1} maps S across itself as in Fig. 3.

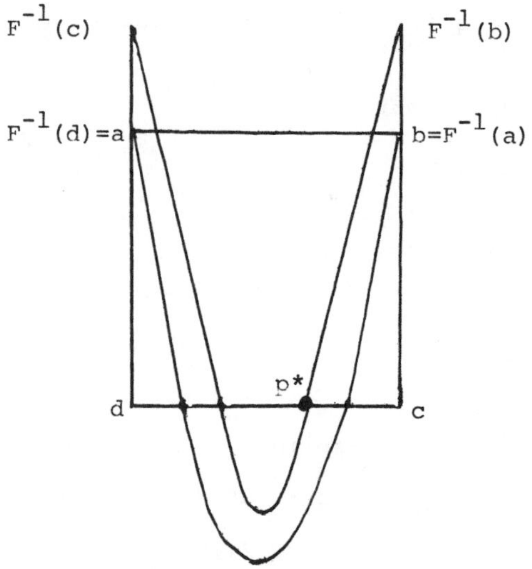

Fig. 3 Geometric description of F^{-1}.

Let p^* be the point on the right-hand intersection of the inner parabolic boundary of $F^{-1}(S)$ and the line $y = -R$ as in Fig. 3. One checks easily that the slope of the tangent line to the parabola at this point is larger than 1 if $A > 3(1+B)^2$. It then follows that the slopes of all tangent lines to vertical boundaries of $F^{-1}(S) \cap S$ are greater than one in absolute value. Hence these curves are vertical curves in the language of the previous section. Similarly, if $A > 3(1+B)^2$, it is easily checked that all horizontal boundaries of $F(S) \cap S$ in Fig. 2 are actually horizontal curves. This verifies Axiom 1 for Hénon's mapping.

To verify Axiom 2, first suppose that $|x| > \frac{1}{2}(\lambda+B)$ where $\lambda > 1$. Then we have, if $|\eta_0| < |\xi_0|$:

$$\begin{aligned} |\xi_1| &= |-2x\xi_0 + B\eta_0| > |2x||\xi_0| - |B\xi_0| \\ &> |2x||\xi_0| - B|\xi_0| \\ &> \lambda|\xi_0| \\ &= \lambda|\eta_1| \\ &> |\eta_1| . \end{aligned}$$

In particular, $|\xi_1| > |\eta_1|$ implies that dF preserves S^+, while $|\xi_1| > \lambda|\xi_0|$ implies that dF expands vectors in S^+ for these values of x. So Axioms 2i and 2ii are satisfied provided $|x| > (\lambda+B)/2$.

Now let x^* be the x-coordinate of p^* in Fig. 3. We claim that, provided A satisfies (1), we have $x^* > (1+B)/2$.

This implies that $|x| > (\lambda+B)/2$ for some $\lambda > 1$ and for all points in $F^{-1}(S) \cap S$.

To prove this, we note that x^* is determined by the equation

$$(x^*)^2 = A - (1+B)R.$$

If we write $A = k(1+B)^2/2$, then we find

$$R = (1 + \sqrt{1+2k})(1+B)/2$$

so that

$$(x^*)^2 = (2k - 2 - 2\sqrt{1+2k})(1+B)^2/4 .$$

For $k > 0$, this quantity increases with k and the left hand side equals 1 when $2k = 5 + 2\sqrt{5}$. Thus for A satisfying (1) we certainly have $x^* > (1+B)/2$. This proves Axioms 2i and 2ii.

Axioms 2iii and 2iv are verified similarly. We leave these proofs to the reader.

We remark that the invariant Cantor set for the Hénon mapping is a subset of the plane of measure zero. Yet this subset contains all of the interesting dynamics of the mapping. One can prove that all points outside of this set wander to infinity under either forward or backward iteration of F. See [DN]. Thus our set forms the entire non-wandering set for F and, as such, contains all of the non-trivial dynamics.

§2.3 The Baker Transformation and a Mapping Associated to the Restricted Three Body Problem.

In the last section we described a nonlinear mapping of the plane which possessed an invariant subset topologically conjugate to a complicated yet well understood model mapping, the shift automorphism. This conjugacy was valid on only a small subset of the plane, in fact, on a set of measure zero. In this section we describe another nonlinear mapping of the plane. Curiously, this mapping was also first introduced by Hénon [H2], this time in connection with his studies of the restricted three body problem. Unlike the previous mapping, chaotic behavior occurs here on the entire plane. Again we can model the dynamic behavior by a complicated yet well understood mapping, the baker transformation. For more details, we refer the reader to [AA].

We will consider the baker transformation of the open square $0 \leq |u|,|v| < 1$. The mapping is given analytically by

$$u_1 = \tfrac{1}{2}(1 + u + 2[v])$$
$$v_1 = 2(v - [v]) - 1$$

where $[v]$ denotes the greatest integer less than or equal to v. Note that $v = 0$ is mapped to $v = -1$, and hence out of the open square. So the domain of the mapping does not include the u-axis. These points are singular points for the mapping. Similarly, one checks easily that no point maps to the v-axis under this mapping.

Geometrically, the baker transformation is described

in Fig. 1.

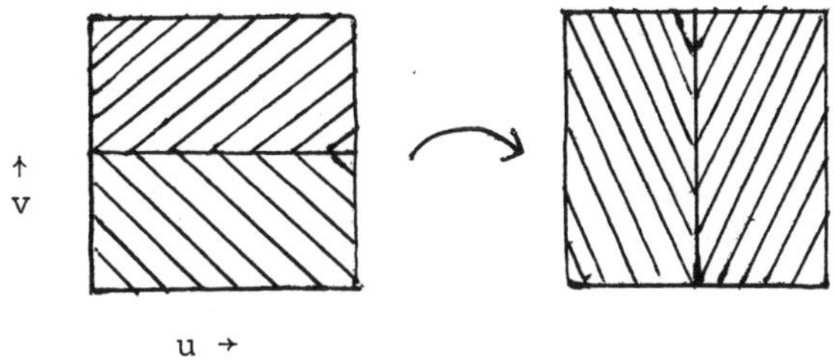

Fig. 1 Geometric description of the baker transformation.

This mapping contracts the upper and lower rectangles in the u-direction and expands them in the v-direction with a cut along the u-axis. As such, the mapping resembles the process of "rolling the dough", hence the curious name.

The u- and v-axes play the role of singularity sets for the baker transformation. Any point which maps to $v = 0$ under forward iteration has an orbit which terminates. Similarly, any orbit which meets $u = 0$ in backward time also terminates. The behavior of collision and ejection orbits in the classical mechanical systems of part one will be mirrored in such singular points.

The baker transformation is intimately related to the shift automorphism introduced earlier. Let $(s) = (\dots s_{-1}\ s_0;\ s_1\ s_2 \dots)$ be a sequence of +1's and -1's, and define the mapping

$$u = \sum_{i=1}^{\infty} s_i/2^i \qquad\qquad v = \sum_{i=0}^{-\infty} s_i/2^{1-i} \tag{1}$$

taking such sequences into points in the closed square. One may check easily that, under this mapping, the shift automorphism on sequences goes over to precisely the baker transformation, and the topology on the set of sequences induces the usual topology on the square.

Note that any sequence which is constant to the right or left of the semi-colon is mapped to the boundary of the square. Any sequence which is eventually constant (in either direction) may be shifted to such a point; hence such points eventually leave the square under iteration of the baker transformation.

Note that such eventually constant sequences are dense in the square. Indeed, they correspond to points in the square one of whose coordinates is a dyadic rational. Note also that sequences of the form $(\ldots s_k +1 -1 -1 -1 \ldots)$ and $(\ldots s_k -1 +1 +1 +1 \ldots)$ are to be identified under the projection above. We will encounter similar identifications later.

Finally note that the baker transformation preserves Lebesgue measure in the square. It can be shown that this mapping is ergodic and even a K-system on the square [AA]. As with the shift automorphism, there is also a dense orbit as well as dense periodic points.

We turn now to a nonlinear analogue of the baker transformation. Consider the mapping $(x_1, y_1) = F(x_0, y_0)$ given by

$$x_1 = x_0 + 1/y_0$$
$$y_1 = y_0 - x_0 - 1/y_0 .$$

In [H2], Hénon introduces this mapping as an asymptotic form of the equations of motion of a special case of the restricted three body problem. One checks easily that F preserves Lebesgue measure in the plane, and its inverse is given by

$$x_{-1} = x_0 - 1/(x_0 + y_0)$$
$$y_{-1} = x_0 + y_0 .$$

F is singular along the x-axis while F^{-1} is undefined along the line $y = -x$.

The interesting fact about F is that it is topologically conjugate to the baker transformation of the open square. More precisely, we have:

Theorem 1. _There is a homeomorphism_ $h: \mathbb{R}^2 \to \{(u,v) \mid 0 \leq |u|, |v| < 1\}$ _which gives a topological conjugacy between_ F _and the baker transformation on their respective domains._

We remark that this result is purely topological; despite the fact that both F and the baker transformation preserve Lebesgue measure, this theorem implies nothing about the ergodicity of F. In fact, it is an interesting open problem whether or not F is ergodic.

The remainder of this section is devoted to sketching a few of the details of the proof of Theorem 1. We refer to [De 6] for complete details.

Define the bundles of sectors in the tangent bundle

$$S^+ = \{(\xi,\eta) \mid \xi\eta \leq 0\}$$
$$S^- = \{(\xi,\eta) \mid \xi\eta \geq 0\} .$$

So S^- consists of all tangent vectors to $\mathbb{R}^2$ which lie in the first or third quadrant. One checks easily that $dF(S^+) \subset S^+$ and $dF^{-1}(S^-) \subset S^-$. Moreover, if we let $\|(\xi,\eta)\| = |\xi| + |\eta|$, we also have for $(\xi_0,\eta_0) \in S^+$, $\|(\xi_1,\eta_1)\|$ min $(2, 1 + 2/y^2)\|(\xi_0,\eta_0)\|$. Also, if $(\xi_0,\eta_0) \in S^-$, then $\|(\xi_{-1},\eta_{-1})\| \geq (1 + 1/(x+y)^2)\|(\xi_0,\eta_0)\|$ where $(\xi_{-1},\eta_{-1}) = dF^{-1}(\xi_0,\eta_0)$. These facts together imply that F is "non-uniformly" hyperbolic; in the sum norm, dF expands vectors in S^+ while dF^{-1} expands vectors in S^-. Note that the expansion coefficient tends to 1 as $|y| \to \infty$, hence the non-uniformity of the hyperbolicity.

We define an <u>unstable curve</u> to be a smooth curve in the plane all of whose tangents lie in the interior of the unstable sector bundle S^+. <u>Stable curves</u> have tangents lying in S^-. It is clear that F maps unstable curves to unstable curves, and, in so doing, increases the length of these curves.

We now describe the conjugacy between F and the baker transformation. Let $(x_n,y_n) = F^n(x_0,y_0) = F^n(p)$ and suppose that (x_n,y_n) is defined for all integers n. Let $s(p) = (\ldots s_{-1}\ s_0;\ s_1\ s_2 \ldots)$ where

$$s_j = s_j(p) = \begin{cases} +1 & \text{if } y_{-j} > 0 \\ -1 & \text{if } y_{-j} < 0 \,. \end{cases}$$

The reversal of signs is confusing, but necessary. So the codin specifies which half-plane the jth iterate of p lies in. We may extend this coding to points whose orbits terminate at one of the singular sets as follows. Suppose $y_{k+1} = 0$ for some $k \geq -1$. Then we let s(p) be the one-sided terminating sequenc $[0\ s_{-k} \ldots s_0;\ s_1\ s_2 \ldots)$. If $y_{-k}(p) = -x_{-k}(p)$ for some $k \geq 1$, then we define $s(p) = (\ldots s_{-1}\ s_0;\ s_1 \ldots s_k\ 0]$. We also allow finite sequences of the form $[0\ s_{-k} \ldots s_j\ 0]$. So to each point $p \in \mathbb{R}^2$, there is assigned a sequence $s(p)$ of the form $[0\ s_{-k} \ldots s_0;\ s_1 \ldots s_j\ 0]$ where k and/or j may be infinite. We assign the sequence $[0;\ 0]$ to the origin. Let Σ denote the set of all such sequences excluding those which begin and/or end with an infinite constant sequence, i.e., excluding sequences of the form $(\ldots \alpha\ \alpha\ \alpha\ s_k\ s_{k+1} \ldots]$ and $[\ldots s_{k-1}\ s_k\ \alpha\ \alpha\ \alpha \ldots)$. The topological conjugacy is then given as follows. To each $p \in \mathbb{R}^2$, we let $h(p)$ denote the point in the square which corresponds to the sequence $s(p)$ under the map (1). Thus one should think of the excluded sequences of the form $(\ldots s_{k-1}\ s_k\ 1\ {-1}\ {-1}\ {-1} \ldots)$ and $(\ldots s_k\ {-1}\ {+1}\ {+1}\ {+1} \ldots)$ as being identified with the allowed sequence $(\ldots s_{k-1}\ s_k\ 0]$.

The following proposition shows why sequences which begin or end with a constant sequence are disallowed.

Proposition 2. <u>Let</u> $p = (x_0, y_0)$. <u>Suppose</u> $y_0 > 0$. <u>Then there are integers</u> $k > 0$ <u>and</u> $j < 0$ <u>such that</u> $y_k \leq 0$ <u>and</u> $y_j \leq 0$. <u>Similarly, if</u> $y_0 < 0$, <u>there are integers</u> $k > 0$ <u>and</u> $j < 0$ <u>such that</u> $y_j \geq 0$, $y_k \geq 0$.

Proof. We prove the first statement; the others follow similarly.

Observe first that if $y_n > 0$ for all n and $x_0 \geq 0$, then we have

$$x_n = x_{n-1} + 1/y_{n-1} > x_{n-1} > \ldots > x_0 > 0.$$

Also

$$y_1 = y_0 - x_0 - 1/y_0 < y_0 - x_0$$

and by induction

$$y_n < y_0 - nx_0 \ .$$

Therefore, if $x_0 > 0$, we must eventually have $y_n \leq 0$.

In case $x_0 < 0$ and $y_0 > 0$ we claim that there exists $k > 0$ such that either $x_k \geq 0$ or else $y_k \leq 0$. Indeed, as before, we have $x_n > x_0$ so that

$$x_2 = x_1 + 1/y_1 > x_0 + 1/y_0 + 1/(y_0 - x_0) \ .$$

Arguing inductively,

$$x_{n+1} > x_0 + 1/y_0 + 1/(y_0 - x_0) + \ldots + 1/(y_0 - nx_0) .$$

Since the series $\sum_{n=0}^{\infty} 1/(y_0 - nx_0)$ diverges, it follows that $x_n \to \infty$, which proves the result. qed

Again we refer the reader to [De 6] for the remainder of the proof.

Part Three: Singularities and Global Dynamics.

The object of this final part of these notes is to combine the techniques of the previous two to show how singularities in the equations of motion of a Hamiltonian system lead to a complicated but nevertheless somewhat understandable orbit structure. Although we cannot completely describe the orbit structure, we do find that a knowledge of the structure of the singular orbits provides in some sense a good overview of the global dynamics.

We deal here with two specific examples: the anisotropic Kepler problem and the isosceles three body problem. We will show that each of these systems admits a "singular" cross-section or surface of section. By this we mean a transverse cross-section for the flow which is crossed by all orbits except a few "special" collision and ejection orbits. Thus the associated Poincaré mapping has singularities, much like the baker transformation and mapping of Hénon of §2.3.

The notion of a singular cross-section has not been discussed very often in the literature, yet this type of cross-section seems to occur much more often in specific Hamiltonian systems than the usual cross-sections.

As with the Hénon mapping, we will relate our two examples to the shift automorphism on certain natural symbol spaces. We prove only that the given system projects onto these symbol spaces. It is an outstanding conjecture whether or not these projections are 1-1 (modulo symmetries). Hence our results give a complete classification of all possible types of motion (classified symbolically) in the given systems. We fail only to show that there is a unique orbit corresponding to each sequence.

§3.1 Singular Cross-Sections.

Let X be a smooth vector field on a manifold M with (not necessarily complete) flow ϕ_t. A singular cross-section S for X is a smooth submanifold of codimension one in M satisfying:

i. X is transverse to S.

ii. There exist smooth submanifolds C_i $i = 0,\ldots,n$ and E_j $j = 0,\ldots,k$ having codimension at least one in S and such that the orbit through any point in each C_i (resp. E_j) does not return to S in forward (resp. backward) time.

iii. The orbit through any point in $S - (\cup C_i)$ (resp. $S - (\cup E_j)$) returns to S in forward (resp. backward) time.

iv. Each orbit of X meets S at least once.

Thus S is almost a traditional cross-section for the flow; only a few points in S fail to return in one direction or the other.

Given a singular cross-section S, there is given as usual a Poincaré mapping F defined on S. Actually F maps $S - (\cup C_i)$ diffeomorphically onto $S - (\cup E_j)$, so we think of $F: S \to S$ as a mapping with singularities. The C_i's play the role of immediate collision orbits, while the E_j's consist of immediate ejection orbits. Of course, points may map onto a C_i after some iterates of F, so many orbits of F eventually reach collision.

Both of the major examples of part one admit singular cross-sections, as we now show. Consider first the anisotropic Kepler problem. From §1.6, the equations of motion are

$$q' = M^{-1}p$$
$$p' = -\nabla V(q)$$

where $M^{-1} = \mathrm{diag}(\mu,1)$ and $V(q) = -1/|q|$. We restrict to a negative energy level $H^{-1}(e)$ given by

$$H(q,p) = \tfrac{1}{2}p^tM^{-1}p + V(q) = e < 0 \ .$$

Let

$$S^+ = \{(q,p) \in H^{-1}(e) \mid p_2 = 0,\ q_2 > 0\}$$
$$S^- = \{(q,p) \in H^{-1}(e) \mid p_2 = 0,\ q_2 < 0\}$$

and $S = S^+ \cup S^-$. We claim that S is a singular cross-section for the flow on $H^{-1}(e)$.

First observe that both S^+ and S^- consist of initial conditions for which the momentum vector is parallel to the q_1-axis or else zero. Thus, over each point in the interior of the Hill's region with $q_2 \neq 0$, there are two initial conditions in S, while over the zero velocity set there is but one (the zero momentum vector).

It is clear that both S^+ and S^- are two dimensional planes in $H^{-1}(e)$. In fact, we may parametrize S^+ by θ and

p_1. Here $0 < \theta < \pi$ and $-\infty < p_1 < \infty$. This second inequality follows from the energy relation restricted to S^+:

$$\tfrac{1}{2}\mu p_1{}^2 = e - V(\theta)/r \;.$$

S^- may be parametrized similarly.

To see that the vector field is transverse to S, we observe that, if $q_2 \neq 0$ and $p_2 = 0$, then $p_2' = -q_2/|q|^3 \neq 0$. So orbits reach a relative maximum q_2-value whenever they cross S^+ and a relative minimum q_2-value when crossing S^-.

Now we claim that a solution beginning at S^+ behaves in one of two possible ways. Either the solution immediately reaches collision with q_2 decreasing to zero, or else q_2 decreases until the solution curve meets S^-. Indeed, $q_2' = p_2$ which is negative just after passing S^+, so q_2 decreases while the orbit remains in the upper half-plane. The solution curve cannot have a limiting q_2-value with $q_2 > 0$, since $q_2'' < 0$ at such points. Hence the solution must meet the origin or else crosses the line $q_2 = 0$. The solution cannot be tangent to $q_2 = 0$; only the homothetic orbits have this property. Therefore any orbit which does not undergo collision after leaving S^+ must pass into the lower half-plane. Then similar arguments as above give that the solution must eventually pierce S^-, proving the claim. Of course, solutions leaving S^- have a similar property.

Now the orbits which leave S and immediately reach collision must lie on the stable manifolds of the rest points on the collision manifold. In fact, for any orbit in these stable manifolds (excluding those along $q_2 = 0$), there is a unique point on the orbit which lies in S and which is the last such point before the orbit reaches collision. These are the immediate collision orbits. They form a finite union of submanifolds since they are given by the transverse intersection of the various stable manifolds with S. Note that they are transverse within the energy surface since the vector field is tangent to the stable manifolds but transverse to S. We will henceforth denote the set of points in S leading to immediate collision by C. Similar remarks hold for points in S leading to immediate ejection orbits. We denote these points by E.

We remark that the homothetic orbits along $q_2 = 0$ do not meet S; but these are the only orbits in each energy level which do not. We will not worry about this slight violation of the definition of a singular cross-section. All other orbits must cross S at least once, since, by the previous arguments, any other orbit must have a maximum or minimum q_2-value, thereby meeting S^+ or S^-. Hence we have proven that S forms a singular cross-section for the flow on a negative energy level in the anisotropic Kepler problem.

Now we discuss a similar singular cross-section for the isosceles problem. Some of the constructions here are similar to those for the anisotropic problem, so these details will be omitted. Actually, there are two different singular cross-sections for the system, each of which will be useful in the sequel. For the first cross-section, we recall the original equations of motion from §1.7:

$$x_1' = 2p_1$$
$$x_2' = (2+\varepsilon/2\varepsilon)p_2$$
$$p_i' = -\partial V/\partial x_i$$

where

$$V(x_1,x_2) = -\frac{1}{x_1} - \frac{4\varepsilon}{\sqrt{x_1{}^2 + 4x_2{}^2}}$$

and $x_1 > 0$. Throughout we will restrict to a fixed negative energy level. Define S_1 to be the set of initial conditions satisfying $p_1 = 0$, $x_1 > 0$. Since $p_1 = 0$ precisely when the primaries reach maximum separation, it is intuitively clear that solutions pass through S_1. More precisely, however, we have

$$p_1' = -\partial V/\partial x_1 < 0$$

along S_1, so solutions do in fact cross S_1 transversely. We postpone the remainder of the proof for a moment.

To define the second singular cross-section S_2, we recall McGehee coordinates (r,v,θ,w) from §1.7. In these coordinates the system is:

$$\dot{r} = \frac{\cos\theta}{\sqrt{W(\theta)}}\,rv$$

$$\dot{v} = \sqrt{W(\theta)}\,[1 - \frac{\cos\theta}{2W(\theta)}(v^2 - 4re)]$$

$$\dot{\theta} = w$$

$$\dot{w} = \sin\theta\,[-1 + \frac{\cos\theta}{W(\theta)}(v^2 - 2re)] - \frac{vw\cos\theta}{2\sqrt{W(\theta)}} +$$

$$+ \frac{W'(\theta)}{W(\theta)}(\cos - w^2/2)$$

and the energy relation is

$$\tfrac{1}{2}w^2/\cos\theta - 1 = \frac{\cos\theta}{W(\theta)}(re - \tfrac{1}{2}v^2)$$

Let S_2 consist of all points in these coordinates with $\theta = \pm\pi/2$. From the energy relation, it follows that $w = 0$ on S_2, while r and v are arbitrary. So S_2 consists of two planes in each energy level. We denote them by S_2^+ and S_2^- depending on the sign of θ.

Now $\dot{w} = -\sin(\pm\pi/2)$ along S_2, so we have that solutions cross S_2 transversely. Again intuitively speaking, S_2 consists of the points of (regularized) double collision, so one expects solutions to pass through S_2.

Now to show that both S_1 and S_2 form singular cross-sections, we first observe:

Proposition 1. <u>Every solution curve</u> $(x_1(t), x_2(t))$ <u>in Jacobi coordinates satisfies</u>: <u>there exist times</u> $t_0 < t_1 < t_2$ <u>such that</u>

i. $x_1'(t_1) = 0$

ii. $x_1(t) \to 0$ as $t \to t_0$ and t_2.

The proof of this proposition is entirely analogous to the corresponding statement in the anisotropic problem, and hence is omitted.

Now $x_1 = 0$ goes over to $\theta = \pm\pi/2$ in McGehee coordinates, so solutions oscillate back and forth between S_1 and S_2. There are only two possibilities for an orbit leaving S_1: either it eventually meets S_2 (and passes through it), or else it reaches triple collision. In the former case, the orbit must return again to S_1 by Proposition 1, while in the latter case, the orbit is an immediate collision orbit lying on one of the stable manifolds of a rest point in Λ. It follows that the set of immediate collision points in S_2 form a finite union of submanifolds as before. This proves that S_1 is a singular cross-section for the flow. The proof in the case of S_2 is equally easy.

§3.2 Symbolic dynamics and a Theorem of Gutzwiller.

The goal of this section is to set the stage for the proof of a theorem of Gutzwiller which classifies all possible types of orbits in the anisotropic Kepler problem. We begin with some symbolic dynamics and end with their relation to the differential equation.

Let Λ' denote the set of all doubly infinite sequences of positive integers, that is

$$\Lambda' = \{(s) = (\dots s_{-1}\ s_0;\ s_1\ s_2 \dots) \mid s_j \in Z^+\} .$$

We will augment Λ' by allowing certain terminating sequences as in the case of the Hénon mapping of §2.3. Namely, we will allow one-sided terminating sequences of the forms

$$[\infty\ s_{-k} \dots s_0;\ s_1\ s_2 \dots)$$
$$(\dots s_0;\ s_1 \dots s_j\ \infty]$$
$$[\infty\ s_{-k} \dots s_0;\ s_1 \dots s_j\ \infty]$$

with $j,k \geq 1$. So $s_0 \neq \infty$. Let Λ denote Λ' augmented by these terminating sequences.

We topologize Λ as follows. If $(s^*) = (\ldots s^*_{-1}\ s^*_0;\ s^*_1\ s^*_2 \ldots)$ is a non-terminating sequence, then we take as a neighborhood basis for (s) the cylinder sets of sequences (s) satisfying

$$s_j = s^*_j \quad \text{if } |j| \leq k$$
$$s_j \quad \text{arbitrary if } |j| > k$$

On the other hand, if $(s) = (\ldots s^*_{-1}\ s^*_0;\ s^*_1 \ldots\ s^*_j\ \infty]$, then we take as a neighborhood basis the sets consisting of sequences of either of the following types:

$$s_i = s^*_i \qquad -K \leq i \leq j$$
$$s_{j+1} \geq K$$

and all other s_i arbitrary, or else

$$s_i = s^*_i \qquad -K \leq i < j$$
$$s_j = s^*_j - 1$$
$$s_{j+1} = 1$$
$$s_{j+2} \geq K$$

and all other s_i arbitrary. The rationale for this strange topology will be given below. We define similar neighborhood bases for the other terminating sequences.

The shift automorphism is defined in this case by

$$\sigma(\ldots\ s_{-1}\ s_0;\ s_1\ s_2\ \ldots) = (\ldots\ s_{-1}\ s_0\ s_1;\ s_2\ \ldots),$$

i.e., σ shifts the semi-colon to the right. We find it more natural to shift in this direction for this problem. Consequently, points of the form $(\ldots s_{-1}\, s_0; \infty]$ are not in the domain of σ, whereas points of the form $[\infty\, s_0; s_1\, s_2 \ldots)$ are not in the range. These points will represent the immediate collision and ejection orbits respectively. Hence we denote them by $\hat{C}$ and $\hat{E}$.

We now return to the anisotropic Kepler problem. Our goal is to associate a sequence in Λ with every trajectory in the system in some meaningful way. This is accomplished as follows.

Let p be an initial condition in S and suppose that $F^j(p)$ is defined. We define the $\underline{j^{th}\text{ passage}}$ of p to be the segment of orbit containing $F^j(p)$ beginning at the first prior crossing of $q_2 = 0$ and ending at the next crossing of the q_1-axis. We include the endpoints of the orbit segment, even if one or both is the origin, i.e., even if this segment of orbit begins and/or ends in collision.

Now let $s_j = s_j(p)$ denote the number of times the orbit through p crosses the q_2-axis during the jth passage. We count collision and ejection as a crossing.

Proposition 1. <u>Suppose p does not lie on the homothetic orbit along the q_2-axis. If $F^j(p)$ is defined, then</u> $1 \leq s_j <$

Proof. The proof that $s_j < \infty$ is trivial. If $s_j(p) = 0$, then there is a point during the jth passage at which $\dot{\theta} = 0$. For definiteness, let us assume that $0 < \theta < \pi/2$ at this point. Now $\ddot{\theta} = -V'(\theta) > 0$ at any such point, so $\theta(\tau)$ has a minimum there. It follows, then, that the orbit must cross $\theta = \pi/2$ before reaching $\theta = 0$. The other cases are handled similarly. qed

So to each point p in S we have associated a sequence of positive integers $s(p) = (\ldots s_{-1}\, s_0;\, s_1\, s_2\, \ldots)$. The sequence terminates to the right at j if $F^j(p) \in C$ and on the left at $-k$ if $F^{-k}(p) \in E$. We associate to such points terminating sequences in Λ of the form $(\ldots s_{-1}\, s_0;\, s_1 \ldots s_j\, \infty]$ or $[\infty\, s_{-k} \ldots s_0;\, s_1 \ldots)$. Therefore, to each point p in S there is associated a sequence $s(p)$ in Λ of the form $[\infty\, s_{-k} \ldots s_0;\, s_1 \ldots s_j\, \infty]$ where k and/or j may be infinite. See Figs. 1 and 2.

So s provides a mapping $s\colon S \to \Lambda$ and we obviously have the commutative diagram

$$\begin{array}{ccc} S - C & \xrightarrow{\;F\;} & S - E \\ {\scriptstyle s}\big\downarrow & & {\scriptstyle s}\big\downarrow \\ \Lambda - \hat{C} & \xrightarrow{\;\sigma\;} & \Lambda - \hat{E} \end{array}$$

where $\hat{C}$ denotes the set of "collision" sequences of the form $(\ldots s_{-1}\, s_0;\, \infty]$ while $\hat{E}$ contains sequences of the form $[\infty;\, s_1\, s_2 \ldots)$. Clearly, the sequence $s(p)$ gives a good qualitative picture of the behavior of the solution through p. Gutzwiller's theorem [Gu 2] mainly asserts that the mapping s is surjective. Thus every conceivable type of motion is possible in the anisotropic Kepler problem. Given a possible sequence of numbers of crossings, there is at least one orbit of the differential equation which has the prescribed behavior.

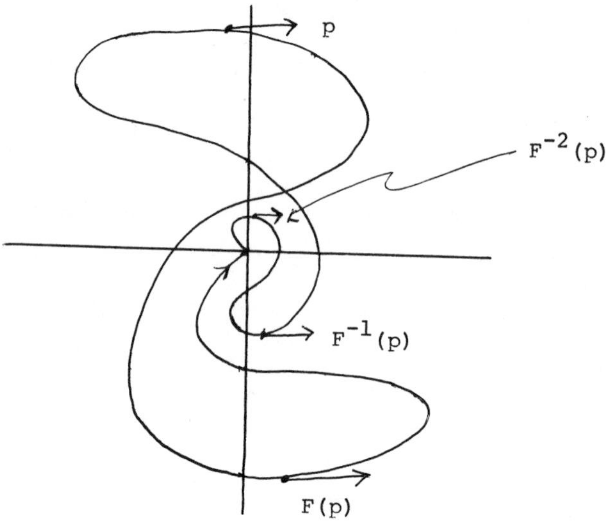

Fig. 1 The orbit through p is a collision/ejection orbit with s(p) = [∞ 2 2 3; 3 ∞].

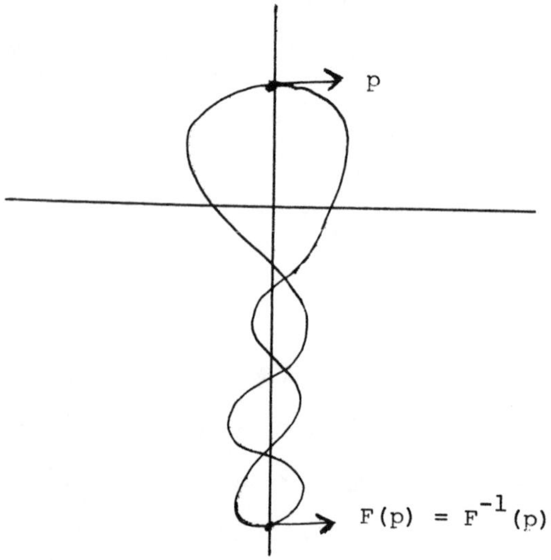

Fig. 2 The orbit through p is periodic with s(p) = (... 7 1; 7 1 7 ...).

One can in fact say more than just surjectivity of s:

Theorem 2. (Gutzwiller [Gu 2]) _The mapping_ $s: S \to \Lambda$ _is a continuous surjection. Moreover, corresponding to any periodic sequence, there is at least one actual periodic solution of the system._

The mapping s is definitely not 1-1 as we have defined it, since the various symmetries of the problem give rise to distinct orbits with the same sequences. However, Gutzwiller has conjectured that, up to these symmetries, there is a unique solution corresponding to each sequence in Λ. If true, this provides a topological conjugacy between the anisotropic Kepler problem and the shift automorphism, a rather unexpected isomorphism!

We might add that Gutzwiller has ample numerical evidence to support his conjecture [Gu 3]. On the other hand, a single generic elliptic closed orbit in the system, however long, would, by the Moser Twist Theorem, serve to disprove the conjecture. No such orbit has been found, however. Thus one might try to prove that the Poincaré mapping F is hyperbolic in some sense. Gutzwiller has shown numerically that the short periodic solutions are hyperbolic [Gu 4], while the author has shown that some of the longest periodic solutions are also hyperbolic [De 1], thus lending some additional credence to the conjecture. Later, we will present a model mapping similar in many respects to F for which one can verify the conjecture. Whether this mapping is conjugate to F, however, is speculation at best at this time.

§3.3 Proof of Gutzwiller's Theorem.

In this section we prove that the mapping $s: S \to \Lambda$ is both continuous and surjective. Our proof is basically geometric and differs completely from Gutzwiller's proof. It also has the advantage of extending to other mechanical systems such as the isosceles three body problem.

We first prove continuity. Let $p \in S$ and suppose first that $s(p)$ is a non-terminating sequence. We show that points close enough to p in S have associated sequences which agree with $s(p)$ arbitrarily far to the right and left of the semi-colon. Indeed, let $\gamma_j(p)$ denote the jth passage of the orbit through p, i.e., the orbit segment connecting $F^j(p)$ to the q_1-axis. $\gamma_j(p)$ is compact and is never tangent to the q_2-axis. Hence $\gamma_j(p)$ intersects the q_2-axis transversely at a finite number of isolated points which are disjoint from the origin. This follows since, if $\gamma_j(p)$ were tangent to the q_2-axis, then uniqueness of solutions would force it to lie for all time along this axis, thereby agreeing with the homothetic orbit already lying on this axis. By continuity of solutions with respect to initial conditions, nearby solutions meet the q_2-axis transversely at nearby points, proving that their associated sequences agree with $s_j(p)$. Hence, given any integer k, for q close enough to p, we have $s_j(q) = s_j(p)$ for all j such that $|j| \leq k$. So s is continuous at such points.

For collision and/or ejection orbits, we argue as follows. Suppose $s(p) = (\ldots s_0^*; s_1^* \ldots s_j^* \infty]$; the other cases are handled similarly. We show that nearby solutions have associated sequences which take one of two forms, agreeing with the topology imposed on Λ. For definiteness, suppose that the orbit through p reaches collision at $\theta = 0$, say from above. See Fig. 1. Now nearby solutions either reach collision,

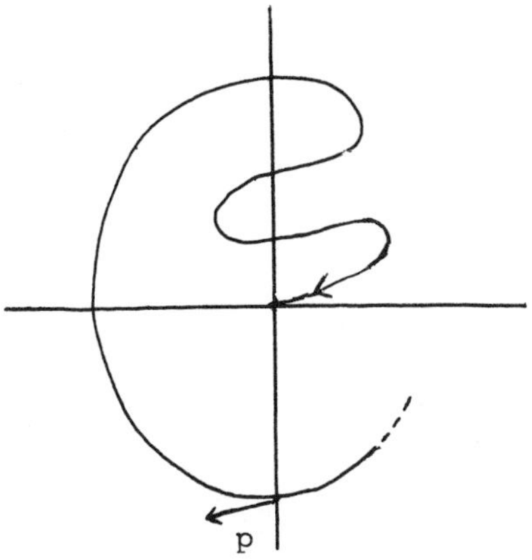

Fig. 1 The initial condition p leads to a typical collision orbit.

or else follow one of the two branches of the unstable manifold emanating from the appropriate rest point in the collision manifold. This was discussed in §1.7. See Fig. 4 in that section. The nearby behavior here is illustrated in Fig. 2 below. For q, we have the associated sequence $(\ldots s_0^*; s_1^* \ldots s_{j-1}^* \; 1 \; k \ldots)$ where k is large. For q', we find $s(q') = (\ldots s_1^* \ldots s_j^* \; k \ldots)$ again with k large. Both of these sequences are close to s(p) in the topology we have chosen on Λ. Hence s is continuous at these points, also.

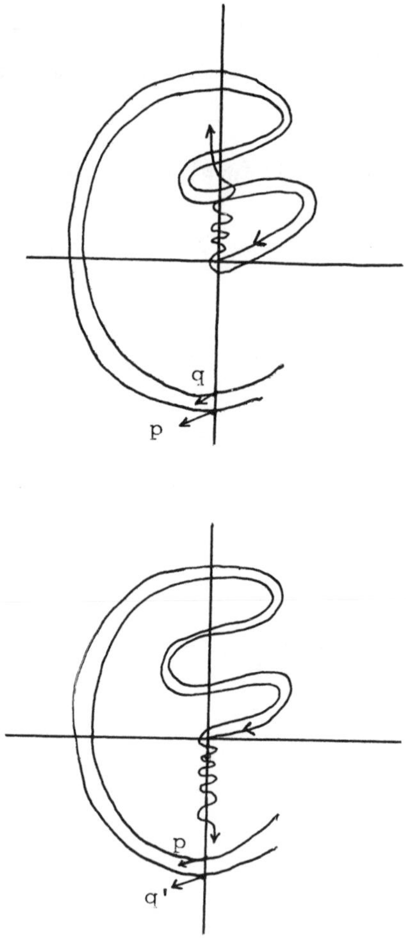

Fig. 2 The two types of nearby behavior for solutions beginning close to p.

We now prove surjectivity. Using θ and p_1 as coordinates, the singular cross-section may be represented as the strip $0 < \theta < \pi$, p_1 arbitrary. The zero velocity set in S^+ is given by $p_1 = 0$. We denote the special central point $p_1 = 0$, $\theta = \pi/2$ by q^+; this is the point where the homothetic orbit along $\theta = \pi/2$ meets S^+. One checks easily that the points of intersection of the other immediate collision and ejection orbits with S^+ form a pair of topological spirals converging to q^+. By this we mean that $C \cap S^+$ consists of two smooth curves which spiral infinitely often about q^+ as they connect $\theta = 0$ or $\theta = \pi$ to q^+. These curves are only continuous at q^+. See Fig. 3. This can be verified

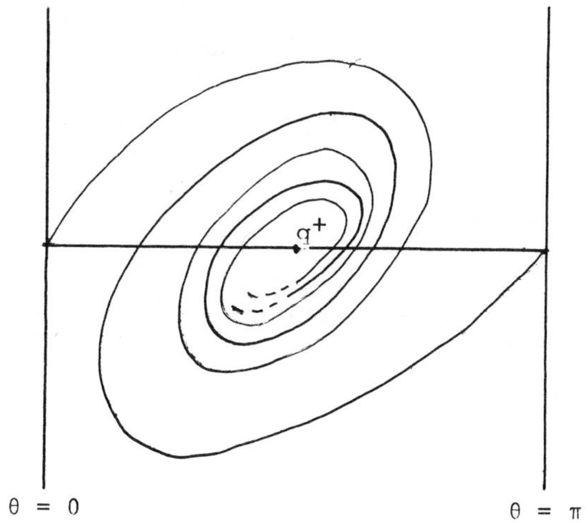

Fig. 3 The set of immediate collision or ejection orbits form a topological spiral in S^+.

using local analysis about the sink at $\theta = \pi/2$ in the collision manifold, together with the fact that one branch of the unstable manifolds at the saddles falls into this sink.

Note that one branch of C can be obtained from the other via the reflection $\theta \to \pi - \theta$, $p_1 \to -p_1$. Also, E can be obtained from C via the reversing reflection $p_1 \to -p_1$.

The full picture of the immediate collision and ejection orbits in S^+ is given in Fig. 4.

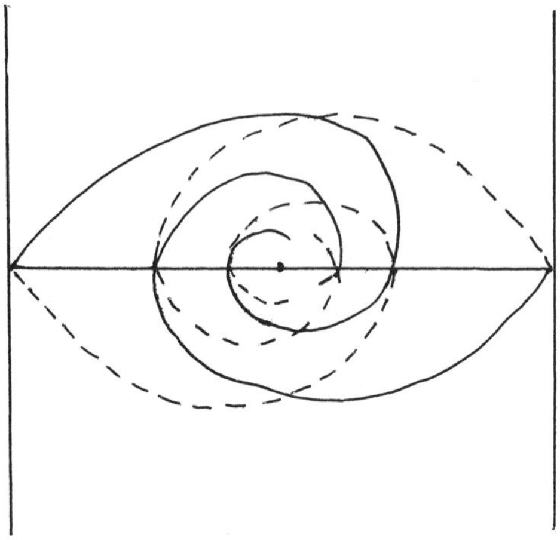

Fig. 4 The immediate collision orbits in S^+ are represented by solid lines, while the immediate ejection orbits are give by broken lines.

One has a similar picture in S^-. In fact, the reflection $q_2 \to -q_2$, $p_2 \to -p_2$ sends the picture in S^+ to that in S^-.

We remark that C and E need not be logarithmic spirals; they may in fact wander quite erratically in S^+, although numerical evidence indicates that this does not happen. All that is necessary for our purposes is that C wind about q^+ infinitely often and separate both S^+ and S^- into two components. Since each spiral in $C \cap S^+$ lies in the unstable manifold of a distinct rest point, it follows of course that these two curves cannot intersect.

Note that E and C intersect infinitely often along the zero velocity set and elsewhere, a fact we have already noted in §1.7.

We henceforth denote $C \cap S^+$ by c^+ and $C \cap S^-$ by c^-. Similarly, e^+ and e^- represent $E \cap S^+$ and $E \cap S^-$ respectively. The subset of c^+ asymptotic to the equilibrium point at $\theta = 0$ will be denoted by c_0^+; c_π^+, e_0^+, e_π^+, etc. are defined similarly.

Now recall that, for any $p \in S$, $s_0(p)$ denotes the number of times the orbit through p crosses the q_2-axis between the crossings of the q_1-axis just prior to and just after p along the orbit. Note that $s_0 = 1$ or 2 near $\theta = 0$ and $\theta = \pi$. This follows from the local description of solutions near the homothetic orbits along $\theta = 0, \pi$ given in §1.7. See Fig. 5. Also, $s_0(p) \to \infty$ as $p \to q^+$.

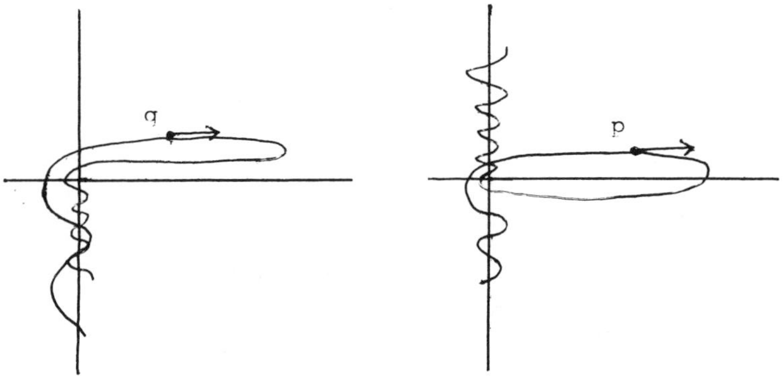

Fig. 5 In these diagrams, $s_0(q) = 2$ and $s_0(p) = 1$.

Now s_0 is continuous on S, except at points in C or E. At such points, s_0 increases or decreases by 1, again by the results of §1.7.

We wish to subdivide S^+ and S^- into a collection of "allowable strips" on which s_0 is constant. Once this is accomplished, we will show that these strips are mapped across each other just as the horizontal and vertical strips were mapped giving the Smale horseshoe. This implies that any positive integer can follow any other in a given sequence. We first need several preliminary lemmas.

Lemma 1. Suppose $p \neq q^+$. Then

i. If $p \in c_0^+ \cap e_0^+$, then $s_0(p)$ is even.

ii. If $p \in c_0^+ \cap e_\pi^+$, then $s_0(p)$ is odd.

iii. If $p \in c_\pi^+ \cap e_\pi^+$, then $s_0(p)$ is even.

iv. If $p \in c_\pi^+ \cap e_0^+$, then $s_0(p)$ is odd.

Proof. Each of these statements follows from a simple counting argument, since the orbit through p can never be tangent to the q_2-axis.

Lemma 2. Suppose $p \in c_0 \cap e_\pi$ and $s_0(p) = 2k + 1$. Then there exist closed intervals in both c_0 and e_π ending at p on which $s_0 = 2k+1$. Also, there exist open intervals abutting p in c_0 and e_π on which $s_0 = 2k$.

Proof. By the local behavior of solutions near an ejection orbit, a solution near p in e_π assumes one of two forms.

Either we have $s_0 = 2k$ as in Fig. 6, or else $s_0 = 2k+1$, as in Fig. 7. Similar considerations hold for collision orbits, thus proving the result. qed

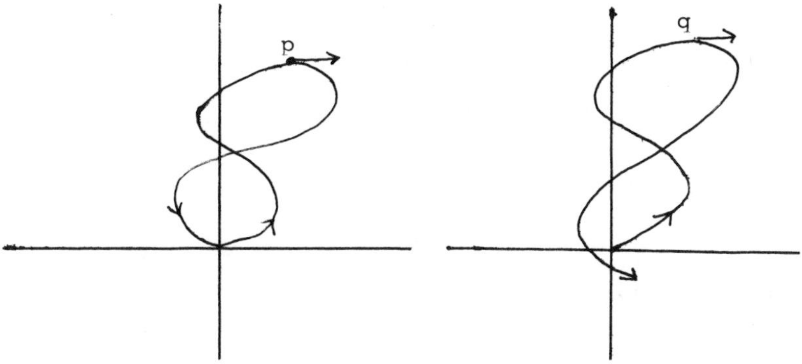

Fig. 6 For the collision/ejection orbit through p, we have $s_0(p) = 2k+1$, while $s_0(q) = 2k$.

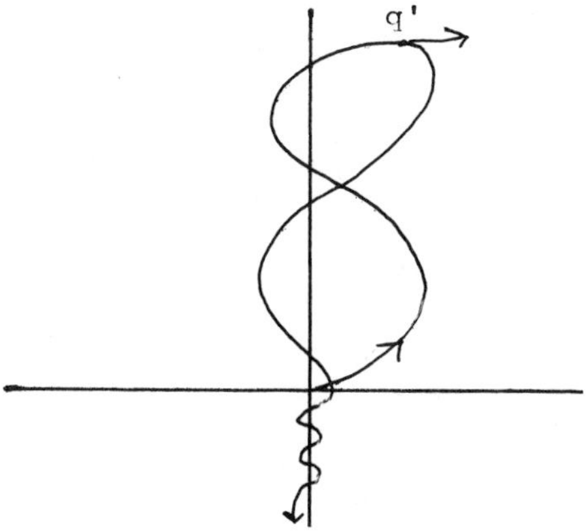

Fig. 7 Here we find $s_0(q') = 2k+1$.

<u>Remark</u>. This lemma shows that, in S, near a point in $c_0 \cap e_\pi$, we have the local picture illustrated in Fig. 8:

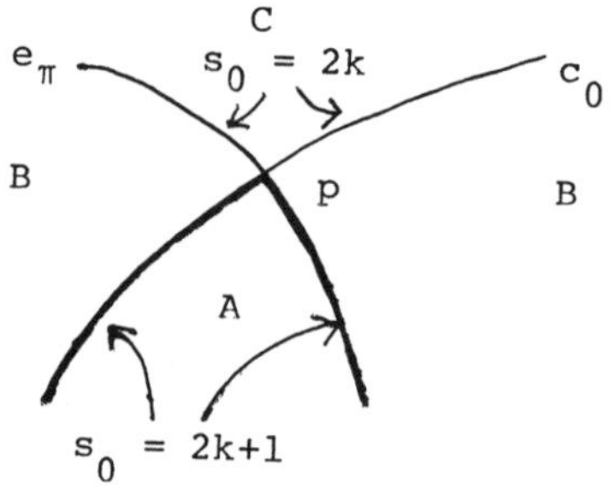

Fig. 8 $s_0(p) = 2k+1$, while $s_0 = 2k$ or $2k+1$ on abutting arcs in C and E. In A, we have $s_0 = 2k+1$, while $s_0|B$ $2k$ and $s_0|C = 2k-1$.

The typical solution in the region A is depicted in Fig. 9.

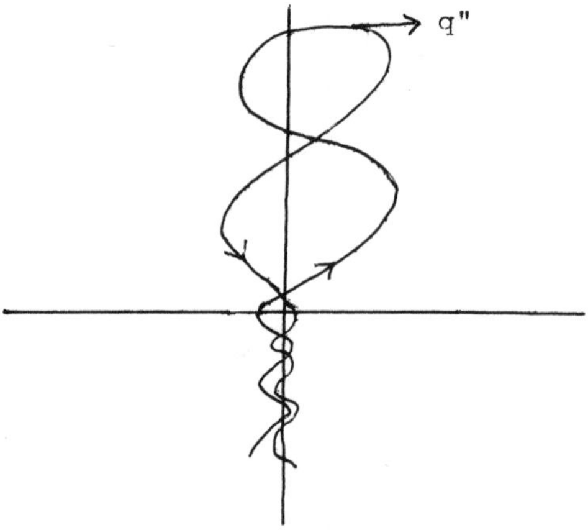

Fig. 9 $s_0(q") = 2k+1$, and the initial condition $q"$ is close to p in Figs. 6 and 8.

Remark. The lemma and following remark are obviously true if $p \in c_\pi \cap e_0$.

Lemma 3. _Let_ $p \in c_0 \cap e_0$ _and suppose_ $s_0(p) = 2k$. _Then there exist closed intervals in_ c_0 _and_ e_0 _ending at_ p _on which_ $s_0 = 2k$. _Also, there exist open intervals abutting_ p _in_ c_0 _and_ e_0 _on which_ $s_0 = 2k-1$.

Proof. The proof is similar to that of Lemma 2 and hence is omitted. qed

Again we remark that a similar result holds for $c_\pi \cap e_\pi$, and that the appropriate local picture is as given in Fig. 10.

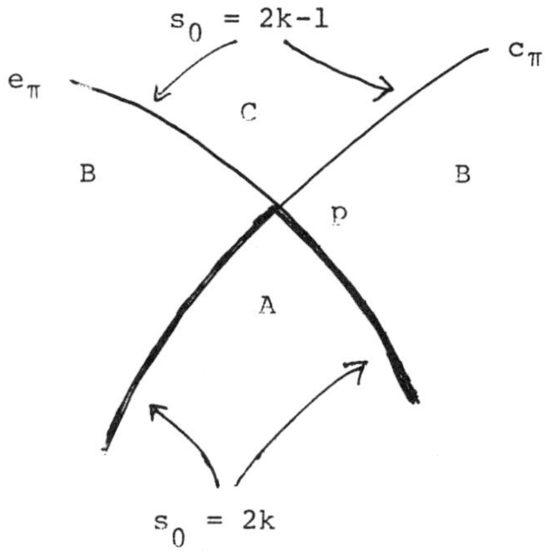

Fig. 10. $s_0(p) = 2k$, while $s_0|A = 2k$, $s_0|B = 2k-1$, $s_0|C = 2k-2$.

Now consider one of the components of $S^+ - E$. This is an infinite spiralling strip winding down to q^+ in S^+ as shown in Fig. 11.

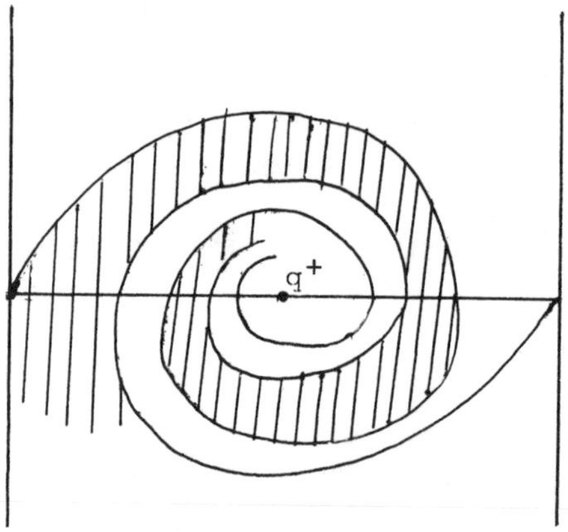

Fig. 11 The components of $S^+ - E$ form infinite spiralling strips converging to q^+.

The function s_0 changes its values in this strip only at points in c_0 or c_π, and there s_0 either increases or decreases by 1. Now C meets the boundary of this strip in one of two ways: either C cuts completely across the strip or else C meets one of the two boundaries twice in succession. See Fig. 12. In the former case, we call the arc in C cutting across the strip an <u>admissible boundary</u>. In the latter case, C bounds a two dimensional disk in the strip which we call an <u>extraneous disk</u>. Note that two or more of these disks may be nested. Now consider the strip with all extraneous disks removed. The remaining admissible boundaries

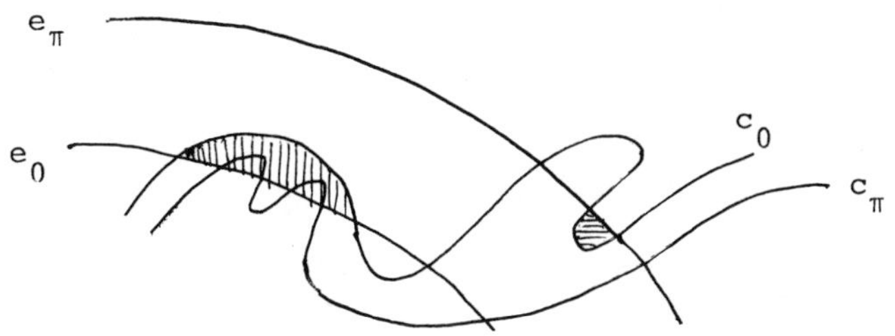

Fig. 12 The shaded regions represent extraneous disks.

cut the strip into a series of substrips on which s_0 is necessarily constant. If $s_0 = k$ on such a strip, we call it a <u>k-strip</u>. Since s_0 is onto the positive integers and changes by 1 upon crossing an admissible boundary, it follows that, for each integer k, there exists at least one k-strip in each component of $S - E$. A typical k-strip is depicted in Fig. 13.

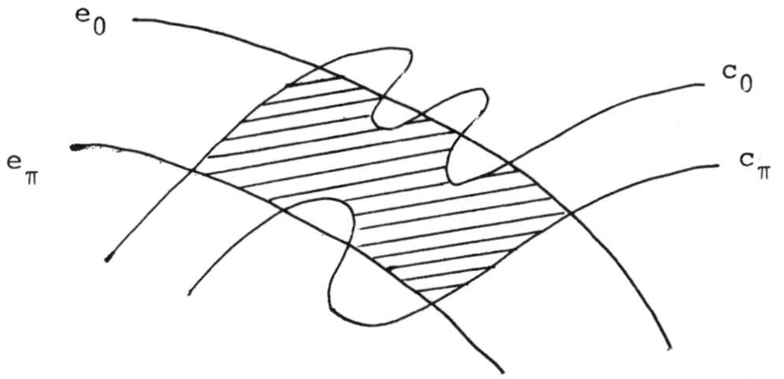

Fig. 13 The shaded region represents a typical k-strip.

So a k-strip is bounded by two admissible boundaries in C as well as several arcs in C (composing boundaries of extraneous disks) and several in E. Each pair of consecutive admissible boundaries in C cuts off a pair of arcs in E, and together these four arcs bound a rectangular region in S which consists of exactly one k-strip together with finitely many extraneous disks.

Lemma 4. Suppose the admissible boundaries of a k-strip both lie in c_0 (or in c_π). Then s_0 assumes the same value on each abutting strip.

Proof. The proof involves repeated applications of lemmas 2 and 3 to take care of successive extraneous disks. Hence we merely present a sketch which gives the relevant details. Suppose we have a 2n-strip whose boundaries lie in c_0 and which is abutted by a (2n+1)-strip. See Fig. 14. We claim that

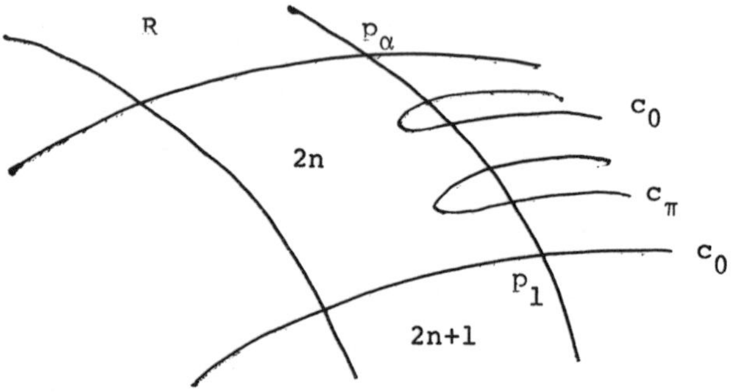

Fig.14.

$s_0(p_\alpha) = 2n+2$, so that, by Fig. 10, $s_0|R = 2n+1$ also. To see this, observe that $s_0(p_1)$ must equal $2n+2$ by lemma 2. Lemmas 1, 2, and 3 can be used repeatedly to fill in the other s_0 values as in Fig. 15:

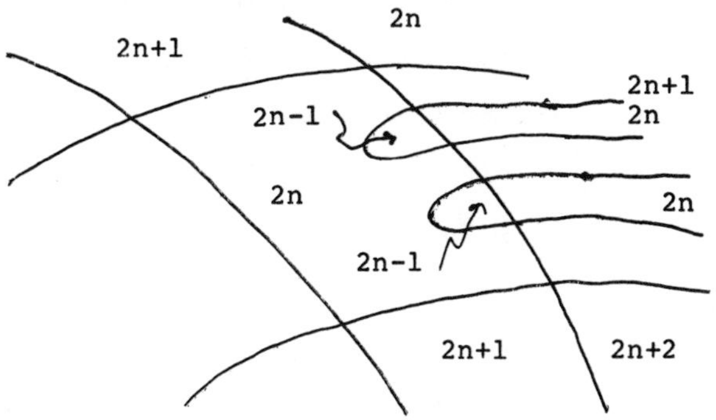

Fig.15.

No matter how many disks meet the 2n-strip, we always have $s_0|R = 2n+1$. All other possibilities may be handled similarly. qed

As a consequence of the preceding lemma, we can prove:

Proposition 5. _In each component of_ $S - E$ _and for each integer_ $k \geq 3$, _there exists at least one k-strip_ R_k _one of whose admissible boundaries lies in_ c_0 _and the other in_ c_π.

Proof. For simplicity, we consider only S^+. For $k \geq 3$, the region in S^+ with $s_0 \geq k$ is compact. In any of the two components of $S^+ - E$, the number of k-strips is finite, and so, given k, we can find the last k-strip in this component, i.e., the "closest" k-strip in this component to q^+. Hence closer n-strips in this component satisfy $n \geq k+1$. In particular, one of the abutting strips must be a $k+1$ strip. By the lemma, if both admissible boundaries lie in c_0 (or in c_π), the other abutting strip must also be a (k+1)-strip. Continuing in this fashion, we must eventually reach a k-strip satisfying the hypotheses of the proposition, or else reach a contradiction since s_0 is onto the positive integers. qed

Remark 1. The proposition does not necessarily extend to the case where $k = 1$ or $k = 2$ since these regions may abut $\theta = 0$ or $\theta = \pi$. In this case, the associated strips must exist (see Fig. 5), but they may be unbounded.

Remark 2. The numerical work of Gutzwiller indicates that, in fact, the k-strips are all bona fide "rectangles" and all extraneous disks are absent. Indeed, in our model mapping, this is the case.

We call k-strips whose admissible boundaries consist of one arc in c_0 and one in c_π admissible k-strips. We now investigate how the Poincaré mapping F affects admissible k-strips. Let R_k be any k-strip in S^+. F maps R_k into one of the components of $S^- - E$, as immediate ejection orbits are not in the range of F. In fact, we claim that $F(R_k)$ intersects every admissible j-strip in this component. To see this, we first prove:

Lemma 6. Let f_0 and f_π denote the boundary components in e_0 and e_π respectively of an admissible k-strip. Then $s_0 = k$ on one of f_0 or f_π, and $s_0 = k+1$ on the other.

Proof. Suppose $s_0 = k$ on the interior of the admissible strip. For each point p in the interior of this strip, the 0th passage begins on the same side of the origin on the q_1-axis. Similarly, the terminal point of the 0th passage for any point p always lies on the same side of the origin. For definiteness, let us assume that the 0th passage of all of these orbits begins to the left of the origin. Then, by deforming orbits to the boundaries of the strip, one checks easily that $s_0 = k$ along f_0 while $s_0 = k+1$ along f_π. See Fig. 16. Similar methods also give the following:

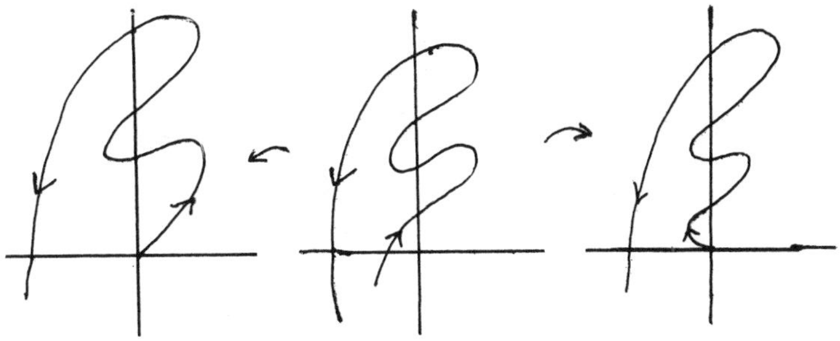

$s_0 = k$ $s_0 = k$ $s_0 = k+1$

Fig. 16 The center orbit is a typical orbit in a k-strip, while the left and right orbits are deformations to the boundaries f_0 and f_π.

Lemma 7. Let g_0 and g_π be the admissible boundaries of an admissible k-strip in S. Then $s_0 = k$ on one of g_0 or g_π, while $s_0 = k+1$ on the other.

We can now finish the proof of Gutzwiller's Theorem:

Proof of surjectivity: Suppose R'_j is an admissible strip in the component of $S^- - E$ intersected by $F(R_k)$. We claim that $F(R_k)$ meets both admissible boundaries of R'_j. Indeed, by Lemma 7, for p on one admissible boundary of R_k we have $s_0(p) = k$, while $s_0 = k+1$ on the other. By the local behavio of near-ejection orbits, we have that $s_1(p) = 1$ for p near the boundary with $s_0 = k+1$. Moreover, as p approaches this boundary, $F(p)$ approaches the collision manifold. On the other hand, $s_1 \to \infty$ as $p \to$ the other boundary, and, in fact, $F(p) \to q^-$. Therefore, $F(R_k)$ meets every admissible j-strip in its component in S^-. Note that, in fact, $F(R_k)$ cannot cross the ejection boundaries of any R'_j. Hence $F(R_k)$ must cross each admissible boundary of R'_j. Moreover, F is undefined on C, so that $F(R_k)$ must be bounded entirely by the image of the ejection orbits. In particular, it follows that $F(R_k) \cap R'_j$ contains a substrip bounded by ejection orbits and connecting the admissible boundaries of R'_j. See Fig. 17.

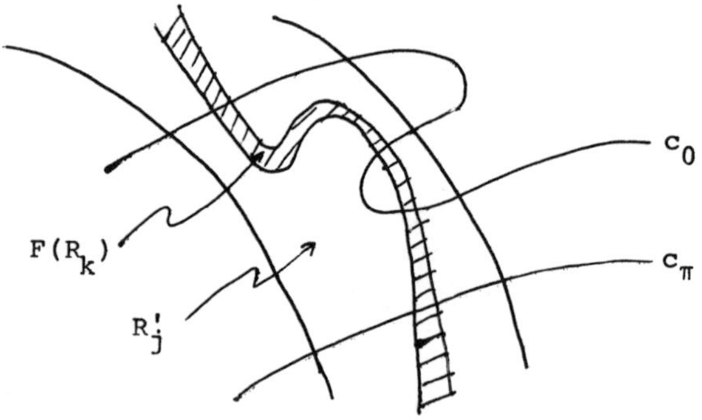

Fig.17 $F(R_k) \cap R'_j$

Of course, this substrip may intersect extraneous disks in R'_j.

Now consider inverse images. Arguing just as before, one finds that $F^{-1}(R_k)$ meets every admissible j-strip in its component of $S^- - C$. Moreover, for any such strip R'_j, we have that $F^{-1}(R_k)$ meets R'_j in a substrip which does not cross any of the collision boundaries of R'_j and which connects both ejection boundaries of R'_j as in Fig. 18. Standard arguments as in the case of the Smale horseshoe in §2.1 then give surjectivity. Note that we do not have hyperbolicity and that prevents us from concluding uniqueness.

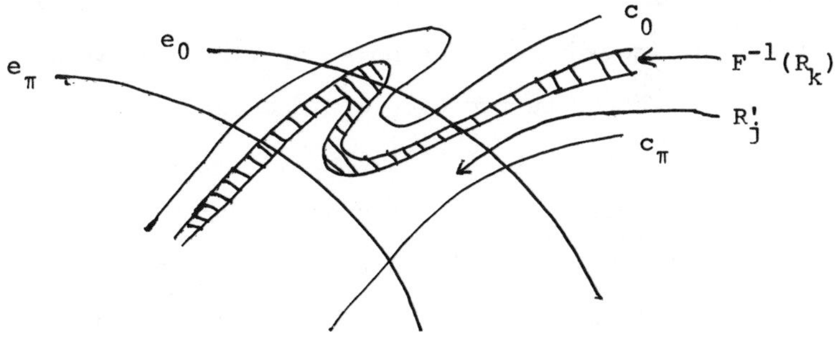

Fig.18 $F^{-1}(R_k) \cap R'_j$

§3.4 <u>A Model Mapping for the Anisotropic Kepler Problem.</u>

In this section we present a mapping which resembles the Poincaré mapping for the anisotropic problem described in the previous section, and for which we can verify topological conjugacy with the shift mapping. Whether this mapping is actually conjugate to F or not, we do not know. We remark that this mapping is similar to Hénon's mapping associated with the restricted three body problem in the sense that the singular points are handled in the same way.

The mapping G is defined on the open annulus $I \times S^1$ parametrized by (x,θ), with $0 < x < 1$ and θ defined mod 1. Via polar coordinates we also think of this mapping as being defined on the punctured two-disk. After removing all symmetries in the anisotropic problem, the Poincare mapping F in this system reduces to a mapping on such a space (the "hole" in the middle corresponds to the homothetic solution alon the q_2-axis). We will use both viewpoints in this section. From an ergodic theorist's point of view, however, the first viewpoint might be preferable since our mapping preserves Lebesgue measure on $I \times S^1$.

We define G in two steps. First let $G_1(x,\theta) = (x,\theta-1/x)$. G_1 is an infinite twist of the annulus. We also define $G_2(x,\theta) = (-\theta,x)$. G_2 interchanges the x and θ coordinates (with a sign change), and is clearly discontinuous along the line $\theta = 0$. In fact, $\theta = 0$ is mapped out of the annulus, and therefore plays the role of the singular set for G_2. Now finally define $G = G_1 \circ G_2 \circ G_1$. So the mapping G may be visualized as a twist followed by a flip followed by another twist. As both G_1 and G_2 preserve Lebesgue measure on $I \times S^1$, it follows that G does also. G is given by

$$x_1 = -\theta + 1/x$$

$$\theta_1 = x + 1/(\theta - 1/x)$$

and its inverse by

$$x_{-1} = \theta + 1/x$$

$$\theta_{-1} = -x + 1/(\theta + 1/x) .$$

So G is singular along the spiral $\theta = 1/x$, while G^{-1} is undefined along $\theta = -1/x$. This first spiral should be thought of as representing the immediate collision orbits in the singular cross-section, while the second represents the immediate ejection orbits. See Fig. 1.

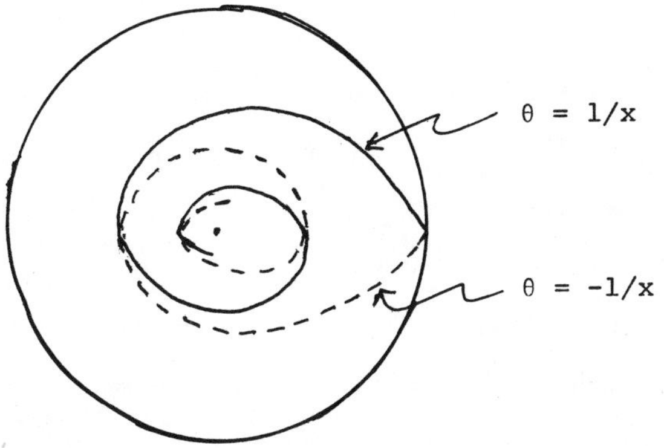

Fig. 1 The solid line represents immediate collision orbits, while the broken line represents immediate ejection orbits.

As in the case of the Hénon mapping, one of the most important features of this mapping is its non-uniform hyperbolicity. Define

$$S^u = \{(\xi,\eta) \mid \xi\eta \geq 0\}$$

$$S^s = \{(\xi,\eta) \mid \xi\eta \leq 0\} .$$

Letting $(\xi_1,\eta_1) = dG(\xi,\eta)$, we have

$$\xi_1\eta_1 = \left(\frac{1}{x^2} + \frac{1}{x^2(\theta - 1/x)^2} - 1\right)\left(1 + 1/(\theta - 1/x)^2\right)\xi\eta$$

$$\geq \left(\frac{1}{x^2(\theta-1/x)^2}\right)\left(1 + \frac{1}{(\theta-1/x)^2}\right)\xi\eta$$

$$> \quad 0 .$$

provided $\xi\eta \geq 0$. Hence dG preserves S^u. Moreover, if $\xi\eta \geq 0$, we have

$$|\xi_1| + |\eta_1| \geq \frac{1}{(\theta-1/x)^2}(|\xi| + |\eta|)$$

so that, in the sum norm, dG expands vectors in S^u.

In a similar fashion, one verifies easily that dG^{-1} preserves and expands S^s. Note that the rate of expansion is $1/(\theta-1/x)^2$, which tends to 1 as $\theta - 1/x$ tends to 1 from below. Hence G is non-uniformly hyperbolic.

Such non-uniformly hyperbolic mappings have been studied recently by Pesin [Pe]. The extension to mappings with singularities has been made by Katok and Strelcyn [KS]. We will not invoke their work, however, since our results are purely topological. Presumably, the work of Pesin and Katok/Strelcyn applies to show that G is ergodic.

Note that the spirals $\theta = 1/x$ and $\theta = -1/x$ partition the annulus into a grid of "rectangles" as in Fig. 2. The rightmost rectangle is actually a "triangle". We enumerate these rectangles from right to left by R_1, R_2, etc. as in Fig. 2.

For technical reasons, we include the two right-hand

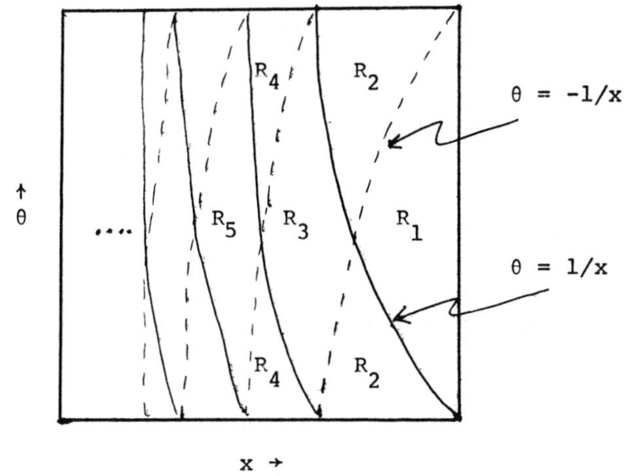

Fig. 2 Construction of the rectangles in $I \times S^1$.

boundaries in each R_j, so that the left hand boundaries of R_{j-1} are part of R_j. One can also view the R_j in the punctured disk as in Fig. 3.

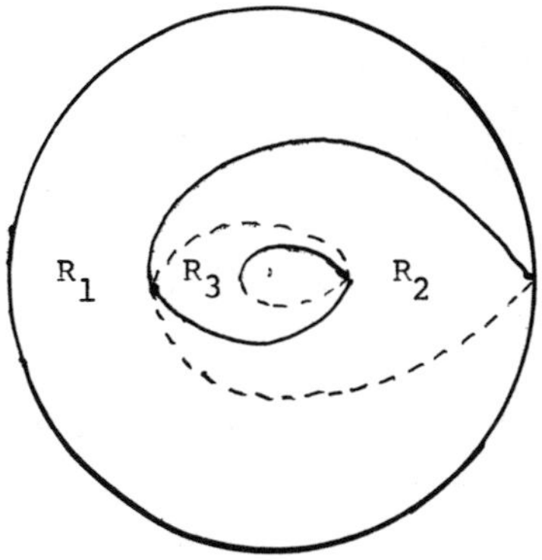

Fig. 3 Construction of the rectangles in the punctured disk.

Each rectangle R_i except R_1 is bounded by a pair of arcs in $\theta = 1/x$ and a pair in $\theta = -1/x$. We call the boundaries in $\theta = 1/x$ <u>stable boundaries</u>, while the boundaries in $\theta = -1/x$ are called <u>unstable</u>. G is undefined on stable boundaries, and G^{-1} is undefined on unstable boundaries.

Lemma 1. *Let γ be a smooth curve in R_j connecting both stable boundaries and having tangents in the unstable sector bundle S^u. Then $G(\gamma)$ is a smooth spiral in $I \times S^1$ beginning at $x = 1$, spiralling down to $x = 0$, and meeting each rectangle R_j.*

Proof. First observe that $G_1(\gamma)$ is a smooth curve connecting $\theta = 0$ to $\theta = 1$. Hence $G_2(G_1(\gamma))$ is a smooth curve connecting $x = 0$ to $x = 1$. Since G_1 is an infinite twist, it follows that $G(\gamma)$ is a smooth spiral connecting $x = 1$ to $x = 0$. Now the spiral $\theta = -1/x$ is not in the range of G. Hence $G(\gamma)$ cannot meet this curve. It therefore follows that $G(\gamma)$ must meet every rectangle. qed

Remark. Since the tangents to $G(\gamma)$ must also lie in the unstable sector S^u, it also follows that $G(\gamma)$ meets each R_i in an unstable curve which is "parallel" to the unstable boundaries.

Corollary 2. *$G(R_j)$ is a spiral band in $I \times S^1$ which meets every rectangle. Moreover, $G(R_j) \cap R_k$ is a sub-rectangle of R_k with two boundaries contained in the stable boundaries of*

R_k <u>and the other two boundaries parallel to the unstable boundaries of</u> R_k.

Proof. The proof is an immediate consequence of Lemma 1. See Fig. 4. qed

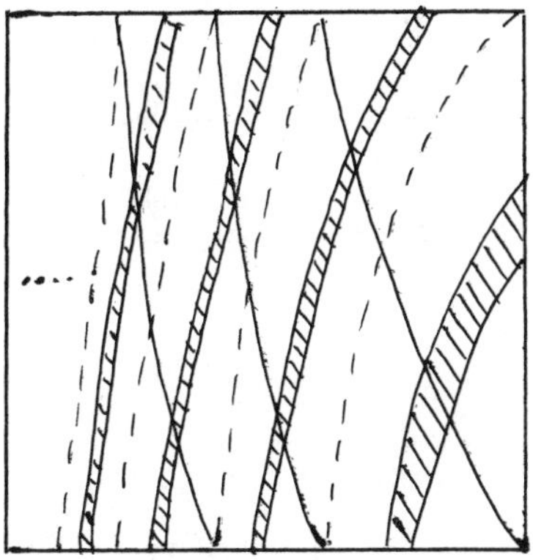

Fig. 4 $G(R_j)$ is a spiral band in $I \times S^1$. Note that $G(R_j)$ meets each R_k in a sub-rectangle.

For G^{-1} we have a similar situation. $G^{-1}(R_j)$ is a spiral band in $I \times S^1$ with boundary parallel to $\theta = 1/x$. See Fig. 5. Hence $G^{-1}(R_j)$ meets each R_k in a substrip with boundaries parallel to the stable boundaries of R_k. This of course gives the familiar Smale horseshoe construction. One verifies immediately that Axioms 1 and 2 of 2.1 hold on the interiors of the R_j, and so there is a subset Ω of $I \times S^1$ on which G is topologically conjugate to the shift automorphism on infinitely many symbols. Any point in Ω has the property that its orbit never meets either of the spirals

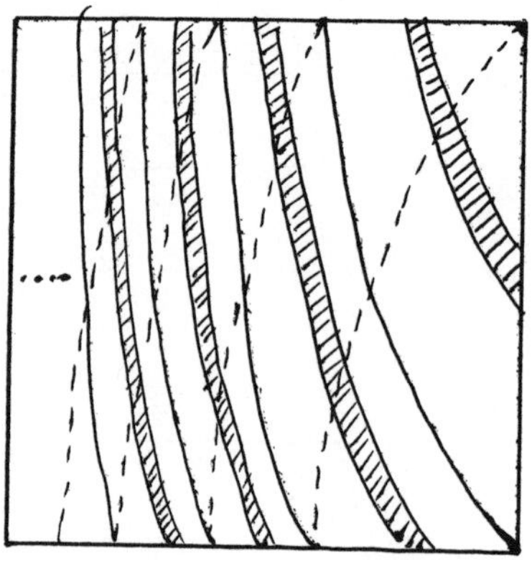

Fig. 5 $G^{-1}(R_j)$ is also a spiral band.

$\theta = 1/x$ or $\theta = -1/x$.

We can in fact extend this conjugacy to the singular orbits. We claim that G is topologically conjugate to the shift automorphism on the augmented symbol space Λ introduced in §3.2. Recall that Λ consists of all sequences of positive integers of the form $[\infty\ s_{-k} \cdots s_0;\ s_1 \cdots s_j\ \infty]$ where $s_0 \neq \infty$ and where j and/or k may be infinite. When j and k are infinite, we have a non-terminating sequence which corresponds to a point in Ω. All other sequences terminate on one or both ends and correspond to singular orbits as follows.

Let $p \in I \times S^1$ and suppose $F^j(p)$ is defined. The sequence $s(p) = (\ldots s_0(p);\ s_1(p)\ s_2(p) \ldots)$ is defined by $s_j(p) = \alpha$ iff $F^j(p) \in R_\alpha$. So $s_0(p)$ specifies which rectangle p originally lies in. We set $s_{j+1}(p) = \infty$ for $j \geq 0$ if $F^j(p)$ belongs to the spiral $\theta = 1/x$. Similarly, we set $s_{-k-1}(p) = \infty$ for $k \geq 0$ if $F^{-k}(p)$ belongs to the spiral

$\theta = -1/x$. So to each point in $I \times S^1$ we have associated a sequence in Λ.

Now recall that the shift automorphism σ on Λ is defined by $\sigma(\ldots s_0; s_1 \ldots) = (\ldots s_0\ s_1; s_2 \ldots)$. So the domain of σ is all of Λ except "collision" sequences with $s_1 = \infty$, and the range of σ excludes "ejection" sequences with $s_{-1} = \infty$. Therefore, s maps points on $\theta = 1/x$ to the collision sequences and points on $\theta = -1/x$ to the ejection sequences. Moreover, we clearly have a conjugacy $\sigma \circ s = s \circ G$ on the respective domains.

We finally claim that s is a homeomorphism. That s is 1-1 and onto follows immediately from hyperbolicity. Hence we confine our attention to continuity of s. For this we need the topology on Λ. If (s^*) is a non-terminating sequence then we take as a neighborhood basis of (s^*) the usual cylinder sets. That is, (t) is close to (s^*) in this topology provided $t_i = s_i^*$ for $|i| \leq K$ with K large. Neighborhoods of terminating sequences of the form $(s^*) = (\ldots s_0^*; s_1^* \ldots s_j^*\ \infty]$ are somewhat different; they consist of two types of sequences. Either

$$s_i = s_i^* \qquad -K \leq i \leq j$$
$$s_{j+1} \geq K$$

with all other s_i arbitrary, or else

$$s_i = s_i^* \qquad -K \leq i \leq j-1$$
$$s_j = s_j^* - 1$$
$$s_{j+1} = 1$$

and all other s_i are arbitrary. Note that this topology on Λ is slightly different from that introduced in §3.2.

Now s is clearly continuous at non-terminating sequences, so we will restrict our attention to sequences of the form $(s^*) = (\ldots s_0^*; s_1^* \ldots s_k^* \infty]$. Suppose $s(p) = (s^*)$. Therefore $F^k(p)$ lies on the stable boundary separating R_k and R_{k-1}. Let N be a neighborhood of p, and let $N_k = N \cap R_k$ and $N_{k-1} = N \cap R_{k-1}$. Note that $G_1(N_{k-1})$ abuts the lower boundary $\theta = 0$, while $G_1(N_k)$ abuts the boundary $\theta = 1$. Hence $G_2(G_1(N_{k-1}))$ abuts $x = 1$, while $G_2(G_1(N_k))$ abuts $x = 0$. Consequently, $G(N_k)$ is a spiral band converging to $x = 0$, and, in particular, meeting every R_ℓ for ℓ large. Similarly, $G(N_{k-1})$ is contained in R_1.

It follows that, if $q \in N_k$, we have $s_{k+1}(q) \geq \ell$, whereas, if $q' \in N_{k-1}$, then $s_{k+1}(q') = 1$. This proves that the conjugacy is continuous.

§3.5 Classification of Motion in the Isosceles Three Body Problem.

The symbolic dynamics approach used in Gutzwiller's Theorem in §3.3 to classify solutions in the anisotropic Kepler problem can also be used in a natural way to classify motion in the isosceles three body problem. The most important ingredient in the proof of Gutzwiller's Theorem was a good understanding of the behavior of solutions which pass close to the singularity. In the isosceles problem, McGehee's coordinates also give a good description of near-collision orbits, so the analogous theorem is easy to prove.

For the remainder of this section, we will assume that ε, the mass of the third particle, is smaller than the numerical value ε_0 computed by Simo [Sim 2]. In particular, this implies that the structure of the stable and unstable manifolds on the triple collision manifold is as in Fig. 7 of §1.7, and that the local behavior of solutions near triple collision is as given in Figs. 8 and 9 of that section.

We now mimic the statement of Gutzwiller's Theorem. Recall that the singular cross-section S_2 consists of all points in a fixed negative energy surface with $\theta = \pm\pi/2$. S_2 is a union of two planes, S_2^+ and S_2^-. Heuristically, S_2 consists of all (regularized) initial conditions for which the primaries are exactly at collision. Let F be the Poincaré mapping defined on S_2.

Let $p \in S_2$ and suppose $F^j(p)$ is defined. We define the jth passage through p to be the orbit segment beginning at $F^j(p)$ and terminating at the next intersection with S_2. So the 0th passage through p is the orbit segment connecting p to the next crossing of S_2 along the orbit.

We can now introduce symbolic dynamics into the problem. If $r > 0$, any point in S_2 leads to an orbit of the system which is not a homothetic solution. Such an orbit must cross the q_1-axis transversely (or not at all), for otherwise the orbit would agree with the homothetic orbit along $\theta = 0$. Since the Hill's region meets this axis in a compact interval, it follows that the number of intersections of the jth passage through any point with this axis is finite. Let $s_j = s_j(p)$ denote the number of such intersections.

As above, one checks easily that $s_j < \infty$. Unlike the anisotropic problem, however, s_j may equal 0. We therefore let the symbol space Σ consist of all doubly infinite sequence of non-negative integers of any of the following types:

$$[\infty\ s_{-k} \dots s_0;\ s_1 \dots)$$
$$(\dots s_0;\ s_1 \dots s_j\ \infty]$$
$$[\infty\ s_{-k} \dots s_0;\ s_1 \dots s_j\ \infty]$$

with $1 \leq j,k \leq \infty$. Again, $s_0 \neq \infty$. Then to each point p in S_2, we assign the corresponding sequence $(s) = (s_j(p))$. If the orbit ends in collision after the jth passage with $j \geq 0$, we let $s_{j+1} = \infty$ and terminate the sequence. If the orbit begins in ejection before the $-k^{th}$ passage with $k \geq 1$, we let $s_{-k-1} = \infty$ and again terminate the sequence. So sequences of the form $(\dots s_0;\ \infty]$ are immediate collision sequences while sequences of the form $[\infty\ s_0;\ s_1 \dots)$ are immediate ejection sequences. See Figs. 1 and 2.

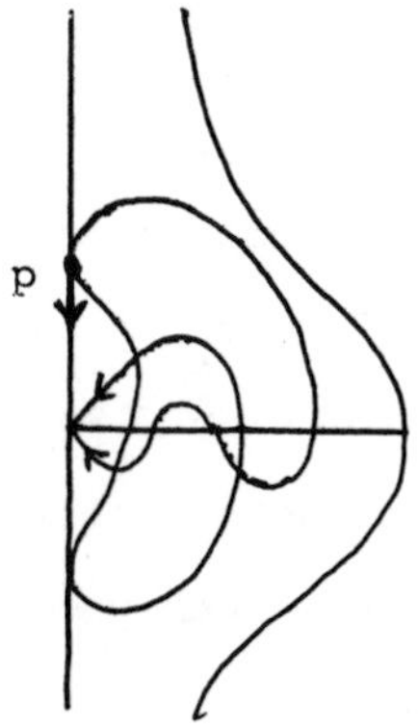

Fig. 1 The associated sequence $s(p) = [\infty\ 4\ 1;\ 2\ \infty]$

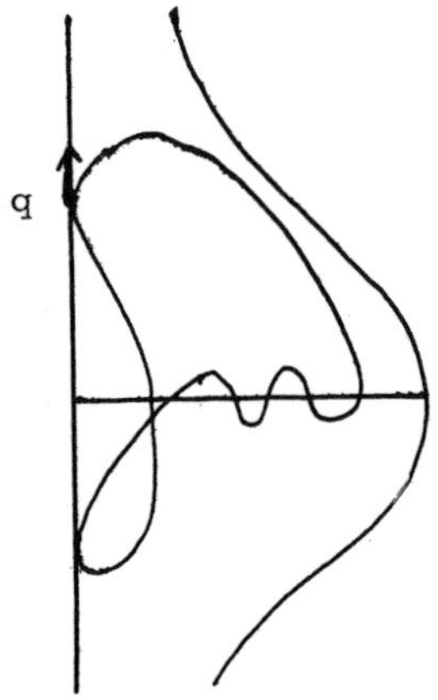

Fig. 2 The associated sequence $s(q) = (\ldots 5\ 1\ 5;\ 1\ 5\ 1 \ldots)$, corresponding to the closed orbit through q.

We topologize Σ as follows. For non-terminating sequences, we take the usual cylinder sets as neighborhood bases. If $(s^*) = (\ldots s_0^*;\ s_1^* \ldots s_j^*\ \infty]$ is a terminating sequence, then we take as neighborhoods $N_{K,L}$ of (s^*) all sequences (s) of one of the following two types. Either

$$s_i = s_i^* \qquad -K \leq i \leq j$$

$$s_{j+1} \geq L$$

or

$$s_i = s_i^* \qquad -K \leq i \leq j-1$$

$$s_j = s_j^* - 1$$

$$s_{j+N} = 0 \qquad 1 \leq N \leq L.$$

Note that these neighborhoods are quite different from the corresponding neighborhoods in the anisotropic problem.

The rationale for this topology rests on our discussion

of the local behavior of solutions near a collision orbit. Recall from §1.7 that near-collision orbits behave in two dramatically different ways depending upon which branch of the unstable manifold in Λ the solution follows. These are illustrated in Figs. 3 and 4.

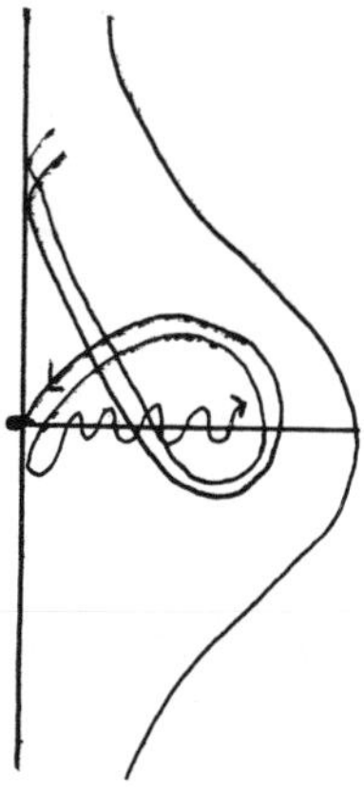

Fig. 3 The nearby orbit has sequence $(\ldots\ s_0;\ \ldots\ s_j\ K\ \ldots)$ with K large.

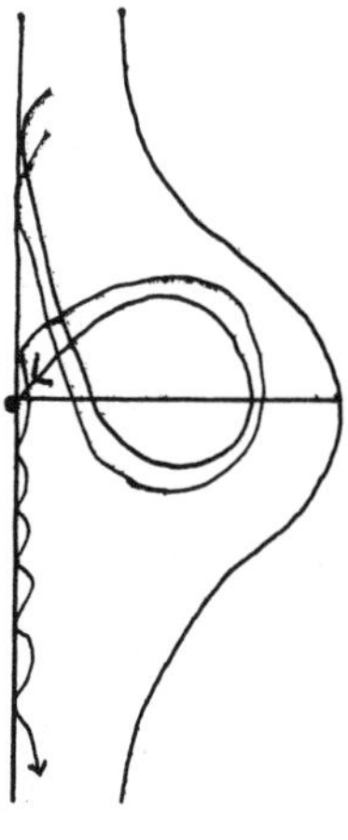

Fig. 4 The nearby orbit has sequence $(\ldots s_0;\ \ldots\ s_{j-1}\ 1\ 0\ 0\ 0.$

Using geometric methods similar to those in §3.3, one can prove the following: let $s: S_2 \to \Sigma$ be the mapping which associates to points in S_2 the corresponding sequence in Σ. On Σ we have the usual shift automorphism $(\sigma(s))_j = s_{j+1}$. The shift is again a homeomorphism defined on $\Sigma - C$, where C represents the immediate collision sequences. Our main result is then

Theorem 1. *The correspondence* s *is a continuous surjection and we have the following commutative diagram:*

$$\begin{array}{ccc} S_2 - C & \xrightarrow{\ F\ } & S_2 - E \\ {\scriptstyle s}\downarrow & & \downarrow{\scriptstyle s} \\ \Sigma - \hat{C} & \xrightarrow{\ \sigma\ } & \Sigma - \hat{E} \end{array}$$

Remark 1. Unlike the anisotropic Kepler problem, the Hill's region for the isosceles problem is non-compact. This means that open sets of orbits of the system can escape (or be captured). At present, the complete structure of these sets is not well understood. However, one can prove that any sequence which terminates to the right (resp. left) with an infinite string of zeroes is an escape (resp. capture) orbit. By the theorem above, these orbits exist.

Remark 2. Again unlike the anisotropic problem, there is no conjecture that the projection s is injective (even up to symmetries). One in fact expects that open sets of orbits escape.

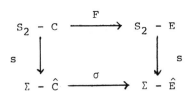

Remark 3. Our work on the isosceles problem is very similar to

the well-known work of Sitnikov and Alekseev [Al]. They study the three dimensional restricted three body problem in which the primaries revolve on Keplerian ellipses and the third mass travels along the axis of symmetry. They classify solutions of this system which oscillate far above and below the center of mass of the primaries. Their methods are similar to ours in the sense that they study a mapping on a cross-section which turns out to be a Smale horseshoe. A nice treatment of their work is contained in Moser's book [Mo].

References

[AA] Arnold, V. and Avez, A. Ergodic Problems in Classical Mechanics. New York: Benjamin, 1968.

[Al] Alekseev, V. Quasirandom dynamical systems, I,II,III. Math. USSR Sbornik 6 (1969), 489-498.

[AM] Abraham, R. and Marsden, J. Foundations of Mechanics. Reading, Mass.: Benjamin/Cummings, 1978.

[Ar] Arnold, V. Mathematical Methods of Classical Mechanics. New York: Springer-Verlag, 1978.

[Bo] Bowen, R. On Axiom A Diffeomorphisms. CBMS Regional Conference Series 38 (1978).

[Br] Broucke, R. On the isosceles triangle configuration in the planar general three body problem. Astron. Astrophys. 73 (1979), 303-313.

[De 1] Devaney, R. Collision orbits in the anisotropic Kepler problem. Inventiones Math. 45 (1978), 221-251.

[De 2] __________. Non-regularizability of the anisotropic Kepler problem. J. Diff. Equations. 29 (1978), 253-268.

[De 3] __________. Transverse heteroclinic orbits in the anisotropic Kepler problem. In The Structure of Attractors in Dynamical Systems. Springer-Verlag Lecture Notes in Math No. 668. New York: Springer-Verlag (1978), 67-87.

[De 4] __________. Morse-Smale singularities in simple mechanical systems. To appear in J. Diff. Geom.

[De 5] __________. Triple collision in the planar isosceles three body problem. To appear in Inventiones Math.

[De 6] __________. The baker transformation and a mapping associated to the restricted three body problem. To appear.

[De 7] __________. Three area preserving mappings exhibiting stochastic behavior. To appear in Classical Mechanics and Dynamical Systems. New York: Marcel Dekker.

[DN] __________, and Nitecki, Z. Shift automorphisms and the Hénon mapping. Commun. Math. Phys. 67 (1979) 137-146.

[Ea 1] Easton, R. Isolating blocks and symbolic dynamics. J. Diff. Equations 17 (1975), 96-118.

[Ea 2] ________. Regularization of vector fields by surgery. J. Diff. Equations 10 (1971), 92-99.

[EM] ________, and McGehee, R. Homoclinic phenomena for orbits doubly asymptotic to an invariant three sphere. To appear.

[Eu] Euler, L. De motu rectilineo trium corporum se mutuo attrahentium. Novi Comm. Acad. Sci. Imp. Petrop. 11 (1767), 144-151.

[Fr] Franks, J. Homology and Dynamical Systems. To appear in CBMS Regional Conference Series.

[Gu 1] Gutzwiller, M. The anisotropic Kepler problem in two dimensions. J. Math. Phys. 14 (1973), 139-152.

[Gu 2] _____________. Bernoulli sequences and trajectories in the anisotropic Kepler problem. J. Math. Phys. 18 (1977), 806-823.

[Gu 3] _____________. Periodic orbits in the anisotropic Kepler problem. To appear.

[H1] Hénon, M. A two-dimensional mapping with a strange attractor. Commun. Math. Phys. 50 (1976), 69-77.

[H2] ________. Etude générale de la transformation. Mimeographed notes.

[KS] Katok, A. and Strelcyn, J.-M. Invariant manifolds for smooth maps with singularities. To appear.

[La] Lagrange, J.L. Oeuvres. Vol. 6. Paris (1873), 272-292.

[L1] Lacomba, E. and Losco, L. Triple collision in the isosceles three body problem. Bull. AMS 3 (1980), 710-714.

[L2] __________. Quadruple collision in the trapezoidal four body problem. To appear in Classical mechanics and dynamical systems. New York: Marcel Dekker.

[MM] Mather, J. and McGehee, R. Solutions of the four body problem which become unbounded in finite time. Lecture Notes in Physics. Vol. 38. Springer-Verlag (1975), 573-597.

[McG1] McGehee, R. Triple collision in the collinear three body problem. Inventiones Math. 27 (1974), 191-227.

[McG2] McGehee, R. Recent Developments in Celestial Mechanics. To appear in CBMS Regional Conference Series.

[Mo] Moser, J. Stable and Random Motions in Dynamical Systems. Annals of Math Studies, No. 77. Princeton Univ. Press, (1973).

[Ni] Nitecki, Z. Differentiable dynamics. MIT Press, (1971).

[Pe] Pesin, Ja. B. Families of invariant manifolds corresponding to non-zero characteristic exponents. Math. USSR Izvestija 10 (1976), 1261-1305.

[Po] Pollard, H. Mathematical Introduction to Celestial Mechanics. Carus Mathematical Monographs, No. 18. Math. Asso. Amer., (1976).

[Sa] Saari, D. Singularities and collisions of Newtonian gravitational systems. Arch. Rational Mech. Anal. 49 (1973), 311-320.

[SM] Siegel, C. and Moser, J. Lectures on Celestial Mechanics. Berlin: Springer-Verlag (1971).

[Sim 1] Simo, C. Analysis of triple collision in the isosceles problem. To appear in Classical Mechanics and Dynamical Systems. New York: Marcel Dekker.

[Sim 2] ______. Masses for which triple collision is nonregularizable. To appear in Celestial Mechanics.

[Sm] Smale, S. Diffeomorphisms with many periodic points. In Differential and Combinatorial Topology. Princeton University Press (1965), 63-80.

[Su] Sundman, K. Mémoire sur le problème des trois corps. Acta Math. 36 (1912), 105-179.

[Sz] Szebehely, V. Instabilities in Dynamical Systems. Dordrecht: D. Reidel (1979).

[W] Waldvogel, J. Stable and unstable manifolds in planar triple collision. In Instabilities in Dynamical Systems. Dordrecht: D. Reidel (1979).

[Wi] Wintner, A. The Analytical Foundations of Celestial Mechanics. Princeton University Press (1941).

Robert L. Devaney
Department of Mathematics
Boston University
Boston, MA 02215